pp. 231
pp. 102–107
pp. 85
pp. 48–51
pp. 24–27
pp. 32
pp. 88/89
pp. 87
pp. 170/171
pp. 174/175
pp. 62/63
pp. 172/173
pp. 176/177
pp. 90–95

westermann

Diercke Praxis

Activity Book 3

Geography | initial level

Herausgeber:
Wolfgang Latz

Autorinnen und Autoren:
Dr. Norma Kreuzberger
Silke Weiß

mit Beiträgen von:
Andreas Bremm
Erik Elvenich
Peter Gaffga
Guido Hoffmeister
Wolfgang Latz
Rita Tekülve

Übersetzer:
John Attfield

unter Mitwirkung
der Verlagsredaktion

C **Choice – Wahlaufgaben**
Hier kannst du wählen. Du bearbeitest nur eine der **C**-Aufgaben.

A **Additional task – Zusatzaufgaben**
A-Aufgaben sind interessante Zusatzaufgaben.

★ **Starthilfen**
Schwierige Materialien sind mit einem ★ gekennzeichnet. Im Anhang findest du zu diesen Materialien Hilfen.

224 **Methodenpfeile**
Blaue Pfeile verweisen auf die Methodenseiten im Anhang. Dort kannst du nachlesen und üben, wie du die Methoden richtig anwendest.

78 **Verweise**
Pfeile in den Farben der Kapitel weisen dir auf den Gewusst? – Gekonnt!-Seiten den Weg zur zugehörigen Arbeitsseite.

schueler.diercke.de | **100800-282-01**
Durch Eingabe des Kartencodes unter ***schueler.diercke.de*** gelangst du auf die passende Seite im Diercke Weltatlas. Dort erhältst du weitere Informationen zu den Karten. Den Code findest du ganz unten auf den Doppelseiten.

M **Medienkompetenz**
Oftmals ist es sinnvoll, einen Computer, einen Laptop oder ein Tablet zur Lösung von Aufgaben zu nutzen. Wie du das machst, zeigen dir die Methoden zur **M**ediennutzung.

Diercke Weltatlas @Diercke
Die Diercke Praxis ist nach Carl **#Diercke** benannt. Er lebte von 1842 bis 1913. Er war Erdkundelehrer und beschäftigte sich mit der Herstellung von Landkarten. 1883 entwickelte er einen Atlas mit einfachen und übersichtlichen Karten: den „Schul-Atlas über alle Teile der Erde“. Er sollte besonders Kinder und Jugendliche ansprechen. Sein Nachfolger ist der **#Diercke** #Weltatlas.

Internet
Die hier angegebenen Suchbegriffe (z. B. Fachbegriffe, Namen von Institutionen) führen nach Eingabe in eine Suchmaschine zu Links mit weiteren Informationen zur jeweiligen Thematik. Dabei kann es sich auch um Videos oder Bilder handeln. **213**

interaktive Karten
Die hier angegebenen Suchbegriffe (z. B. Ortsnamen, Fachbegriffe) führen nach Eingabe in einen digitalen Kartendienst zu Karten oder Satellitenbildern, die die jeweiligen Räume darstellen oder die die jeweilige Thematik an einem weiteren Raumbeispiel veranschaulichen. **191**

Titelfoto:
Zwei Mädchen vor dem Lotustempel in Neu-Delhi, Indien

wordlist – Vokabelliste hilft, das Verständnis der Aufgabenstellung und der Materialien zu erleichtern.

Helpful words and phrases for task 3:
Unter den Arbeitsaufträgen findest du oft Hilfen, mit denen du leichter formulieren kannst.

Wenn du diese Aufgaben erfolgreich bearbeitet hast, kannst du ...
... mithilfe dieses Kastens überprüfen, ob deine Forschung erfolgreich war.
... prüfen, ob du die ***Fachbegriffe*** *der Doppelseite verstanden hast.*

westermann GRUPPE

Druck A[1] / Jahr 2021
Alle Drucke der Serie A sind im Unterricht parallel verwendbar.

Redaktion: Jan-Berent Schmidt, Steffen Stierhof
Umschlaggestaltung: JANSSEN KAHLERT Design & Kommunikation GmbH, Hannover
Layout: JANSSEN KAHLERT Design & Kommunikation GmbH, Hannover
Druck und Bindung: Westermann Druck GmbH, Georg-Westermann-Allee 66, 38104 Braunschweig

ISBN 978-3-14-**115403**-0

A = Addition (Ergänzung – fakultativ) **O** = Optional topic (Wahlthema)

A = Addition (Ergänzung – fakultativ) **O** = Optional topic (Wahlthema)

A = Addition (Ergänzung – fakultativ) **O** = Optional topic (Wahlthema)

A = Addition (Ergänzung – fakultativ) **O** = Optional topic (Wahlthema)

M

Methoden 198

A

Anhang 226

A = Addition (Ergänzung – fakultativ) **O** = Optional topic (Wahlthema)

1. Global population – growth, food supply and migration

Housing in Mexico City

Enough for all? – population growth and food security

Development of the world's population is not the same everywhere

Every minute, the world population grows by about 150 people - about 80 million a year. Some people speak of a "population explosion" - but is that really true? Exactly how fast is the world population growing? What differences are there between the countries and regions of the world?

1. Explain the cartoon. 209

C **2.** **A** Describe the global population growth (M1).
B Write a short newspaper report about the scenarios for global population development (M1).

3. Go to the "Weltbevölkerungsuhr" on the internet pages of the *Deutsche Stiftung Weltbevölkerung* (M7). 210
a) Read the clock precisely every minute for 10 minutes and produce a diagram. 207
b) Calculate what number you would be on the world population clock.

4. Illustrate the population and population development until 2050 in a country of your choice (M7).

5. Describe which continents have particularly high population growth. (M3 – M5).

development	–	Entwicklung
distribution	–	Verteilung
food security	–	Ernährungssicherung
global population	–	Weltbevölkerung
growth	–	Wachstum
newly industrialised country	–	Schwellenland

Helpful words and phrases for task 2:
From (year) to (year) / about (year), the population was about ....
Today / x years ago, y people live / lived on earth...
From (year) to (year), the population grows slowly / constantly / rapidly / very fast / exponentially...
A population of (number) is reached in (year).
Population growth depends on different factors, e.g. life expectancy, numbers of births and deaths.
According to the low / middle / high scenario, there will be (number) people on earth in (year).

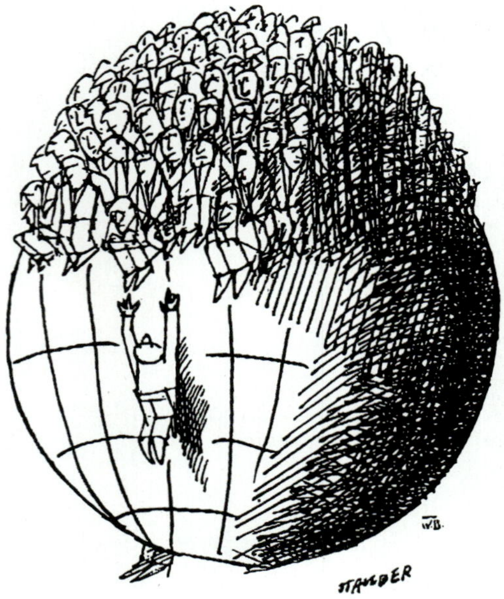

M2 Cartoon

The growth of the world's population is very unevenly distributed. In the so-called **industrialised countries**, the population is growing very slowly or even shrinking. In the so-called **developing countries** and the newly industrialised countries, the population is growing strongly.
The proportion of people living in developing countries has risen sharply in recent years: from 68% in 1950 to 80% in 2000. By 2050, 90% of people will probably be living in developing and newly industrialised countries – most of them in India and China.
Population development depends mainly on births and deaths. In addition, millions of people leave their homes every year and move to another place or country - sometimes even to another continent. This **migration** also contributes to the differences in population distribution worldwide.

M3 Regional differences

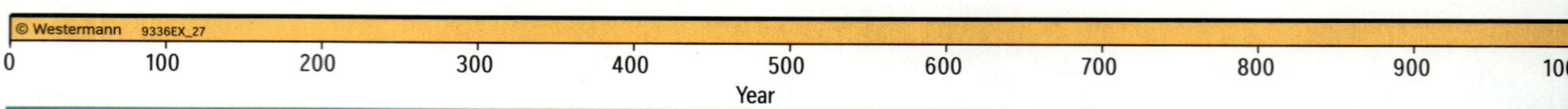

M1 The development of the world population

If the world in 2020 was a village with only 100 inhabitants, there would be ...

	2020	2050
Total inhabitants	100*	125
Africans	17	32
Asians	60	68
Europeans	10	9
Latin Americans	8	10
North Americans	5	5
Oceanians	1	1

*differences due to rounding
Source: UN DESA 2020

M4 **World population by regions**

INTERNET

M7 Numerous organisations offer excellent material on population on the internet: statistical data, scenarios up to 2100, PowerPoint presentations, films and much more. The Deutsche Stiftung Weltbevölkerung also offers the latest scenarios on further population development for download in „Datenreport" and in„Weltbevölkerungsbericht". 210

- Deutsche Stiftung Weltbevölkerung (DSW)
- United Nations (e.g. World Population Prospects, United Nations Population Fund)
- Population Reference Bureau (PRB)
- Statistisches Bundesamt (DESTATIS)

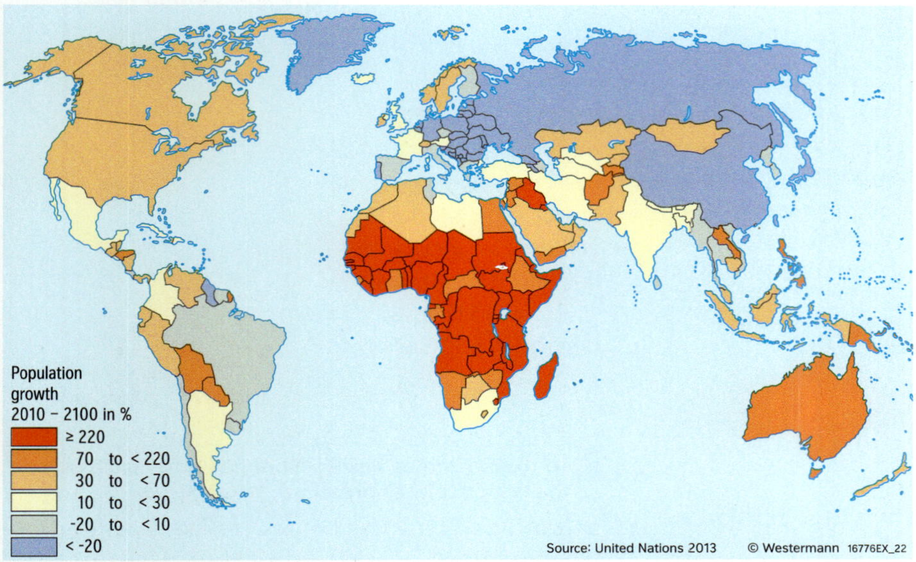

M5 **Regional distribution of population growth to 2100**

M6 To find out how many people will live on Earth in 2050, scientists from the **UNO** regularly create **scenarios** on global population development. Today's population figures and assumptions on the development of births, deaths and migration in the various regions of the world serve as a basis. Different assumptions lead to different scenarios.

INFO

Wenn du diese Aufgaben erfolgreich bearbeitet hast, kannst du ...

- ... *die Bevölkerungsentwicklung und das unterschiedliche Bevölkerungswachstum auf der Erde beschreiben.*
- ... *die Fachbegriffe* ***developing country*** *(****Entwicklungsland****),* ***industrialised country*** *(****Industrieland****),* ***migration*** *(****Migration****),* ***UNO*** *(****UNO****) und* ***scenario*** *(****Szenario****) erklären.*

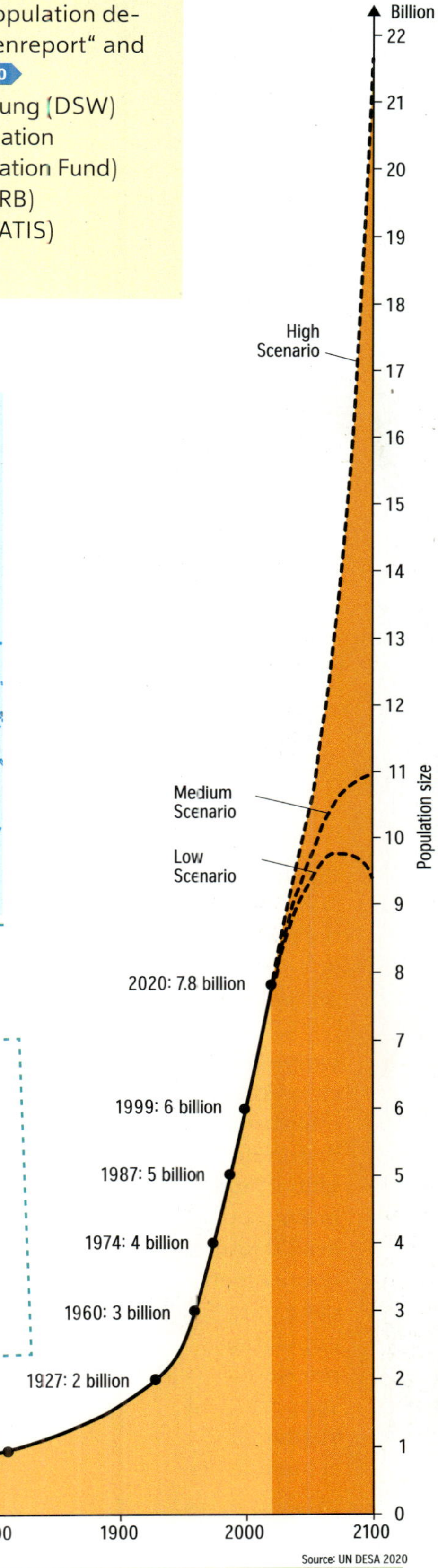

How many people can the Earth support?

The population is rising and rising. But how many people can the Earth support? How many people can it feed? Already, many people suffer from hunger.

1. Compare the calculations of the carrying capacity of the Earth by Malthus and the Club of Rome. (M2, M3).

2. Describe and evaluate the development of global food distribution (M4, M5)

3. a) Present M7 in the form of a diagram. Discuss in class which presentation makes the most sense. 207
b) Using M7 and your diagram, illustrate the term "population pressure".

C **4.** Research the topics of "carrying capacity" in the internet (M8). 213 Report on
A further myths about global hunger (M6, M8).
B the Club of Rome and one of its current scenarios for the future (M8).

5. Name possibilities you personally have to increase the carrying capacity of the Earth. Give reasons for your decision. (M8)

agricultural scientist	–	Agrarwissenschaftler
arable land	–	landwirtschaftliche Nutzfläche
armed conflict	–	bewaffneter Konflikt / Krieg
carrying capacity	–	Tragfähigkeit
chronically malnourished	–	chronisch unterernährt
distribute	–	verteilen
global hunger	–	Welthunger
grain production	–	Getreideproduction
livable future	–	lebenswerte Zukunft
nourishment	–	Nährstoffe
per capita	–	pro Kopf
population pressure	–	Bevölkerungsdruck
support	–	tragen / unterstützen
sustain	–	ertragen
sustainability	–	Nachhaltigkeit

M1 **How many people could the Earth sustain?**

The power of population is so superior to the power of the Earth to produce subsistence for man, that premature death must in some shape or other visit the human race.

Source: Malthus, Thomas Robert: An Essay on the Principle of Population. London 1798, p. 44

Malthus argued that population growth initially increases exponentially, but food production can only be increased linearly. A shortage of food would then lead to a decline in population.

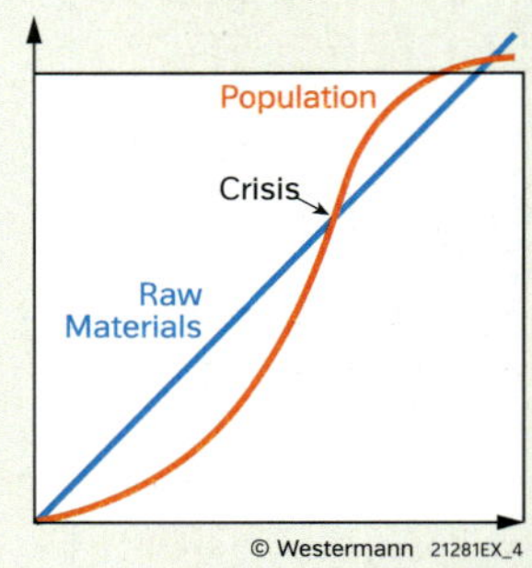

A model of Malthus's theory.

M2 **In 1798, Thomas Robert Malthus wrote "An Essay on the Principle of Population"**

In a complex simulation, the scientists of the Club of Rome calculated scenarios for the **carrying capacity** of the Earth. The Club of Rome is a think tank of about 100 scientists and business leaders. They write: *„Our goal is to actively advocate for paradigm and systems shifts which will enable society to emerge from our current crises, by promoting a new way of being human, within a more resilient biosphere."*

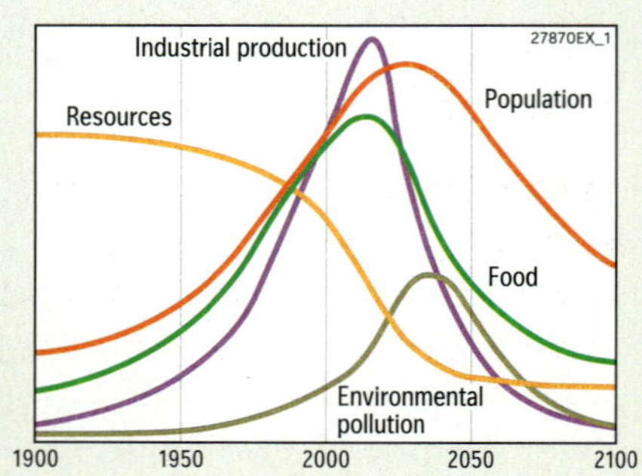

"World3" simulation (2004)

Source: The Club of Rome: About The Club of Rome. www.clubofrome.org, accessed: 03.12.2020

M3 **The study "The New Limits to Growth " was published on behalf of the Club of Rome in 1992**

Agricultural scientists assume that it is technically possible to produce enough food with 0.07 hectares of arable land per person. By 2025, 2.5 billion people could be affected by a scarcity of arable land (in 2010 it was 260 million people). 830 million people are already chronically malnourished.

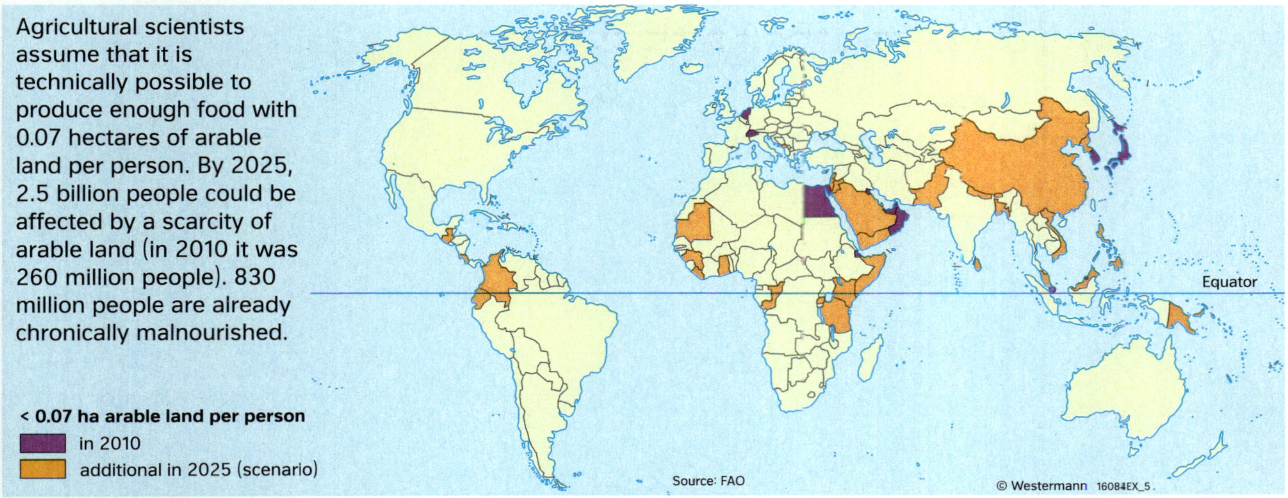

M4 Countries with less than 0.07 hectares of arable land per capita

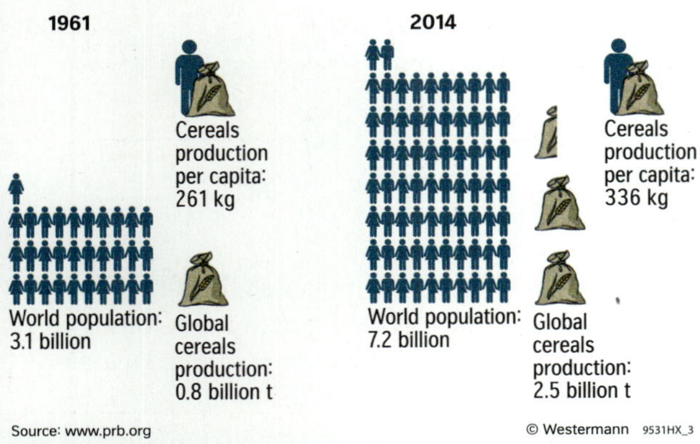

M5 Global cereals production and world population

		China	Germany	India	Peru
1961	Pop.	686	73	460	10
	AA	343	19	175	17
2000	Pop.	1320	81	1057	26
	AA	524	17	180	23
2018	Pop.	1459	83	1353	32
	AA	529	17	180	24
2050*	Pop.	1366	79	1663	40
	AA	?	?	?	?

Pop: Population in millions.
AA (Agricultural area): Arable land + meadow and pasture
*Forecast

M7 When the population grows and the arable land area stays the same, **population pressure** increases

Myth 1: Hunger means not having enough to eat.
Reality: Hunger is more the an empty stomach. There is also “hidden hunger”, which means that the food lacks nutritional value [...]

Myth 3: There will never be enough food to feed everybody.
Reality: At the moment, there is plenty of food to feed everybody on this planet. Earth’s population will increase to 9 Billion by 2050. [...] To ensure that there is plenty of food to go around, we have to change production methods and our behaviour. [...]

Myth 4: Only Droughts and other natural disasters are to blame for hunger.
Reality: Nature is only one of many factors.[...] Armed conflict is often at the heart of today‘s worst food crises. People are displaced, local markets destroyed, and food prices increase. [...]

Source: World Food Program: 10 Mythen über den Welthunger. www.wfp.org, 01.08.2017, accessed: 01.08.2021

M6 Myths about global hunger (selection)

INTERNET

M8 In the blog of the *World Food Programme (WFP)* „Stories“, you can find information on hunger crises and relief measures, and facts about myths on world hunger, malnutrition and sustainability.

The Club of Rome informs on its homepage about scenarios and ideas for a livable future.

Wenn du diese Aufgaben erfolgreich bearbeitet hast, kannst du ...
- *... die Kernaussagen zweier Tragfähigkeitsberechnungen nennen.*
- *... die Entwicklung der Weltbevölkerung und der Nahrungsmittelversorgung vergleichen.*
- *... Mythen über den Hunger bewerten.*
- *... erste Maßnahmen zur Erhöhung der Tragfähigkeit beschreiben.*
- *... die Fachbegriffe **carrying capacity** (**Tragfähigkeit**) und **population pressure** (**Bevölkerungsdruck**) erklären.*

Carrying capacity – what does it have to do with us?

The population is increasing, the carrying capacity of the Earth is limited ... what does this have to do with us? What difference can we make locally?

1. Describe the ecological footprint (M2) and the concept of "Earth Overshoot Day" (M1).
2. Name examples of overshooting the Earth's carrying capacity in the various dimensions (M3)
3. Due to the Corona pandemic, the Earth Overshoot Day, which had moved further forward each year, was pushed back. Explain the phenomenon and comment of the slogan of the initiative "We don't need a pandemic to move the date!". (M1, M2, internet)
4. Analyse the cartoon (M7). 209
5. Discuss the statement by the Club of Rome (M4).
6. How can I contribute to sustainability in all areas of life?
   A Match each Sustainable Development Goal to an opportunity for action. (M5, M6, M8). – Do it!
   B Name another 10 opportunities for action.

accumulate waste	–	Abfälle anhäufen
carbon footprint	–	CO_2-Fußabdruck
demand	–	Nachfrage
ecological footprint	–	ökologischer Fußabdruck
economical	–	wirtschaftlich
fibre	–	Fasern
fossil fuel	–	fossile Brennstoffe
liquidate stocks	–	Vorräte auflösen
locally	–	lokal / vor Ort
pollutant	–	Schadstoff
poverty	–	Armut
primarily	–	vor allem
sequester	–	hier: binden
sustainable development	–	nachhaltige Entwicklung
sustainable development goal	–	Nachhaltigkeitsziel
timber	–	Holz / Bauholz
unanimously adopted	–	einstimmig verabschiedet

INTERNET

M1 On the Earth Overshoot Day website you can find explanations, videos and statistics about Earth Overshoot Day.
- Earth Overshoot Day
  - For Kids and Teachers
  - Steps to Move the Date

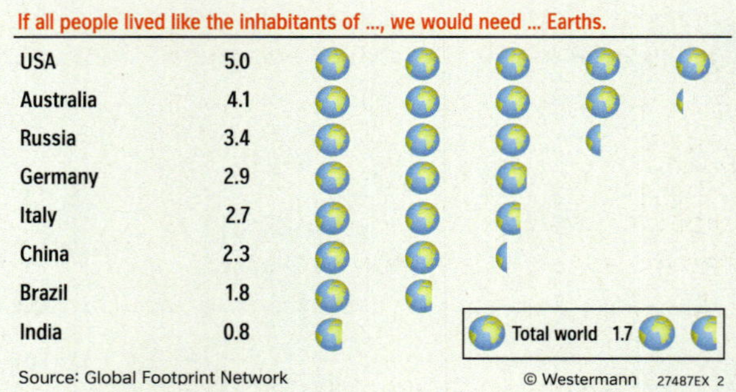

The ecological footprint measures the amount of biologically productive land and sea area required to provide for the demand of population, including the areas for growing the food, fibres, and timber they consume, the space they occupy with their houses and roads, the area needed to sequester the carbon dioxide from burning fossil fuel.

M2 The ecological footprint (2017)

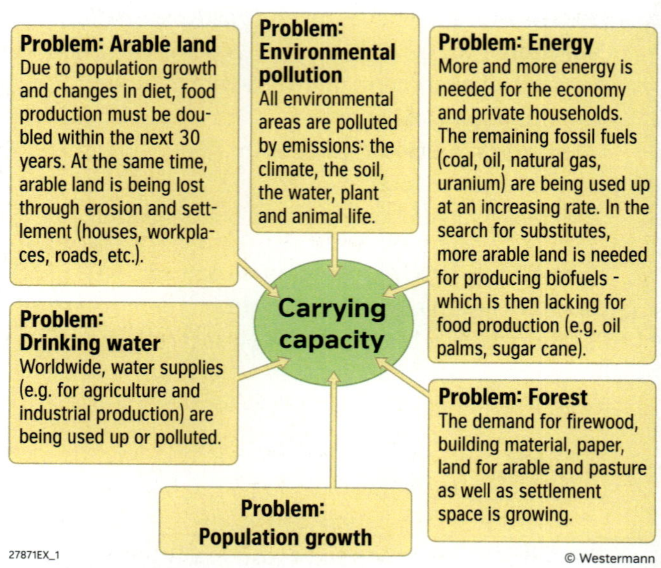

M3 Dimensions of carrying capacity.

New technologies make it possible to release fewer pollutants into the environment. Even mass poverty can be defeated.
The prerequisite is that population growth is slowed down, and that energy and resources are used more efficiently and sparingly and distributed more evenly worldwide. **Sustainability** is the most important challenge for human activity and creativity.
The Club of Rome believes that people today are capable of meeting these challenges. A long-lasting future for humanity is still technically and economically possible. This requires maturity, sharing in partnership and wisdom.

M4 **The Club of Rome's hopes in 1992 – are they justified?**

M7 **Cartoon**

In September 2015, the **2030 Agenda** for Sustainable Development was unanimously adopted by the member states of the United Nations. With the 2030 Agenda, the global community has set itself 17 **Sustainable Development Goals** (SDGs) for socially, economically and ecologically sustainable development.

Source: BMZ: Die Agenda 2030 für nachhaltige Entwicklung. www.bmz.de, accessed:03.12.2020

M8 **Agenda 2030**

Do it!

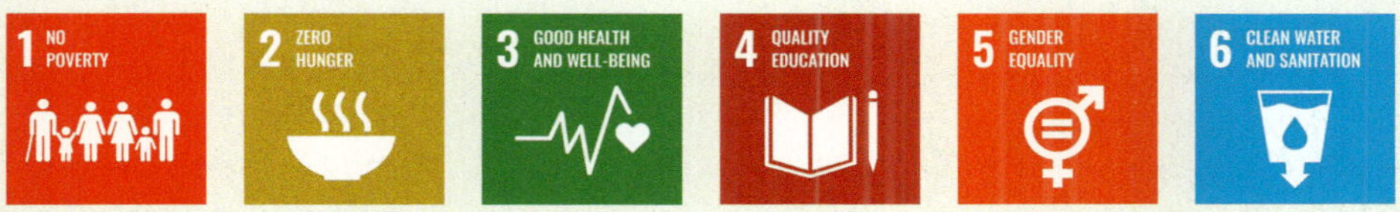

One opportunity for action can fit several goals: A Donate blood. B Introduce an electricity saving day. C Use grit instead of salt on the road. D Campaign for the rights of girls and women. E Donate your old bicycle. F Save food and prevent waste. G Water your plants with rainwater. H Organise and get involved in trade unions, for example. I Visit clothing swaps and second-hand shops. K Eat vegetarian food more often. L Volunteer as a reader with the "VORLESEN" network. M Sign petitions to stop unsustainable urban development projects. N Give language and integration courses. P Consider what invention would make your town or city better. R Put pressure on politicians. S Stop using microplastics. T Go and vote. Tell everyone about your 17 goals!

Source: Engagement Global gGmbH: 17 Ziele. www.17ziele.de, accessed: 03.12.2020, amended

M5 **The goals of the 2030 Agenda with suggestions on what you could do yourself: Do it!**

INTERNET

M6 The *Bundesministerium für Zusammenarbeit und Entwicklung* offers information and suggestions on the 2030 Agenda: BMZ, Ziele für nachhaltige Entwicklung, or on the website "17 Ziele".

Wenn du diese Aufgaben erfolgreich bearbeitet hast, kannst du ...
- ... *unterschiedliche Dimensionen von Tragfähigkeit und neue Ansätze zur Tragfähigkeitsberechnung beschreiben.*
- ... *die Bedeutung von nachhaltiger Wirtschaftsweise für die Tragfähigkeit der Erde erklären.*
- ... *die Fachbegriffe **sustainability (Nachhaltigkeit)**, **Agenda 2030** und **Sustainable Development Goals (SDGs – Nachhaltigkeitsziele)** erklären.*

How can we secure our food supply in future?

According to experts, global food production must be increased by 50 percent by 2050 (base 2012). What is the situation today? What ways do we have to ensure that everyone will be fed in the future?

1. a) Locate two major cultivation areas where arable farming predominates for each continent. Note two crops in each case (M4, atlas).
b) Based on two countries, explain possible risks for cultivation (M4, atlas).

2. Outline, in each case, the methods that lead to an increase in food production with the same area of cultivation:
a) Intensification in the cultivation of lentils (M1, M5),
b) Intensification in the cultivation of rice (M6).

3. Name the most important proposition of the agricultural scientist Joachim von Braun. (M2)

4. Analyse M3. Using the insights from M2 and M6 derive measures that could improve food security worldwide.

arable farming	Ackerbau
cattle	Rinder
climate-related hazard	klimabedingte Gefährdung
crop	Anbauprodukt
crop diversity	Nutzpflanzenvielfalt
crop rotation	Fruchtfolge
fodder	Futtermittel
genetically modified	gentechnisch verändert
harvest	Ernte
hybrid variety	Hybridsorte
large scale	großmaßstäblich
legume	Hülsenfrucht
lentils	Linsen
maize	Mais
minerals	Mineralstoffe
pasture	Weideland
protein	Eiweiß
rapeseed	Raps
seed bank	Saatgutbank
smallholder	Kleinbauer / -bäuerin
soy	Soja
trace elements	Spurenelemente
tropical cyclones	tropische Wirbelstürme

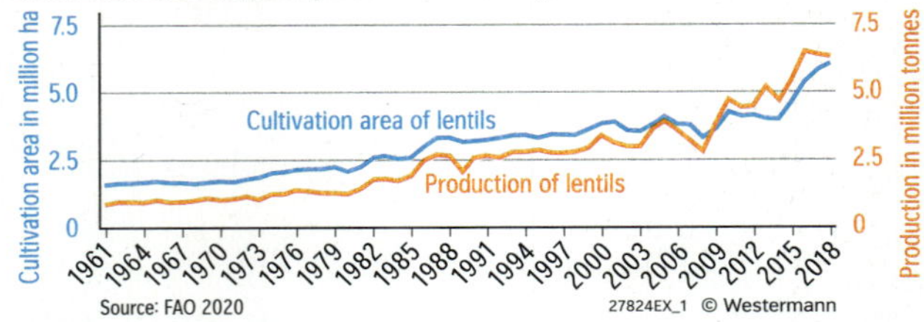

M1 Data on global food production.

Enormous increases in production at the global level will be necessary to meet the demand for food and agricultural raw materials [...]. It is hardly possible to expand the area of land beyond the approximately 1.4 billion hectares that are currently available and used for agriculture. [...] Higher yields are in principle possible through more intensive production and technical progress. It is precisely in developing countries [...] that yields per hectare are lowest. In the longer term, yields will only increase sustainably if new crop varieties and technologies are developed that are adapted to the specifics of the location (for example, drought, heat, soil salinisation, disease and pest pressure) and make better use of scarce resources, especially water and soil.
With regard to the quality of nutrition [...] staple crops with higher micronutrient contents are also important. Plant breeding makes an important contribution to this.

Source: von Braun, Joachim (bpb): Die Bedrohungen der Ernährungssicherheit. www.bpb.de, Bonn, 13.11.2014, accessed: 16.10.2020

M2 Agricultural scientist Professor Joachim von Braun on ways to increase food production globally

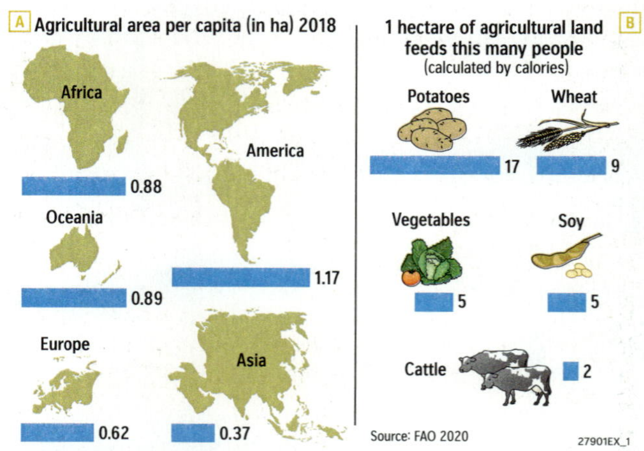

M3 Agricultural land and food

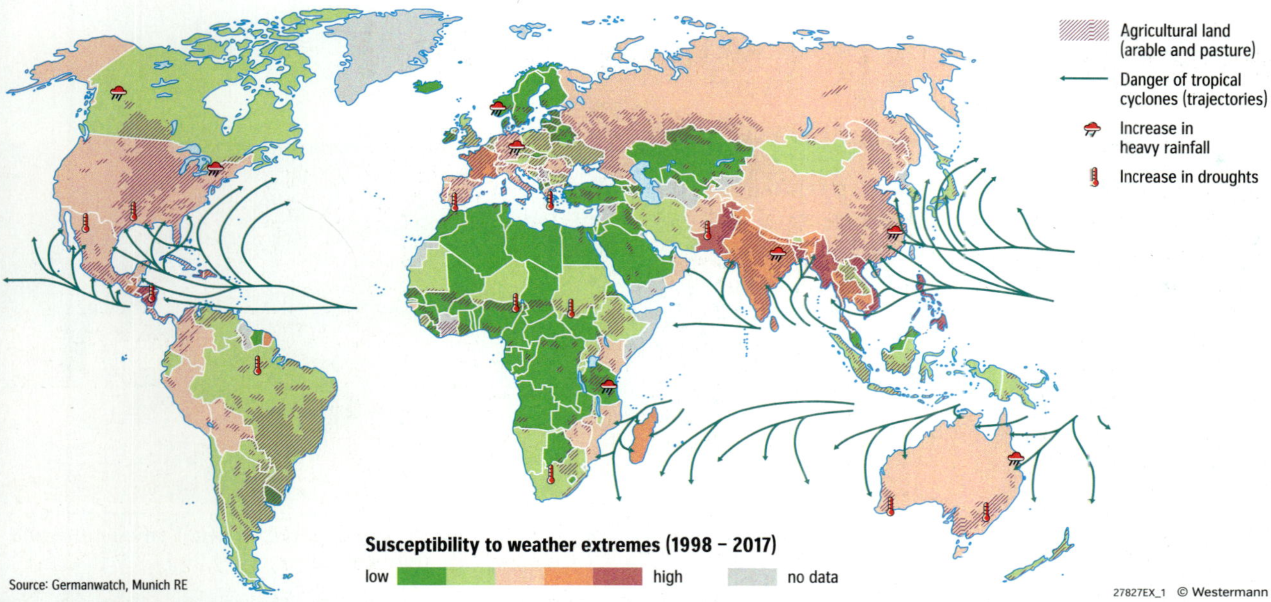

M4 Global agricultural regions and climate-related hazards for food security

Lentils can be a miracle for nutrition. The legume grows in dry regions. The plant can fix nitrogen from the air and store it in its roots - so it fertilises itself and the soil. Lentils are very healthy - rich in protein, minerals, and trace elements.

Ethiopia, Demekesh, smallholder farmer:
"The new varieties must be cheap, give high yields and be resistant to plant diseases. Only then can we harvest enough and sell the surplus."
India:
Here, lentils are grown that contain certain micronutrients (iron, zinc) to reduce malnutrition. Also, the new varieties mature faster, allowing four harvests a year and more variety in crop rotation.
Canada:
Canada is the most important lentil producer, accounting for 65 percent of world production. Here, cultivation is carried out on a large-scale agro-industrial basis. The seeds have been genetically manipulated so that the varieties are resistant to many pests, can be harvested with machinery and are adapted to the climate.
Worldwide:
Teams of researchers from the seed industry and governmental and non-governmental organisations are working to develop hybrid varieties that are adapted to local conditions, high-yielding and resistant to disease and pests.

M5 Lentils as a nutritional miracle: cultivation of adapted hybrid varieties in Ethiopia, India and Canada

I have been training farmers for 3 years. Before, I did not raise fish in my rice fields. After harvesting the rice, I got nothing from the field. But now – in harvest season, we can have both rice, fish and other animals, and this is very good. [...]
We notice that we have a healthier crop, and we don't have enemies of the rice – because fish eat the insects and worms. So we have nice rice and big fish, which enables our family to have enough fish to eat, and we don't have to go far away to find food. We even have some for sale too.

Source: FAO: Training farmers in aquatic resource building. Interview with Bunlom Phantavong, Rice/Fish Farmer and Trainer, Savannakhet Province, Lao People's Democratic Republic. Bangkok 2019, p. 1

M6 Intensive, sustainable agriculture - Mr. Bunlom Phantavong from Laos explains the dual use of his rice fields

Wenn du diese Aufgaben erfolgreich bearbeitet hast, kannst du ...
- ... *die globale Nahrungsmittelproduktion und ihre Risiken erläutern.*
- ... *die grundlegenden Wege zur Produktionssteigerung (kritisch) darstellen.*
- ... *Beispiele (Linsenzüchtungen) zur Intensivierung der Produktion ausführen.*
- ... *den Fachbegriff **hybrid variaty** (**Hybridsorte**) erklären.*

OPTIONAL TOPIC 1

New food options are being developed – on the land

Food or food raw materials come from agriculture (the soil) or from water (mainly the sea). This will continue to be the case in the future. What are the ways in which agriculture can produce more food in the future? What is sustainable agriculture?

1. Describe the different methods of food production (M1 – M4).
2. For three of the examples, analyse their effects on the local population, global nutrition and the climate (M2, M3, M5, M6).
3. Give reasons for the great importance of women and small farmers for global nutrition.
4. Ⓒ Research on the internet. Report on
   A an "Edible City" near you. 213
   B sack gardening in Nairobi. 213
5. Summarise in keywords the idea and objectives of the vegetable farm on the roof of the exhibition centre in Paris. Use the Deutschlandfunk podcast "Europa heute" ("Städtische Landwirtschaft: Urban Farming auf Pariser Messehalle").

artificial	–	künstlich
beef	–	Rinfleisch
consumption	–	Verbrauch
edible	–	essbar
farmscraper	–	Farmhochhaus
fattening	–	Mast
greenhouse	–	Gewächshaus
irrigation	–	Bewässerung
majority	–	Mehrheit
multistory	–	mehrstöckig
nutrition	–	Ernährung
stem cell	–	Stammzelle
tissue sample	–	Gewebeprobe
urban farming	–	urbane Landwirtschaft
utopian	–	utopisch

Helpful words and phrases for task 2:

The project tries to improve...
The most important measures are...
People are shown / trained in
To increase the productivity, ...
The project is small scale / large scale...
To save energy / time / space ... people use / build / apply...
It is good / harmful for the environment / climate / local people, because ...

M1 Using machinery is expensive, but it saves time and also avoids child labour.

Smallholder farmers with less than 2 hectares of land are very important for the world's food supply. They manage 60 percent of the world's arable land. 85 percent of all farms and 80 percent of agricultural production in Asia and Africa.
Yet many of these farms produce little more than their own needs. Due to a lack of reserves, many small farmers quickly find themselves in need during crises, e.g. droughts.
These farms have great potential to increase productivity, as they could intensify their production in various areas with often simple means – e.g. irrigation, seeds, fertilisers, or storage.
80 percent of small farms are managed by women. They have traditional knowledge about local resources, plants, soils, animal husbandry and breeding. Through training, they can combine this with modern knowledge and thus increase yields. Therefore, development projects often focus on water management, small animal husbandry, crop rotation, agricultural technology, and self-managed micro-credits for investments.

M2 Innovations and support for female smallholders

M3 Clearing savannah in Ethiopia, where a Saudi Arabian agribusiness group is creating arable land for rice cultivation

a) **Vertical Farming**: Plants grow in vertical beds, often on large shelves or in small spaces in supermarkets. In urban agglomerations, “farm-scrapers” or high-tech greenhouses could be worthwhile. There, lettuce and vegetables grow in a nutrient solution instead of soil, under LED lights instead of in the sun. Light, water, and nutrient supply are computer-controlled. Several harvests per year are possible. Delivery traffic with trucks can be avoided because the food is produced where it is needed. However, energy supply is a problem as long as electricity does not come from renewable sources.

Sack gardening in Kibera slum, Nairobi

b) Urban gardening plays an important role in African towns and cities: gardens in sacks and bags between wooden or tin huts, with small livestock in between. This micro-agriculture contributes to food security and can be a source of income.

c) Multi-storey pigsties south of Guigang in South China.

Multistorey pig fattening pens south of Guigang in Southern China

d) In the “Edible City” of Andernach, food is grown on fallow land. In Germany, there are such projects in over 100 cities – the majority in Berlin.

e) In Detroit, about 20,000 residents produce food on former industrial sites. They produce for their own consumption. Some also sell their fruit and vegetables.

M4 Urban farming: various forms of industrial and small-scale agriculture in the city. This is not utopian. In 2050, 80 percent of people will live in cities.

With rising standards of living, meat consumption is also increasing in many countries of the world. More and more agricultural land is needed to grow animal feed. To avoid this, there are many ideas to replace meat.
Companies in California, Israel and the Netherlands are leading the way. Muscle cells are grown in the laboratory from tissue samples from donor animals and stem cells. The first burger made in London from lab-grown meat cost 250,000 euros in 2013. Since then, research has moved on and prices are falling. The space and water requirements of in-vitro meat will be smaller, but the energy consumption higher than for “real” meat.
Worldwide, more than two billion people regularly consume insects, especially in Asia.
Recently, insect products have also started to appear in our shops – e.g. in noodles or biscuits. Insects are very rich in protein. Compared to beef, insect production requires much fewer resources of water, energy, or space, 10 times less feed, 100 times less time, 100 times less greenhouse gases.

M5 Two alternatives to increasing meat consumption

AMAZING

M6
- 800 million people worldwide practise urban farming.
- In China, four high-rise pig farms are expected to fatten around 840,000 pigs per year (see M6).
- If all people wanted to eat as much meat as we do in Germany, we would need three Earths just to grow enough feed.

Wenn du diese Aufgaben erfolgreich bearbeitet hast, kannst du ...
- ... *klassische und neue Wege erläutern, mit denen die Nahrungsmittelproduktion der Zukunft an Land sichergestellt werden kann.*
- ... *die Vor- und Nachteile neuer Wege zur Nahrungsmittelproduktion erläutern und im Hinblick auf ihre nachhaltige Wirkung in Teilbereichen bewerten.*
- ... *die Fachbegriffe* ***vertical farming*** *(****Vertical Farming****) und* ***urban farming*** *(****Urban Farming****) erklären.*

OPTIONAL TOPIC 2

New food options in the water

Never before have people consumed so much fish and seafood as they do today. What are the potentials and ways to obtain more food from the water in the future?

1. Describe the global global distribution of fish production and fish consumption (continents and oceans) (M4, atlas).
2. Describe the ways of producing food shown in M1, M5 and M6. Compare the three methods in terms of their sustainability. 217
3. Contrast the advantages and disadvantages of aquacultures in a table (M5, M7). 207
4. Ⓒ Research on the internet. Describe
   A sustainable aquaculture for fish and plant production (aquaponics).
   B the development of fishing yields off the coast of Peru.
   C possibilities of reducing the by-catch in Industrial fishing.
5. Demand for fish and seafood is increasing. Industrial fishing supplies about 80 million tonnes of fish and creates some 500,000 jobs. Compare the importance and the sustainability of industrial fishing with that of aquacultures.

M1 Industrial fishing with factory vessels in the Arctic: The fish is caught with large nets or longlines and processed and frozen on board. By-catch – unwanted fish that are thrown back dead into the sea – is a problem.

Seafood:
Crustaceans: **e.g., lobster, prawns, scampi**
Molluscs: **squid, snails, mussels**
Edible fish: **e.g., salmon, herring, cod, plaice**
Sea vegetables: **seaweed, vegetable algae**: raw as a salad or dried as a snack, especially in Japan; **Nori seaweed leaves** – used for wrapping maki sushi; **Agar-Agar** – gelatine substitute from **seaweed**

Ecological balance: CO_2 equivalent of salmon 1.5 kg, beef 9.5 kg

M2 Perspective: More food from the sea. Do you know what it is?

algae – Algen
aquaculture – Aquakultur
by-catch – Beifang
catch rate – Fangquote
coastal – küstennah
environmental and animal welfare requirements – Umwelt- und Tierschutzauflagen
factory vessel – Fabrikschiff
fish consumption – Fischkonsum
fish meal – Fischmehl
fish oil – Fischöl
fish production – Fischproduktion
fish stocks – Fischbestände
fishing fleet – Fangflotte
industrial fishing – industrielle Fischerei
inland water – Binnengewässer
net pen – Netzgehege
non-compliance – nicht-Einhalten
overfishing – Überfischung
seafood – Meeresfrüchte
tank – hier: Becken

M3 Production of algae (spirulina) on a rooftop in Bangkok. This plant production is an example of urban farming. Food is produced where it is needed. Today, 55 percent of the world's population already lives in cities. By 2050, it will be 80 percent.

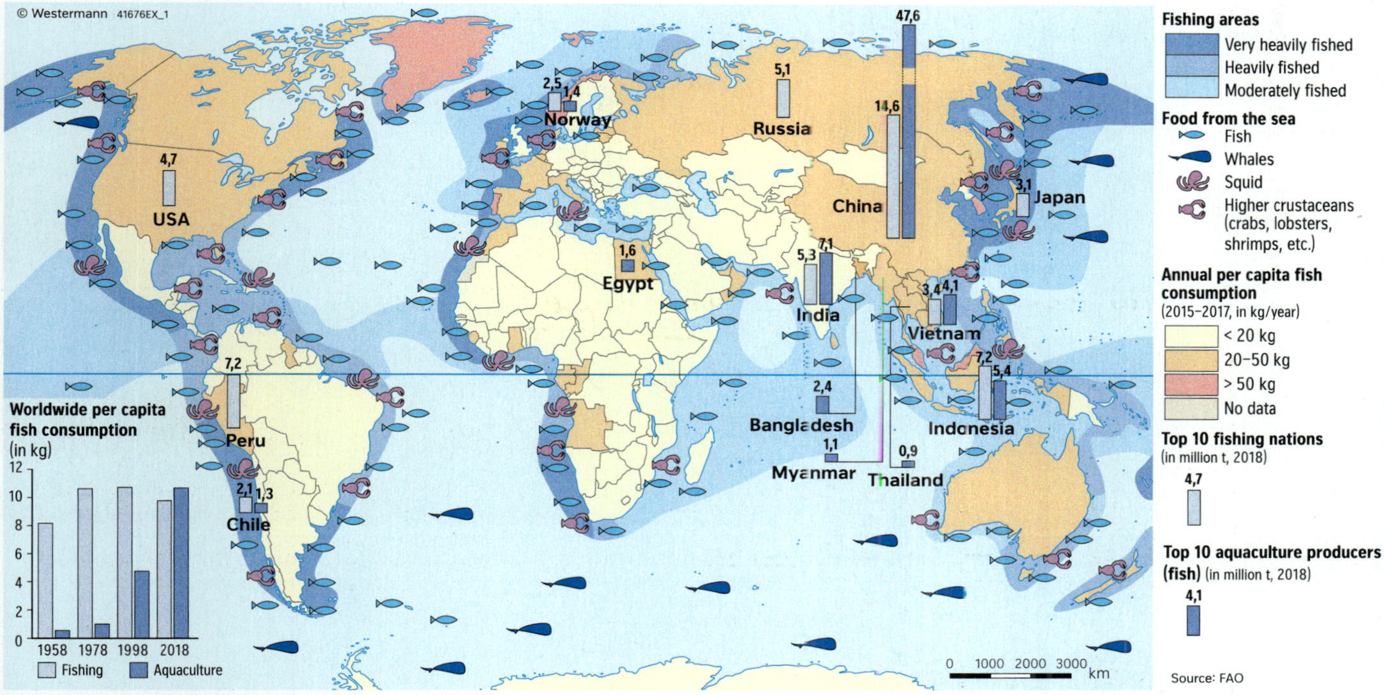

M4 Global fisheries: fish stocks, fish production and fish consumption

Aquaculture is food production in water – in the sea, in inland waters or in tanks. The most important products are fish, shrimps, algae, and seaweed. Coastal net pens are often 25 – 50 m deep and 30 – 50 m in diameter. Salmon are fattened for about 2 years with fish meal, fish oil and soy. The fish excrements and the antibiotics used in mass fish farming end up in the sea and affect the ecosystem. For every kilo of salmon, 1.2 kg of fish meal is fed. CO_2 equivalents of 1.5 kg are produced (in comparison: beef 9.5 kg). Aquacultures help to supply the growing world population with inexpensive fish and seafood. They can reduce overfishing. Aquacultures create 19 million jobs worldwide. In Southeast Asia, aqua farms produce protein-rich food, income, employment, and foreign exchange. Often, mangroves – valuable ecosystems and shelter from cyclones and tsunamis – are cut down for their construction.

M5 On the importance of aquaculture

M6 **12 million small-scale fishermen catch 35 million t. of fish annually worldwide. fish for the local market. They are very important for the food supply of many developing countries. However, traditional, small-scale fishing (as here in Sri Lanka) is severely threatened by large fishing fleets which systematically overfish the waters, and curtailed fishing rights.**

AMAZING

M7
- Over 30 percent of the world's fishery yields come from illegal fishing (i.e., non-compliance with environmental and animal welfare requirements and protection zones).
- According to the WWF, two thirds of the mangrove forests in the Philippines alone have been cleared for shrimp production.

Wenn du diese Aufgaben erfolgreich bearbeitet hast, kannst du …
- … *neue Wege erläutern, mit denen die Nahrungsmittelproduktion der Zukunft durch Ernährungsmöglichkeiten aus dem Wasser sichergestellt werden kann.*
- … *die Vor- und Nachteile der neuen Wege zur Nahrungsmittelproduktion erläutern und im Hinblick auf ihre nachhaltige Wirkung in Teilbereichen bewerten.*
- … *den Fachbegriff* ***aquaculture*** *(**Aquakultur**) erklären.*

M1 **Biologists experiment with rape plants at the Leibniz Institute in Gatersleben. They simulate the climate of the future and look for high-yielding plants adapted to climate change and new cultivation methods.**

WHAT WE HAVE LEARNED

Enough for all? – population growth and food security

The development of the world population is not the same everywhere

10 The world population has been increasing very rapidly for decades. Although birth rates are falling worldwide, **United Nations (UNO) scenarios** predict a population of over nine billion by 2050. The distribution of the population depends on natural population development (births and deaths) in the individual countries and regions of the world and on **migration**. Overall, the population in the **industrialised countries** will remain the same or decrease. In developing countries and emerging economies, it is growing strongly.

How many people can the Earth support?

12 The first theories on the carrying capacity of the Earth were developed over 200 years ago.
Food production was seen as the limiting factor for the world's population. Later, it was recognised that other factors besides **population pressure**, e.g., lifestyles and meat consumption, limit the **carrying capacity** of the Earth.
The question remains: is there enough arable land to produce food for a growing population and increasing expectations?
So far, global food production has been able to keep pace with population growth, yet millions of people suffer from hunger for a variety of reasons.

Carrying capacity – what does it have to do with us?

Our **ecological footprint** is clearly too high. We are 14
"overloading" the Earth in the various dimensions of carrying capacity. The "Earth Overshoot Day" is moving forward year after year.
The goal must be not to exceed the Earth's carrying capacity in all areas – the key word here is **"sustainability"**. Almost all the countries of the world have committed themselves, with **Agenda 2030**, to pursuing 17 **Sustainable Development Goals** (SDGs). Should this succeed, it would be an important step towards preserving the carrying capacity of our planet. For this to happen, however, everyone must examine their own way of life and, if necessary, change it. Do it!

How do we achieve food security in the future?

To answer this question, one must consider several fac- 16
tors: The world population is growing, as is the standard of living in many countries. At the same time, demand for land to grow agricultural commodities (e.g., animal feed) is increasing and soil is at risk from salinisation and overexploitation as well as drought and heavy rainfall.
This means that large global increases in yields and sustainable land management are needed, as it is hardly possible or ecologically and socially justifiable to increase the amount of land under cultivation.
It is necessary to increase food production on land and in water in a sustainable way to achieve **food security** in the future.

New food options on the land

18 New agricultural land is created by clearing savannah or rainforest, for example, or by **urban farming**, high-rise farm buildings or shelves (**vertical farming**) or on fallow land. Less meat consumption also frees up land for the cultivation of food.

Intensification and thus higher productivity can be achieved through agricultural technology, adapted varieties, irrigation, and training for small farmers.

New food options in the water

Industrial fishing with factory ships works very efficiently for the world market, but it causes overfishing and is often not sustainable. Traditional fishing is less productive but conserves natural resources and provides the food base for countless people in developing countries. 20

Aquacultures reduce overfishing but are usually a burden on local ecosystems. In large cities, the production of food from water is being tested, e.g., on rooftops.

Borneo, Indonesia: To establish rice farms on peat soils in the rainforest, the government has the rainforest cleared and the peatlands dried out. This leads to high greenhouse gas emissions.

Bali, Indonesia: Farmers keep ducks in rice fields. They eat pests and weeds. This saves farmers pesticides, fertiliser and duck feed.

Inle Lake, Myanmar: Floating beds to adapt to changing water levels and protect against flooding. This secures the harvest.

M2 **Three different cultivation methods for rice. Rice is the staple food for over three billion people. Over 90 percent of them live in Asia.**

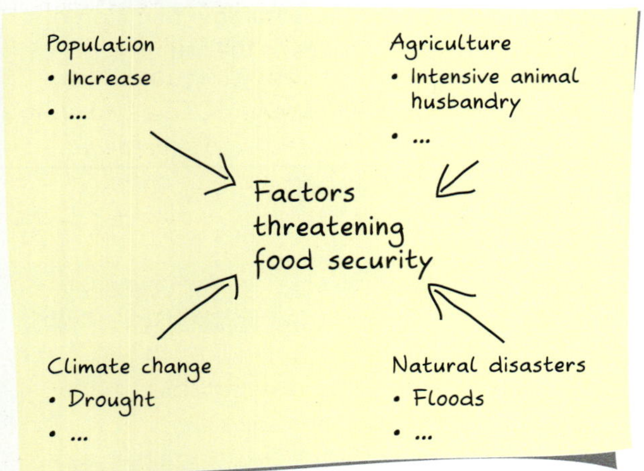

© Westermann 27902EX_1

M3 Idea for a causal network on the subject of "risks to global food security".

carrying capacity – urban farming – sustainable farming – biodiversity – aquaculture – industrial farming – industrial fishing – overfishing – food security – population pressure – meat consumption – food production – ecological footprint – earth overshoot day

M4 Technical terms – do you know them?

1 Prepare a causal network from M3. Use terms from M4.

2 Explain to your partner three terms from M4 in your own words.

3 Describe the three types of rice cultivation (M2). Compare them in terms of their sustainability.

C **4** Record a "three-minute talk" for a podcast. Topics:

A Threats to food security.

B Possibilities for increasing food security.

What factors determine population development?

What does the birth rate depend on?

Worldwide, 273 children are born every minute, 246 of them in developing and emerging countries. In India alone, 28 million children are born every year. That is more than the entire population of Australia. India will soon be the most populous country with 1.5 billion people. Why are so many people born in poor countries?

1. Name the reasons for the high birth rates in developing countries (M1 – M9). 210
2. Name ten countries around the world in which the birth rate is particularly high (atlas).
3. "Education is the first step towards a lower birth rate". Explain. (M3, M4)
4. Outline why teenage pregnancies are especially problematic. Include in your answer the future of the children and their young mothers (M5, M8, M9).
5. In your opinion, what should be done to reduce birth rates? Give reasons for your decision.

abortion	–	Schwangerschafts-abbruch
blessing	–	Segen
contraception	–	Verhütung
degree	–	hier: (Schul-)Abschluss
demographic development	–	Bevölkerungs-entwicklung
health insurance	–	Krankenversicherung
illiteracy rate	–	Analphatebenrate
life-threatening	–	lebensbedrohend
pregnancy	–	Schwangerschaft
primary school	–	Grundschule
sex education	–	Aufklärung
unemployment insurance	–	Arbeitslosen-versicherung

Helpful words and phrases for task 4:

There are different measures...
To reduce birth rates, it makes sense to ...
In my opinion ... / I think...
The most efficient / human / widely accepted way to reduce birth rates is ...
If people / women are provided with ... / taught how to... / given the chance to ...
There is a close connection between ... and ...
Teenage pregnancies must be reduced because ...

M1 Family outing to Kerala (India)

In developing countries, poor families in particular have many children. Children are important for life planning: in many developing countries there is still no state pension scheme, no unemployment insurance and no health insurance. In old age, people must be provided for by their families. Children are needed to care for their parents when they are sick or old. They must then work in the fields, for example, and earn money for necessary medicines. Up to 50 million children even have permanent, usually extremely poorly paid, jobs in industrial enterprises. **Child labour** is an important contribution to the family income of poor families. Thus, a high number of children is important for poor families. Moreover, in many regions of the world, a large number of children, especially sons, contributes to the family's reputation.

M2 Many children – a blessing!

	Proportion of illiteracy	Average no. of children per woman
Germany	c. 0 %	1.6
Japan	c. 0 %	1.4
USA	c. 0 %	1.7
China	5 %	1.7
Brazil	7 %	1.7
Peru	8 %	2.3
South Africa	14 % (2017)	2.4
India	34 %	2,2
Niger	73 %	6.9
Mali	74 %	5.9

Source: World Bank, 2020

M3 Illiteracy and number of children per woman (2018)

In many countries, women's lives are still very much focused on the role of wife and mother. Parents are therefore often unwilling to invest in their daughters' education. Many are not even allowed to attend primary school. In Africa, this is the case for 40 percent of all girls. They are therefore **illiterate** and can neither read nor write. However, a rule of thumb says: the longer the schooling and vocational training takes and the higher a woman's educational qualification, the lower the number of children. In the African country of Uganda, for example, women without a school education have an average of 7.8 children, while those with a higher education have only 3.9. It is therefore not uncommon for girls to be married off at a childhood age, about 10 million per year worldwide. For these children, it is taken for granted that they will have become mothers at least once before the age of 18. This early motherhood also contributes to keeping the birth rate high and the population growing rapidly.

M4 **The role of women.**

Every woman should be able to decide for herself when, if, with whom and how many children she wants to have. Sounds logical, doesn't it? But 218 million women in middle and low-income countries do NOT have this decision in their own hands. One in four women in developing countries cannot use contraception, even though they would like to. That is why 111 million unwanted pregnancies occur every year. Of these: 30 million unplanned births, 35 million often life-threatening abortions performed by untrained staff and in secret due to prohibitions or lack of access to safe abortion.

Source: Deutsche Stiftung Weltbevölkerung: Freiwillige Familienplanung. www.dsw.org, accessed: 13.11.2020

M5 **Sex education and contraception are not available everywhere**

INFO

M6 The **birth rate (br)** is the number of births per 1,000 inhabitants. It is twice as high in developing countries (20.1 ‰) as in industrialised countries (10.6 ‰). Germany, for example, has a birth rate of 9.4 per thousand, whereas Niger has 46.3 per thousand. Instead of per thousand (‰), the birth rate is often given as a percentage (%).

M7 **Nasoin Akhter, 15, and her husband, 32**

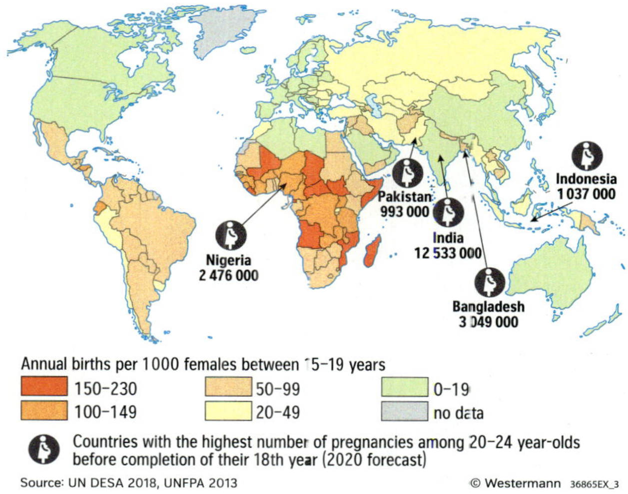

M8 **Teenage pregnancies (2018)**

Every year, 7.3 million teenagers have a child, two million of whom are under the age of 15. [...] "95 percent of teenage pregnancies are in the developing world. 20,000 girls become mothers every day. 90 percent of these pregnancies occur in partnerships or within marriage. Teenage pregnancies pose a particularly high health risk to mothers and their children." [...] Every year, around 70,000 girls die worldwide from complications during pregnancy or childbirth.

Source: APA/CSM: Kurier: Handlungsbedarf bei Teenagerschwangerschaften. www.kurier.at, Wien, 30.10.2013, accessed: 13.11.2020

M9 **Need for action on teenage pregnancies**

Wenn du diese Aufgaben erfolgreich bearbeitet hast, kannst du ...
- ... *die hohen Geburtenraten in Entwicklungsländern begründen.*
- ... *Vorschläge machen, wie man durch die Stärkung und Förderung von Frauen die Geburtenrate senken kann.*
- ... *die Fachbegriffe **child labour (Kinderarbeit)**, **illiterate (Analphabet)** und **birth rate (Geburtenrate)** erklären.*

Falling death rates...

While the birth rate in many countries is falling only very slowly, it has been possible to reduce the death rate significantly in recent decades. How has this been achieved?

1. List the measures by which the death rate could be reduced in many countries (M2 – M7). 210

(C) **2.** Show the development of the population growth rate (M1) in 217
A a line diagram.
B a bar chart.
Explain the development.

3. "There is still much to do to reduce the death rate further", says V. Kumar. Make suggestions on how the death rate in India could be further reduced (M2, M6, M7).

4. Draw a mind map on the subject of: "The causes of population growth" (pages 24 – 27).

(A) **5.** Investigate the course of the Ganges. What is its source? Where does it flow into the sea? Which countries and which large cities lie on the Ganges (atlas)?

achieve	–	erreichen
body	–	hier: Leiche
burial site	–	Begräbnisstätte
bury	–	bestatten
child mortality	–	Kindersterblichkeit
cholera	–	Cholera
disease	–	Krankheit
epidemic	–	Seuche / Epidemie
famine	–	Hungernot
fatal	–	tödlich
hunger	–	Unterernährung
living conditions	–	Lebensbedingungen
maternal mortality	–	Müttersterblichkeit
measure	–	Maßnahme
medical care	–	medizinische Versorgung
medical research	–	medizinische Forschung
polio	–	Kinderlähmung
previously	–	vorher / vormals
sewage system	–	Abwasserentsorgung
typhoid	–	Typhus

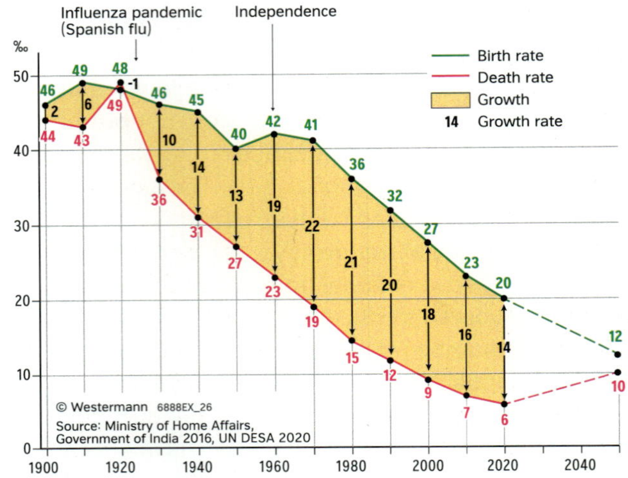

M1 Development of the birth, death and population growth rates in India (in per thousand)

Mr. Kumar, you have been working on demography, population science, for some years now. In contrast to the birth rate, the death rate has fallen very quickly in many developing countries. Why is this?
Well, reducing the birth rate is a long and difficult process. For example, one has to introduce pension and health insurance, improve education and break with old traditions. The death rate, on the other hand, could and can be lowered with a few measures.
Yes, medical research has made it possible to develop vaccines and medicines for many formerly deadly diseases, for example polio, typhoid or cholera. And while hospitals and doctors used to be available only in the big cities, today there is also medical care in the countryside.
Hygiene has also been improved. More and more villages have access to clean drinking water and a regulated sewage disposal system.
What is the status in the fight against AIDS?
Worldwide, more than 30 million HIV-infected people have died of their disease so far. Countries in southern Africa are particularly affected. But thanks to better education, the number of new infections is declining in most countries.
People are still dying of malnutrition. How do you assess the food supply worldwide?
The nutrition situation has improved considerably. But unfortunately, hunger is still a pressing problem in many regions. We must not rest until every human being is fed.

M2 Interview with Vimal Kumar, a scientist from Delhi

Between 1891 and 1901, India's population declined by 41 million due to famine, epidemics, and plagues, and in 1918, 18.5 million people died during an influenza epidemic. It was not until the beginning of the 1920s that living conditions improved. There were better medicines, and doctors also settled outside the cities.
In 1991, India had already caught up with Germany in terms of its mortality rate (12 out of 1000). The average **life expectancy** for men in 2020 was 68 years (1971: 44 years) and for women 70 years (1971: 46 years). In Germany, for comparison, the average life expectancy is 78 years for men and 83 years for women.

M3 From an internet encyclopaedia

INFO

M4 **Death rate (dr)**
The death rate is the number of deaths per 1,000 inhabitants. It is also given as a percentage (%) instead of per thousand (‰).

Growth rate (gr)
The growth rate results from the birth rate and the death rate. When there is a high birth rate in combination with a declining death rate, it leads to a high growth rate.

br - dr = gr

	br - dr	= gr
World	19‰ - 7‰	= 12‰ (1.2%)
Industrialised countries	10‰ - 10‰	= 0‰ (0%)
Developing countries	20‰ - 7‰	= 13‰ (1.3%)

Data: PRB 2020

	Death rate 2020	Child mortality 1990	Child mortality 2020	Maternal mortality per 100,000 births 2017	Life expectancy in years (at birth) 2020
Germany	12	10	4	7	81
China	7	54	12	29	77
India	6	134	39	145	69
Nigeria	12	112	102	917	54
Peru	5	94	16	88	76
Industrialised countries	10	16	5	–	79
Developing countries	7	106	44	–	71

Source: UN DESA 2020, UNICEF 2020

M5 Death rates and life expectancy in selected countries

M6 The Ganges: sacred river, drinking water reservoir, sewer, and burial site

INCREDIBLE

M7 Every 13th person in the world lives along the Ganges, for 330 million it is the only source of drinking water. At the same time, four billion litres of untreated sewage flow into the river every day. 40,000 corpses are buried in it every year - and they are not always cremated first. This is how epidemics break out again and again. On average, at least one person dies of diarrhoea every minute, more than 100,000 children alone every year. For 900 million Hindus, the Ganges is above all a sacred river. 60,000 pilgrims visit it every day and bathe in it. In 2013, during the great Hindu festival Kumbh Mela, 30 million pilgrims bathed in the holy water within 24 hours.

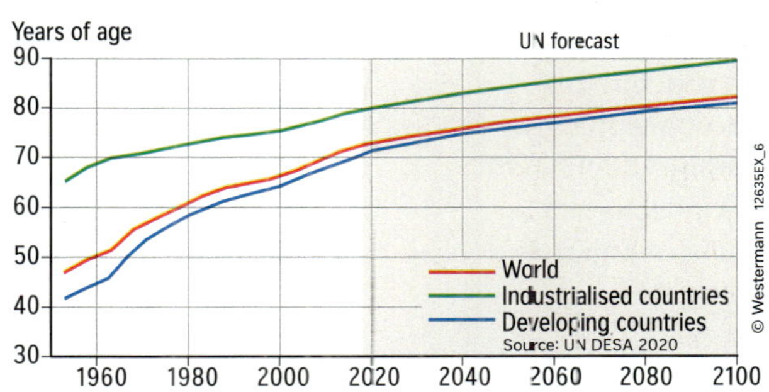

***M8** Life expectancy at birth

Wenn du diese Aufgaben erfolgreich bearbeitet hast, kannst du ...
- *... begründen, warum die Sterberate in einigen Staaten stark zurückgegangen ist.*
- *... die entscheidenden Ursachen des Bevölkerungswachstums erläutern.*
- *... die Fachbegriffe **death rate** (**Sterberate**), **growth rate** (**Wachstumsrate**), **life expectancy** (**Lebenserwartung**) erklären.*

The model of demographic transition

Population development can give an indication of a country's level of development. What connections are there between demographic development and a country's overall level of development? What conclusions can be drawn about the economic and social situation of the people?

1. Compare the demographic development of India and Germany. Name similarities and differences (M2, M3).
2. Allocate the factors influencing the birth and the death rate to the different phases of the demographic transition (M4, M7).
3. Summarise the key statements of the model of demographic transition (M4, M6).
4. Select two countries from M1 and determine their birth, death and growth rates.
   a) Allocate the country to a phase of the demographic transition.
   b) Research the socio-economic conditions of the countries (see. pp. 240/241, internet).
   c) Assess whether the model of demographic transition makes it possible to make statements about the development status of the country in this case.
5. Compare the advantages and disadvantages of a model (M1, M4, M6, M7) to those of diagrams of individual countries (M2, M3).

210

applicable – anwendbar
childcare – Kinderbetreuung
contraceptives – Verhütungsmittel
demographic transition model (DTM) – Modell des demographischen Übergangs
social scientist – Sozialwissenschaftler/in
social standards – gesellschaftliche Normen

Helpful words and phrases for task 3:

The demographic transition model shows...
Three different rates are shown ...
The lines represent the development of ... over time ...
The area between the lines shows...
Different stages can be identified...
In stage ... the birth rate / death rate is high / low / fluctuating / increases / decreases ...
The population growth in stage ... is negative / fast / slow ...

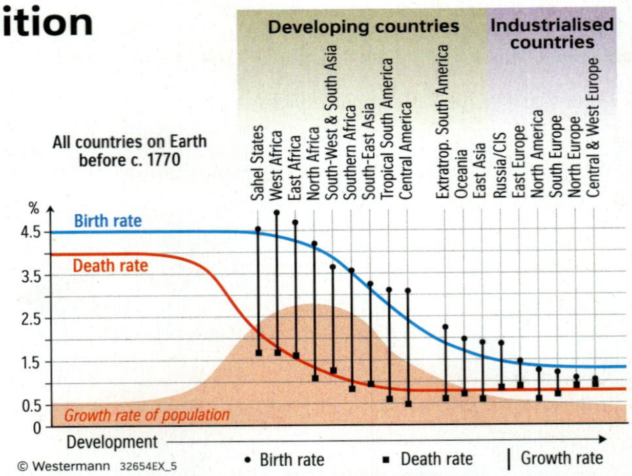

M1 Regions of the Earth in demographic transition

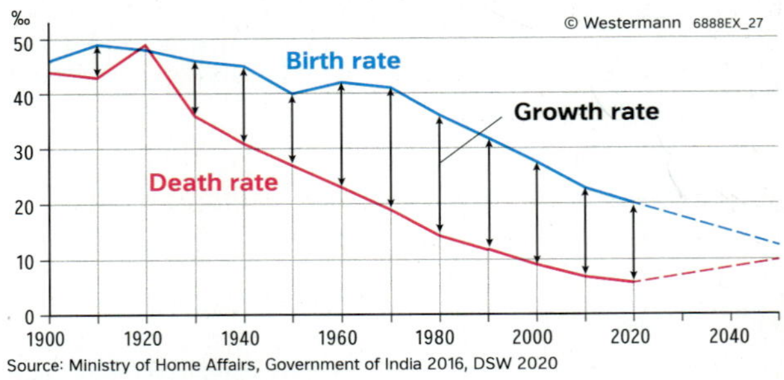

M2 The demographic development of India

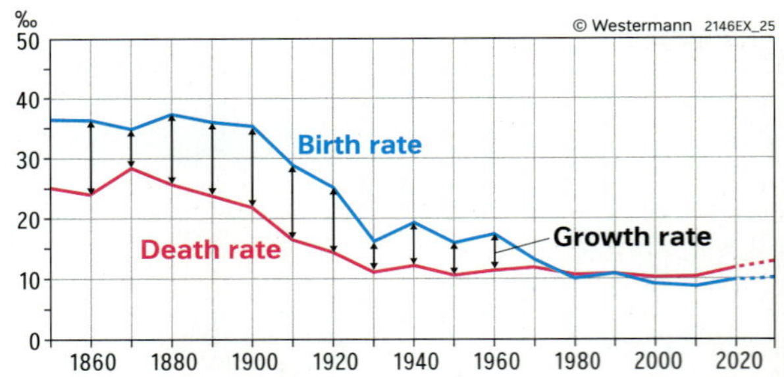

M3 The demographic development of Germany

INFO

M4 **The model of demographic transition ...**

In the 1920s and 1930s, American social scientists investigated whether regularities could be found in the population development of countries. They analysed the birth and death rates of European countries.

They used their results as a basis for creating a generally applicable model of population development – the **demographic transition model**. It is still used today to illustrate the status of a country in the development process.

M5 East Africa and West Europe: In the second and the fifth phase of the demographic transition

Factors influencing the birth and death rate:

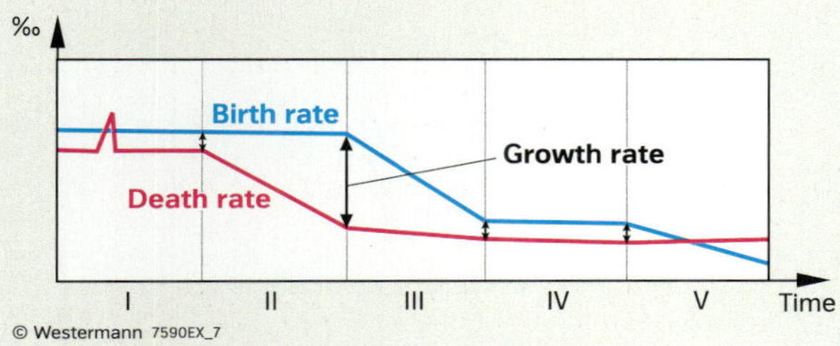

Birth rate:
Children as a necessary labour force, children as social and health insurance as well as old-age provision, importance of the number of children for the family's reputation, information about family planning options, access to contraceptives, children as a (high) cost factor, age at marriage/age at the first birth, generation gap, care services for infants and children, opportunities for women to work, social position of women.

Death rate
Health care, hygiene, food supply, spread of epidemics, plagues and famines, wars, environmental factors

M6 The model of demographic transition

INFO

M7 **... with limited validity**
However, the validity of the model is limited: it is derived exclusively from the development of European countries. Many social and cultural factors (e.g., different social standards, religious values, the position of women) that are important for population development are not included. Therefore, it cannot be assumed that this development will repeat itself in all countries of the world in exactly the same way as it did in Europe 100 - 150 years ago. In each region, therefore, the specific factors must be examined.

INTERNET

M8 The *Deutsche Stiftung Weltbevölkerung (DSW)* provides demographic information on all countries.
UN World Populiation Prospects and the-*Kreditanstalt für Wiederaufbau (KfW)* offer key data on population development and interactive world maps.

Wenn du diese Aufgaben erfolgreich bearbeitet hast, kannst du ...
- ... anhand der Stellung eines Landes im demographischen Übergang Hinweise auf seinen Entwicklungsstand geben.
- ... die Aussagekraft des Modells vom demographischen Übergang kritisch beurteilen.
- ... den Fachbegriff **demographic transition** (**demographischer Übergang**) erklären.

Regions with a shrinking population

A growing population? Germany and other countries wish it were so. Because in some industrialised countries the population has been declining for years.

Why is this so? And what are the consequences?

1. Draw a diagram in which you present and compare the birth rates and death rates in three countries of your choice (you will find sources of data on pages 240/241). 217
2. Illustrate the reasons for the fall in population in some industrialised countries (M6).
3. Comment on the reasons for the decline in births in Germany. Which reasons do you find understandable? And which ones not? (M6)
4. Interpret the cartoon M1. 209
5. List possible consequences for Germany of the low birth rate (M1, M2, M4).
6. a) Describe the development of family sizes in Germany (M7, M8).
   b) Carry out a survey in your class. Ask about the family sizes of your parents and grandparents and present the results for the class in graphical form (diagram / table). 207
7. Work in pairs. Explain to each other how historical events have influenced the birth and death rates in Germany. Use examples (M5, internet).

bourgeois	–	bürgerlich
compatibility of famliy and work	–	Vereinbarkeit von Familie und Beruf
headline	–	Schlagzeile
old-age poverty	–	Altersarmut
shrinking population	–	schrumpfende Bevölkerung
working age	–	arbeitsfähiges Alter

Helpful words and phrases for task 3:

There are different reasons for ... for example ...
In my opinion it is understandable / clear that...
The most important factors for the decline in births are ...
Many people decide against parenthood because ...
Lifestyles have changed and so ...
If we do not.... / We should ... / The state should ...

Der deutsche Rentenzahler 2050 — *Horst Haitzinger*

M1 Cartoon

We will have an **ageing population**. There will be fewer younger people of working age between 15 and 65, and many older people over 65. In Germany in 1900, there were 12.4 people of working age for every elderly person. Today, the ratio is 1:4. Scientists fear that there may not be enough workers in the future, which would lead to an economic crisis. At the same time, there would be a lack of money for pensions. The result would be widespread old-age poverty.

M2 Alarming scenarios

	2015	2050*
Germany	81.1	76.4
Italy	62.5	58.9
Japan	126.9	96.9
Russia	144.3	134.2

*Forecast

M3 Population development in selected industrialised countries (in millions of inhabitants)

Demografische Zeitbombe
Japan schrumpft im Rekordtempo

Source: ntv.de, mmo/rts: Japan schrumpft im Rekordtempo. ntv Nachrichtenfernsehen www.ntv.de, Köln, 05.07.2017.

Italy, Spain, Japan and Thailand will see populations halved by 2100, the study by Washington University's School of Medicine says

Source: The SUN: MANKIND ‚TO SHRINK'. World's population will start shrinking within 45 years, experts claim. 05.07.2020.

Geburtenrate: Warum bekommen wir immer weniger Kinder?

Source: RTL: Geburtenrate: Warum bekommen wir immer weniger Kinder? RTL interactive, www.rtl.de, Köln, 17.07.2018.

M4 Headlines

M8 A bourgeois family in Berlin around 1900

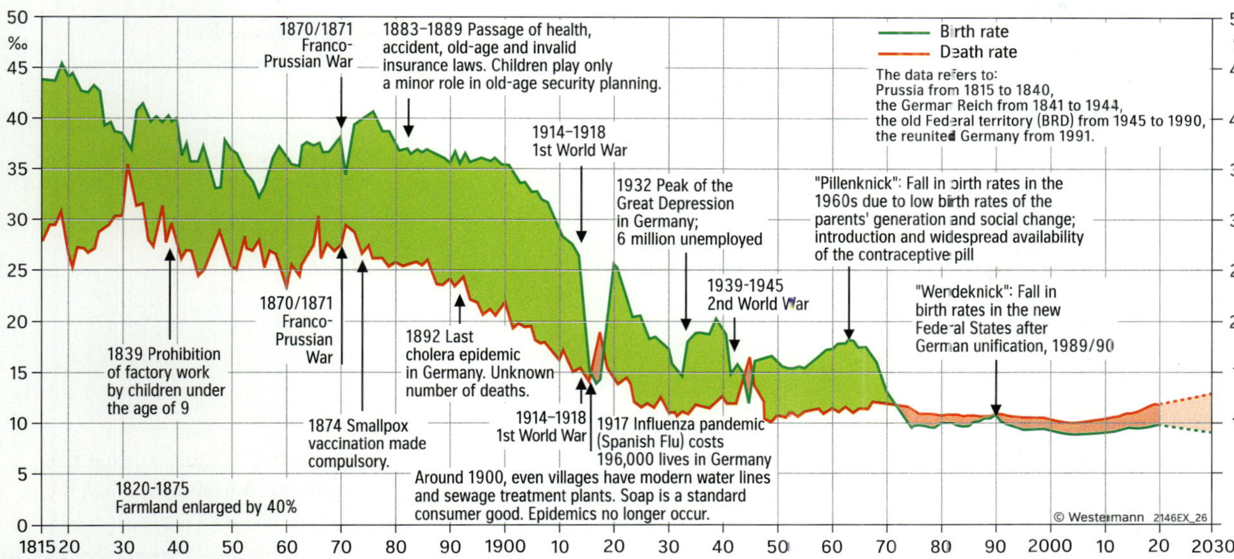

M5 The development of birth, death and growth rates in Germany

- The age of marriage and first motherhood has risen.
- Many women choose to complete vocational training or studies before becoming mothers and then pursue a career.
- A large number of children no longer means higher prestige.
- Children result in high costs. In Germany, a child costs its parents an estimated 200,000 to 300,000 euros by the time it reaches adulthood.
- Possibilities for contraception are very good.
- Family, job, and career are often hard to combine: For example, there are not enough kindergarten places.

M6 Reasons for the fall in the number of births in industrialised countries

Year of marriage	Number of children per marriage (figures per 100 marriages) 0 children	1 child	2 children	3 children	4 and more children
1910 – 1912	12	17	22	17	32
1951 – 1955	13	25	31	17	14
1970 – 1974	19	29	40	10	2
2000 – 2004	44	26	22	6	2

***M7 Family sizes in Germany**

Wenn du diese Aufgaben erfolgreich bearbeitet hast, kannst du ...
... begründen, warum die Geburtenrate in einigen Staaten sehr stark zurückgegangen ist.
... mögliche Folgen des Geburtenrückgangs nennen.
*... den Fachbegriff **ageing population** (**Überalterung**) erklären.*

The age structure of the population

A high birth rate means that there are many children in a country. A low death rate and a high life expectancy means that there are many old people.
How does the change in birth rate and death rate affect the age structure of a population?
How can the age structure of a population be represented graphically so that the structures can be seen at first glance?

1. Describe the three basic forms of population pyramids (M4).

C **2.** Historical events influence the age structure. Explain this
A based on M7.
B based on the population pyramid of 1950 (M3) with the aid of M6 on page 31.

3. Analyse the population pyramid of India in 2020 in three steps: (M1, M2) 209
a) Describe the population pyramid and assign it to one of the three basic forms,
b) Explain possible differences from the ideal form,
c) Point out future developments.

4. Work with the interactive population pyramid (M6). Report on the scenarios that you have created. 209

abort	–	hier: abtreiben
bell	–	Glocke
cohort	–	Jahrgang/Alterskohorte
contraceptive	–	Schwangerschafts-verhütungsmittel
foetus	–	Fötus
global economic crisis	–	Weltwirtschaftskrise
life expectancy	–	Lebenserwartung
jobseekers	–	Arbeitssuchende
surplus	–	Überschuss
urn	–	Urne
wide base	–	breite Basis

Helpful words and phrases for task 1:
Differently shaped population pyramids represent ...
The expansive / stationary / constrictive pyramid shows a shrinking / stationary / growing population ...
It indicates a young / old / ageing / stable population ...
Most countries can be matched to

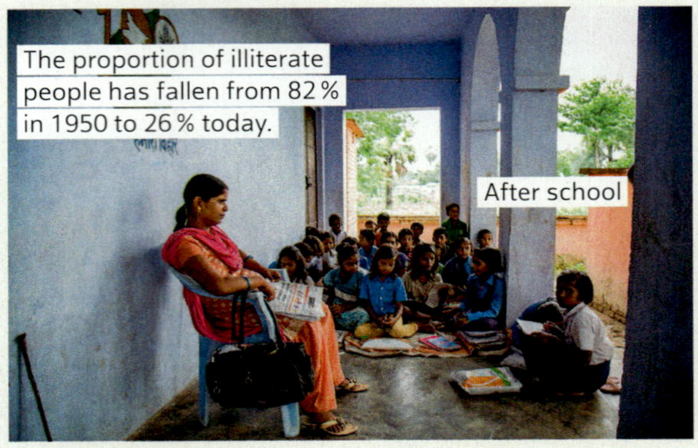

India (2020) – a land of children
The pyramid has a wide base. 10 percent of the population is younger than five and 26 percent younger than 15. To enable all six-year-olds to attend school, 1,200 new schools would have to be built annually and 300,000 additional teachers would have to be qualified. Only 6 percent of the population is older than 65.

India (2020) – a land of job seekers
67% of the population is of working age. They would be able to work. But already today there are not enough jobs, and every year 13 million young people leave school in search of a job or a place at university.

India (2020) – a land of males
Because, in the tradition of many Indian families, sons are worth more than daughters, female foetuses are often aborted. The surplus of boys and men can be seen in the age-gender pyramid.

M1 Facts about India's population, 2020

1950, 2020 and **2100** (forecast)

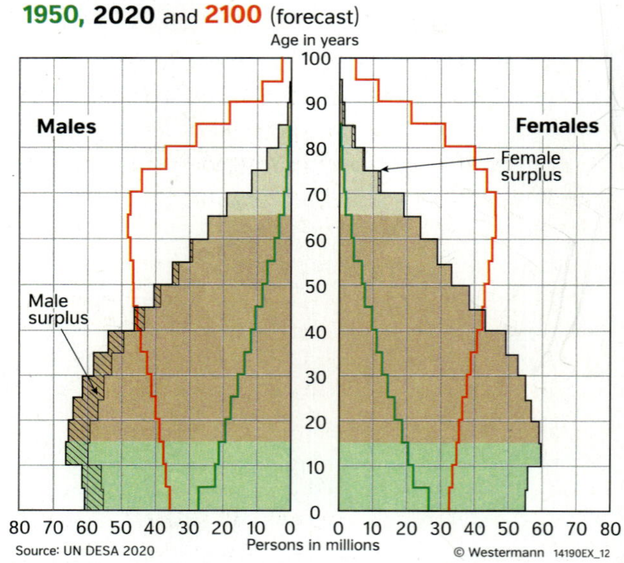

M2 India's population pyramids

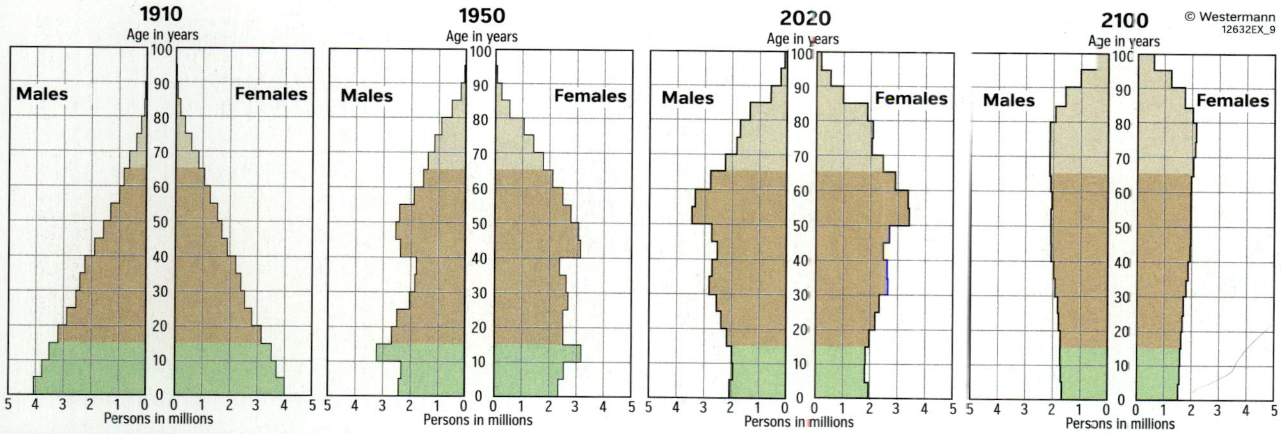

M3 Population pyramids in Germany over time: In 1910, the population pyramid resembled that of a developing country. By 1950, major differences were apparent. Low birth rates during the World Wars and the global economic crisis of 1929 – 1932 lead to substantial changes.

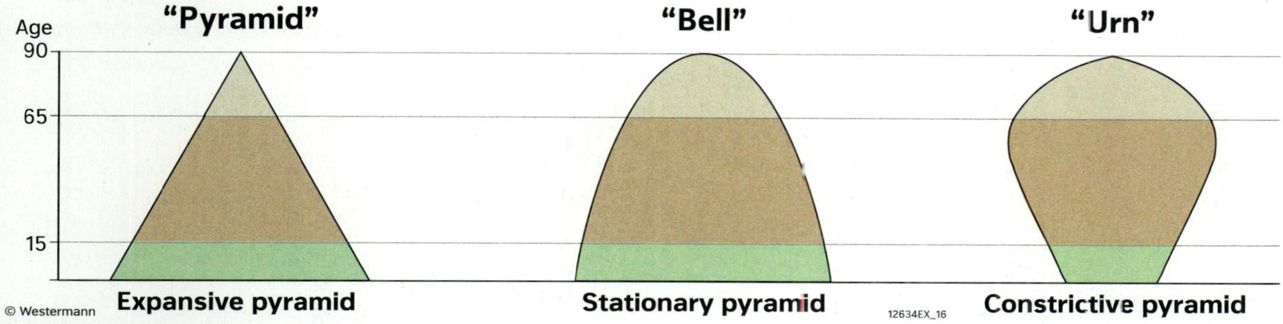

M4 There are three simplified basic forms of population pyramids. Of course, the actual developments in each country are different, but most can be assigned to one of these three forms.

INFO

M5 Not all age cohorts are equally represented in the population of a country or region. In some cohorts many people are born, in others few. In some cohorts more people die, in others fewer. In addition, people migrate in and out of the region. Thus, every population has a certain **age structure**. This can be well represented in an **age-gender pyramid**. It shows the age structure of a population and the proportion of women and men in the individual cohorts.

INTERNET

M6  Unter *www.destatis.de/bevoelkerungs-pyramide* das Statistische Bundesamt eine interaktive Bevölkerungspyramide Deutschlands. Hier kannst du dir die Veränderung der Pyramide bis 2060 verdeutlichen, Altersgruppen einfärben und abgrenzen sowie verschiedene Szenarien simulieren.

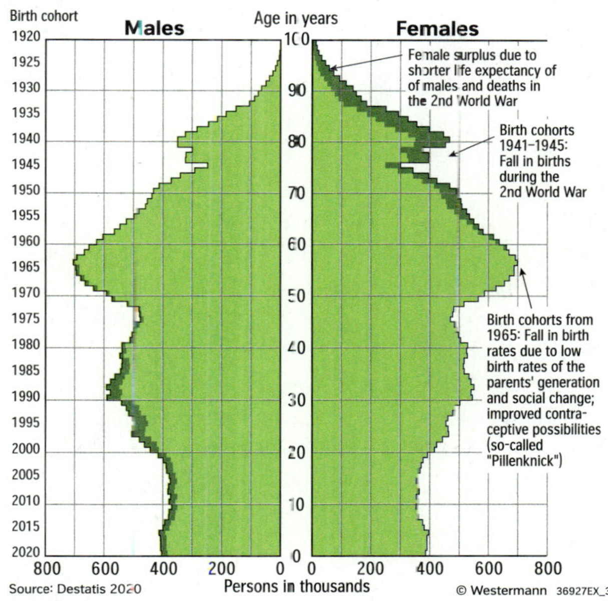

M7 Historical events influence the age structure of the population (Germany, 2020)

Wenn du diese Aufgaben erfolgreich bearbeitet hast, kannst du …
- *… den Altersaufbau der Bevölkerung eines Landes oder einer Region erläutern.*
- *… unterschiedliche Typen der Altersstruktur einer Bevölkerung beschreiben.*
- *… die Fachbegriffe **age structure** (**Altersstruktur**) und **age-gender pyramid** (**Bevölkerungspyramide**) erklären.*

Family planning: Fewer children! – More children!

In India, Nigeria or Indonesia, the birth rate is very high. In Japan and Germany, the birth rate is very low. Politicians and scientists are in favour of influencing the birth rate to improve the future prospects of particular countries. What are the possibilities for lowering the fertility rate? What measures can increase it?

1. Describe the concrete measures and the social framework conditions that should lead to a reduction in the birth rate. (M2 – M8).
2. Name three measures or framework conditions, that seem to you to be particularly important or meaningful (M1, M2).
3. Point out which framework conditions could lead to an increase in the birth rate. (M2).
4. Draw a mind map on the subject of population policy / family planning. 199

affordable housing	–	bezahlbarer Wohnraum
childcare facilities	–	Kinderbetreuungsstätten
contraceptive	–	Schwangerschaftsverhütungsmittel
demographer	–	Bevölkerungswissenschaftler/in
equal	–	gleich
financial support	–	finanzielle Unterstützung
fine	–	Strafe/Strafgebühr
general conditions	–	Rahmenbedingungen
one-child-policy	–	Ein-Kind-Politik
population policy	–	Bevölkerungspolitik
publicly awarded	–	öffentlich ausgezeichnet
sexually transmitted disease (STD)	–	sexuell übertragbare Krankheit
sterilise	–	sterilisieren
street vendor	–	Straßenhändler/-in
treat	–	behandeln

Helpful words and phrases for task 3:

How can the birth rate be increased ...
Framework conditions for a higher birthrate are e.g. ...
If women / people / families are offered ...
In order to change the attitude of young people / women / the state must / society must...
More money should be spent on ...
To combine career and family, people must be offered / given ...

Vimal Kumar, demographer (India)

The goal should be to reduce the birth rate! People must be convinced to practise ***family planning****. In Indonesia, large posters advertise: "Two children are enough"; here in India, puppet shows parade through the countryside to convey ideas of family planning. But the conditions must also be right: the state must provide old-age pensions, health and unemployment insurance, women and men must be treated equally – with a legal right to education. And contraceptives must be freely available everywhere. Until 2015, China applied very strict measures. One-child families received an annual premium. The only child was also given preference in health care, kindergarten, school selection and job allocation. Couples who had themselves sterilised were publicly rewarded. These measures worked: the birth rate dropped drastically. Since 2021, however, three children per family have been allowed again in China - for fear of an ageing population.*

M1 Influencing the birth rate – statement 1

Lena Schriedendahl, politician (Germany)

The goal should be to increase the birth rate! Good framework conditions must be created for families with children: for example, affordable housing as well as sufficient day-care places and childcare facilities so that mothers can also pursue a career. In addition, families must be supported financially. For most countries, it would be sufficient if each couple had an average of 2.1 children. Then at least the population would not decrease but remain the same.

M2 Influencing the birth rate – statement 2.

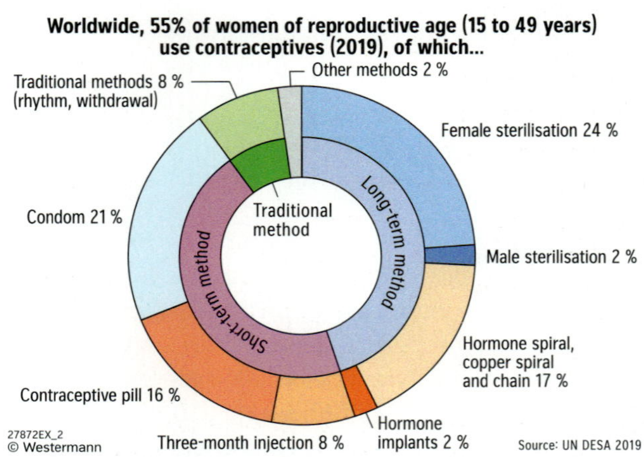

M3 Access to contraceptives for women, 2020

M4 **In Kenya (East Africa), women and men inform themselves about family planning options**

M7 **Mozambique: Sale of condoms ("JeitO") by street vendors**

The card symbols are assigned to thematic areas:

- ♣ **Clubs** indicates the dangers of sexually transmitted diseases / STDs;
- ♠ **Spades** names the family planning laws;
- ♥ **Hearts** deals with health and hygiene during sex;
- ♦ **Diamonds** presents contraceptive methods.

M5 **China: A card game used for family planning: The first 6,000 were sold out in the family planning offices in Beijing after a few hours**

"Governments worldwide must ensure that young women and men have access to education and sex education, as well as health care and modern contraceptives. [...] In addition, there must be investment in research and development for new methods of contraception that are better adapted to the needs of young women in developing countries in particular. Important decision criteria are that they must be safe, discreet, easy to use, inexpensive, widely available and with as few side effects as possible. [...] Current population studies emphasise for good reason that education and access to contraceptives are the most important factors in slowing population growth. They show that there is a clear correlation between formal educational attainment, access to contraceptives and the average number of children per woman."

Source: DSW: Weltverhütungstag 2020: Fast jede zweite junge Frau in Entwicklungsländern kann nicht verhüten. www.dsw.org, 25.09.2020, accessed: 13.11.2020

M6 **Jan Kreutzberg, from the German Foundation for World Population (DSW) on the occasion of World Contraception Day 2020 (annually on 26 September)**

AMAZING

M8

- If there were free access to contraceptives, there would be only 35 million unintended pregnancies a year instead of 111 million.
- In 2020, the COVID pandemic led to more than seven million more unintended pregnancies worldwide due to more difficult access to contraception.

Wenn du diese Aufgaben erfolgreich bearbeitet hast, kannst du ...

... Maßnahmen erläutern, die die Geburtenrate senken, und Maßnahmen, die sie steigern.

*... den Fachbegriff **family planning** (**Familienplanung**) erklären.*

WHAT WE HAVE LEARNED

What factors determine population development?

What does the birth rate depend on?

24 Especially in developing countries, and there especially among poor families, there are numerous reasons to have as many children as possible. They can help with work and are often **child labourers**. They provide for the parents' old age and replace health insurance. In addition, in many regions a large number of children ensures high prestige. However, the **birth rate** is also very much dependent on the role of women in the respective country. For many countries with a high birth rate, it is typical that women marry very early (teenager pregnancies), that many are insufficiently educated (often **illiterate**). They also often have poor access to contraceptives.

Falling death rates...

26 In contrast to the birth rate, the **death rate** has fallen rapidly in many countries and average **life expectancy** has increased. This has increased the **growth rate**, which means that the population of these countries is now growing faster. The decrease in the death rate has been achieved mainly through better medical care, vaccinations, improved hygienic conditions and an improved food supply. Nevertheless, there are still many deficits in these areas worldwide. If these are eliminated, the death rate will fall even further, and life expectancy will increase.

The model of demographic transition

28 The typical development of birth and death rates is reflected in the **demographic transition** model. Although this model is not entirely uncontroversial, the birth, death and growth rates can serve well as indicators of a country's development status.

M1 Cartoon – "Demographic Balance"

Where the population is shrinking

In some industrialised countries, the population is even falling. The birth rate there has fallen below the death rate. The reasons for this are manifold: the age of first motherhood is high because many women first learn a profession, complete their education and then start work. It is often difficult to combine having a family with children and working at the same time. For these reasons, the number of children in families is also lower than in the past. Thus, the proportion of old people in the population is increasing. The result is an **ageing population**. As a result, fewer and fewer young people must care for more and more old people. 30

The age structure of the population

The **age structure** of the population of a country or region can be well represented in a **population pyramid**. This special type of diagram shows how many people there are within a cohort and how many of them are male and female. In this way, one can draw conclusions about the birth rate or life expectancy. Historical events also influence the birth and death rates, and this can also be seen in the population pyramid. 32

Family planning: Fewer children! – More children!

Most countries try to control the birth rate with different measures: either to promote births or to reduce the birth rate. 34

In the framework of **family planning**, the main aim in countries with a rapidly growing population is to reduce the number of children per family. To achieve this, the reasons that lead to a high birth rate must be eliminated as far as possible.

In countries with shrinking populations, incentives are created to start a family. The most important thing is that governments in all countries create the appropriate framework conditions, e.g. laws and financial support.

M2 Information event on possibilities for family planning in Uganda

TRANSFER

Population development in figures

1 Obtain an overview of the goals of the respective organisation and the information it offers on the topic of population.

DESTATIS
Statistisches Bundesamt

→ *Themen* → *Bevölkerung* → *Bevölkerungsvorausberechnung*: you will find the animated population pyramid for Germany.

2 Name three years in which the population pyramid can ideally be assigned to one of the three basic forms.

C 3 Different scenarios: Open the animated population pyramid. Create three scenarios each (low, moderate, high) and report on the main changes in the population pyramid if you make different assumptions for the following factors
A Birth rate.
B Life expectancy.
C Net migration.

DSW youth can
Deutsche Stiftung Weltbevölkerung

The country database is a GIS that allows you to display desired population data for each country on the map or in the form of a table.

C 4 Make a chart for one country per continent with data on the following indicators:
A *"Bevölkerung aktuell", "Bevölkerung 2050", "Einwohner pro km² Anbaufläche"*
Calculate the number of inhabitants per km² of cultivated land in 2050 (assuming an unchanged cultivated area)
B *"Lebenserwartung m/w", "Sterberate", "Säuglingssterblichkeit"*
Name possible reasons for the values.

INTERNET

M3
- Deutsche Stiftung Weltbevölkerung (DSW)
- Population Reference Bureau (PRB)
- Statistisches Bundesamt (Destatis)
- United Nations (z. B. World Population Prospects, United Nations Population Fund)

UN DESA

→ Interactive Data:

C 5 Open a chart for selected countries for the years 1950 to 2100 in ten-year steps, for
A population development (total population, both genders).
B life expectancy (total, both genders)
C fertility rate (number of children per woman).

→ Graphs and Profiles → Demographic Profiles → Population Pyramids

C 6 Open an (interactive) population pyramid. Report on the changes between 1950 and 2100.
A China and Japan.
B Nigeria and Rwanda.
C India and Singapore.
D Brazil and Italy.

→ Probabilistic Projections → Population →Total Population/ Age 0 – 14/ Growth Rate

C 7 Open graphs of the projected developments. Report on the different scenarios from 1950 – 2100.
A World.
B India and a country of your choice.
C Germany and a country of your choice.
Report on the changes.

ORIENTATION

Population growth and megacities

Population growth is very unevenly distributed worldwide. The same applies to the growth of large cities and agglomerations.

1. Name ten countries each with high, low and negative population growth (M2).
2. Describe the cartoon M5 and assign it to a group of countries. Give reasons. 209
3. Name five regions in Germany that are growing and five that are shrinking. (M4, atlas)
4. Make a chart with the data for 1970, 2000 and 2035 showing the ten largest cities, their ranking, their number of inhabitants, their country and their continent. 201
5. (C) **A** Describe how the number of megacities worldwide has changed (M1, M2). 201
   B Describe how the distribution of megacities worldwide has changed (continent / regions). 201
   C Compare the development of the number of inhabitants of two megacities in countries with differing rates of population growth.
6. Describe the shares of different city sizes in worldwide urban growth (M3). Assess the assertion: "Urban growth mainly takes place in small cities".
7. (A) Prepare a profile of a megacity of your choice.

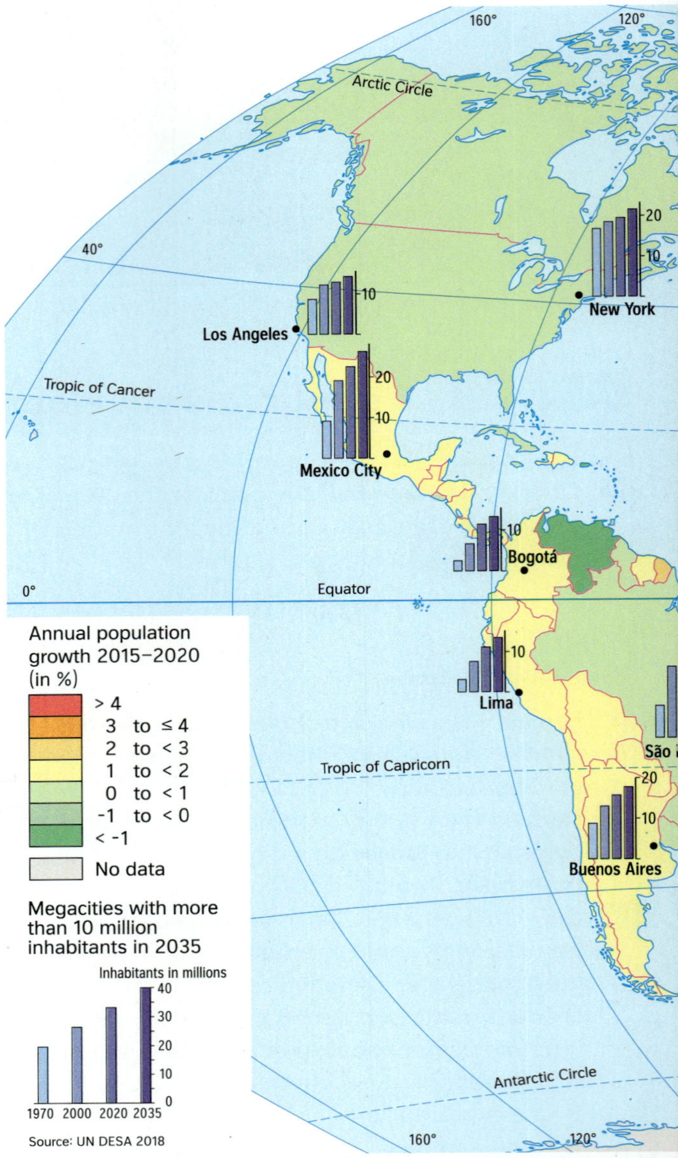

M2 Average population growth in the countries of the world

agglomeration	–	Agglomeration/ Ballungsraum
megacity	–	Megastadt
urban growth	–	Städtewachstum
urban hinterland	–	städtisches Umland
urban population	–	Stadtbevölkerung

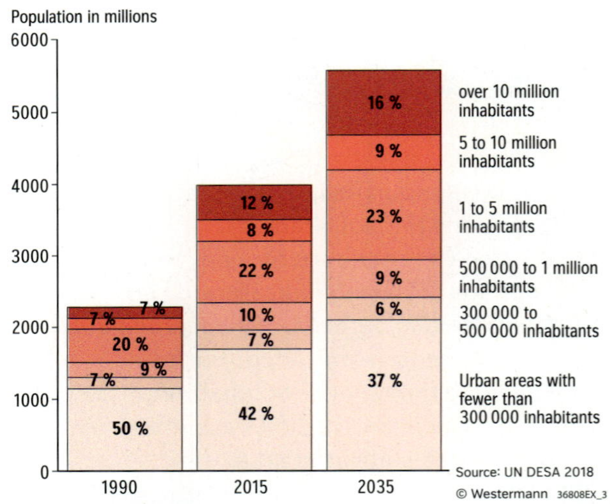

M3 Urban population of the world and shares by city size

INFO

M1 Megacities

Cities with more than 10 million inhabitants are called megacities. As the population grows, so does both the population density and the area of the cities. They spread further and further out into the urban hinterland.

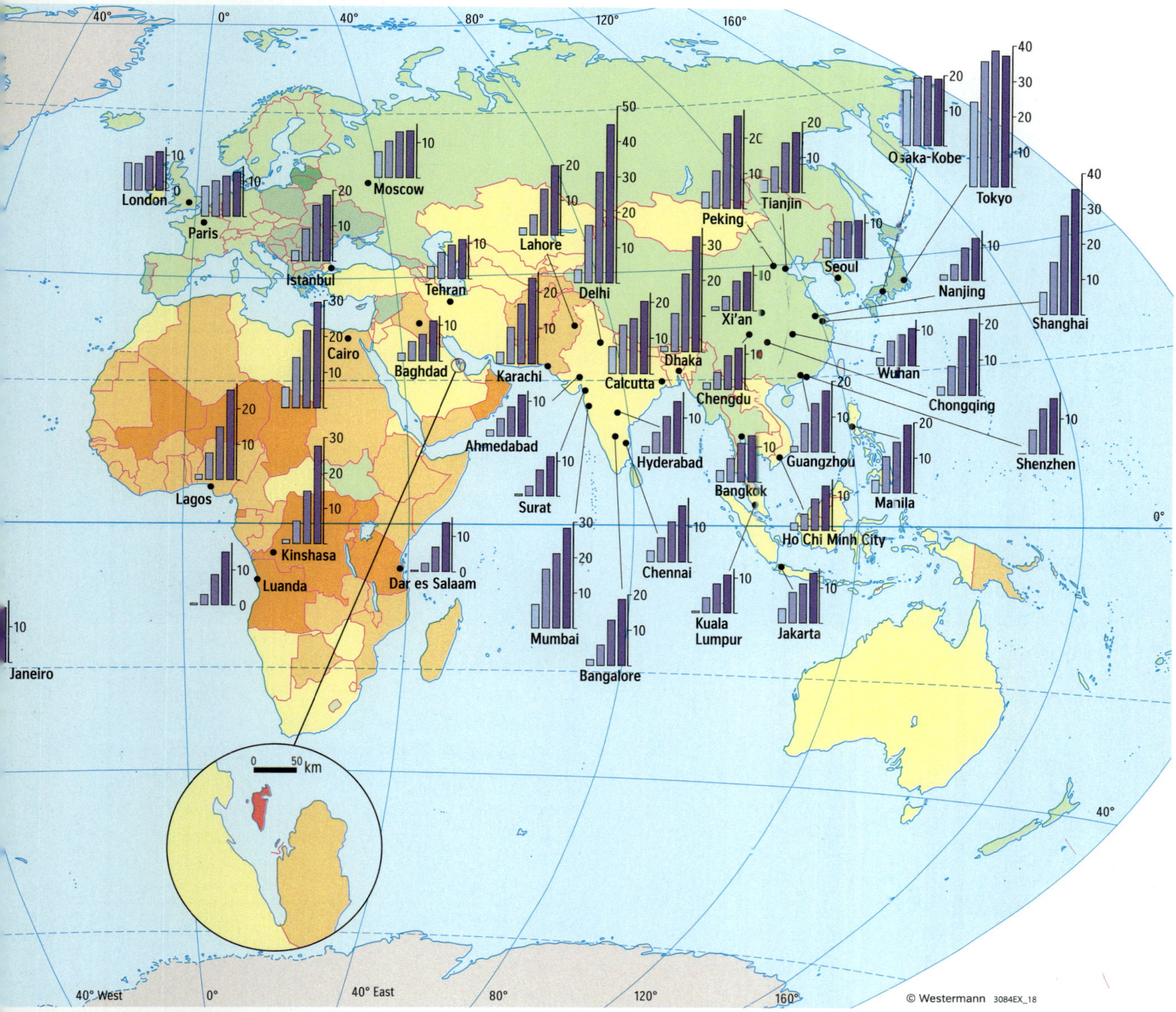

in the years 2010 to 2015 and megacities that will have ten million inhabitants and more in 2035

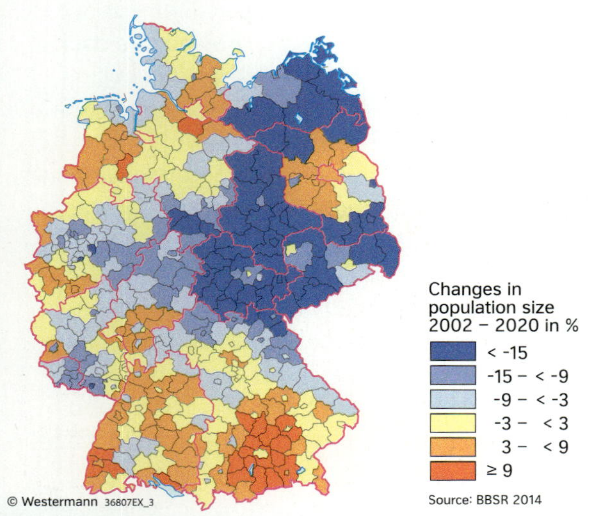

M4 Regions with growing and shrinking populations in Germany

M5 Cartoon

Wenn du diese Aufgaben erfolgreich bearbeitet hast, kannst du ...
- ... weltweite Unterschiede im Bevölkerungswachstum beschreiben.
- ... die Veränderung von Zahl und Verteilung von **megacity** (**Megastadt**) beschreiben.

To go or to stay? – the significance of migration for regions of origin and destination

Migration is normal

Migration has always been part of human life: for example, in the 19th century millions of Europeans emigrated to the USA and said good-bye to Europe forever. Refugees in small boats risking their lives in the Mediterranean to get to Europe are only a small part of migration. Most people leave their place of residence by ordinary means of transport, planes or buses, and a great many go to the next town or neighbouring country. Why do people change their place of residence? What are the different forms of migration?

1. Compare the different examples of migration (M2, M4 – M11):
   a) Make a list summarising the reasons that are mentioned in the examples, and find headings.
   b) Point out differences between the examples.

2. Describe the migration to Germany in the years 2000 to 2019 (M1).

3. Work with the atlas (map: Erde. Migration). Make lists:
   a) Countries and regions with high inward migration, 201
   b) Countries and regions with high outward migration, 201
   c) Countries and regions that cut themselves off from migration. 201

4. Explain some migration flows using examples from the map (atlas).

ancestors	–	Vorfahren
backward	–	rückständig
destination	–	Zielregion
emigrant	–	Auswanderer/-in
geriatric nurse	–	Altenpfleger/-in
homeland	–	Heimatland
immigrant	–	Einwanderer/-in
internal migration	–	Binnenmigration
migrant	–	Migrant/-in
origin	–	Herkunft
persecution	–	Verfolgung
protection	–	Schutz
refugee	–	Flüchtling
resettler	–	Aussiedler/-in
skilled workers/ professionals	–	Fachkräfte

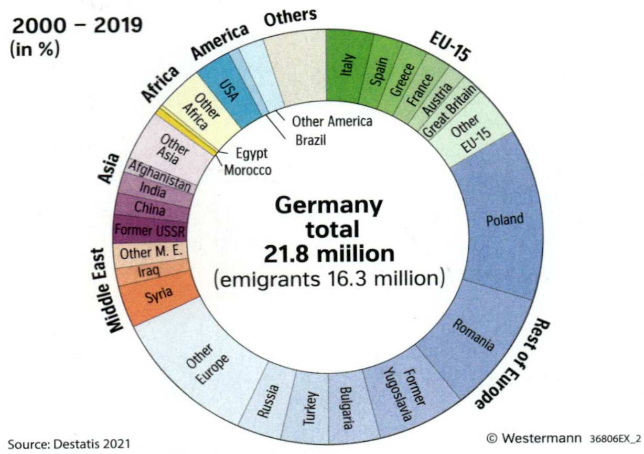

Source: Destatis 2021

© Westermann 36806EX_2

M1 Countries and regions of origin of new immigrants to Germany in the years 2000 – 2019

I grew up in Anatolia. I came to Germany in 1970 as a so-called guest worker. At that time, the offer to work in Germany came just at the right time for me, because at that time I felt that life in my birthplace was backward, and you couldn't earn much. My wife came later, and our children were born in Essen. I first worked in construction, then in an industrial company. Now I am retired and already have two grandchildren, one of whom goes to kindergarten and the other to grammar school.

M2 Umut P., 71 years old

INFO

M3 **Migration** is the movement of people with the aim of changing their place of residence. Emigration means leaving a country or region, immigration means moving to a country or region. Today, more than 230 million people live outside their home country. Many more people are on the move within countries. Internal migration is migration within a country, international migration is migration across borders.
Refugees are people who have had to leave their country or region because of war, political or religious persecution, for example. Refugees enjoy special international protection. In general, the term is understood to mean people who have been forced to leave their homeland.

I fled from Aleppo (in Syria) with my parents in 2015 to escape the war. We came to Germany via Greece. At first, we lived in a refugee home. I am still at school, hardly need any German lessons anymore, and will graduate - I'm proud of that!

M4 **Rafat H., 16 years old**

My father works for a big Japanese company and has been sent to Europe. My sister and I go to the Japanese school here in Düsseldorf. Next year we're going back to Osaka.

M5 **Masaki T., 13 years old**

In 1981, I came to Germany with my parents as so-called "Aussiedler" (resettlers) from what is now Russia. My parents say today that we are better off in Germany than we were in Russia. My family has German roots.

M6 **Franz P., 49 years old**

I am a marine architect. Two years ago I got an offer to move to Norway and work for a company that builds oil platforms. My salary is here is high, and they take great care of me – a language course, housing, etc. It was made very easy for me to get started. I plan to stay for about five years and then return.

M7 **Petra O., 38 years old**

After my studies in Alicante, I had no chance of finding work, despite my good degree. I am a telecommunications engineer. Now I live in Lünen and work for a small electrical company that urgently needed skilled workers. It's getting better and better with the language! I like it here and maybe my boyfriend from Spain will join me soon.

M8 **Nina G., 29 years old**

I went to primary school in New Zealand. At that time, my mother had taken a particularly good job there. She is a doctor and works in the hospital. New Zealand had been my parents' dream island for a long time. Actually, we only wanted to stay for three years. It turned out to be five, and when we got back, I started school in grade 6.

M9 **Ella B., 15 years old**

I graduated from high school in Bucharest last year and am now spending a year as an au pair in Dortmund. I was very curious about my German family. I want to improve my language skills because I probably want to study in Germany.

M10 **Ioana R., 19 years old**

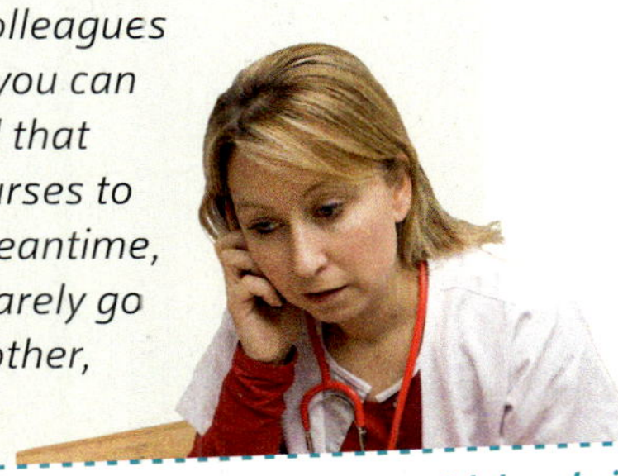

In Poland, I last lived with my mother in Warsaw. I am a trained nurse. My colleagues and I quickly realised that you can earn more in Germany and that there is a great need for nurses to care for the aged. In the meantime, I have settled in and only rarely go back to Poland with my mother, maybe three times a year. The separation is very hard every time!

M11 **Beata D., 36 years old**

Wenn du diese Aufgaben erfolgreich bearbeitet hast, kannst du ...

- ... Gründe für Migration erläutern und Arten von Migration unterscheiden.
- ... weltweite Migrationsbewegungen beschreiben.
- ... die Fachbegriffe **migration** (**Migration**) und **refugee** (**Flüchtling**) erklären.

Causes and consequences in the countries of origin

What are the causes of migration? What are the effects of migration in the migrants' countries and regions of origin?

1. Reasons for migration are referred to as push and pull factors:
a) Explain either factor by using the example of Senegal (M1 – M3),
b) Allocate your results from task 1a the push and pull factors in M4.

2. Locate the destination countries and regions of Senegalese migrants. List possible reasons for choosing these countries and regions (M7).

3. In a chart, arrange the arguments (M6) as "positive" and "negative" for the country of origin. 207

4. Present the effects of brain drain and remittances in a causal network.

Ⓐ **5.** Work with a partner. Write a dialogue between two schoolchildren – one with relatives abroad and one without. The effects of migration on the families should be made clear.

abroad – im Ausland
brain drain – Braindrain (wörtlich: Gehirn-Abfluss)
cause – Ursache
consequence – Folge
country of origin – Herkunftsland
creditor – Gläubiger/-in
fishing vessel – Fangschiff/Fischerboot
higher professional education – höherer Berufsabschluss
large landowners – Großgrundbesitzer/-in
nursing staff – Pflegepersonal
remittance – Rücküberweisung
small scale agriculture – kleinbäuerliche Landwirtschaft
support – Unterstützung

Helpful words and phrases for task 5:

I come from a family that ...
My brother / sister / father / mother ... went abroad ... to work in ...
None of my family has gone abroad...
We miss him / her ... / Going abroad can be dangerous ...
He / She / They send money that we use for ...
I can go to school / we have a better house because ...
With the money he / she / they send, we can ...
We have decided to stay ... because ...

M1 **Large fishing vessels from China, Russia and European countries fish directly off the coast of Senegal. Fewer and fewer fish remain for the local fishermen with their traditional boats.**

Living abroad:
24 % of Senegalese people with a higher professional qualification, 51 % of doctors trained in Senegal and 27 % of nursing staff.
70 % of households have a relative living abroad.

- Remittances from Senegalese migrants:
  2005 – US$ 789 million
  2010 – US$ 1479 million
  2015 – US$ 1758 million
  2020 – US$ 2286 million (= 9.4 % of GDP)
- Senegalese people living in Germany remit an average of around 200 euros per month to their families,
- In Dakar, 85 % of households with migrants have at least 343 euros per capita and year at their disposal. But this is only true for 69 % of households without migrants in the family,
- In rural areas, there is often little improvement in income. This is explained by the fact that the rural population often relies on migrants for support.

M2 **Consequences of migration: the example of Senegal**

INFO

M3 **Brain drain** – means the loss of well-educated people (skilled workers, academics) for the country of origin or the home region due to migration to regions with better working, living or earning opportunities.

Remittances are monetary payments made by migrants, for example, to their family back home. These mostly smaller amounts add up to huge capital flows.

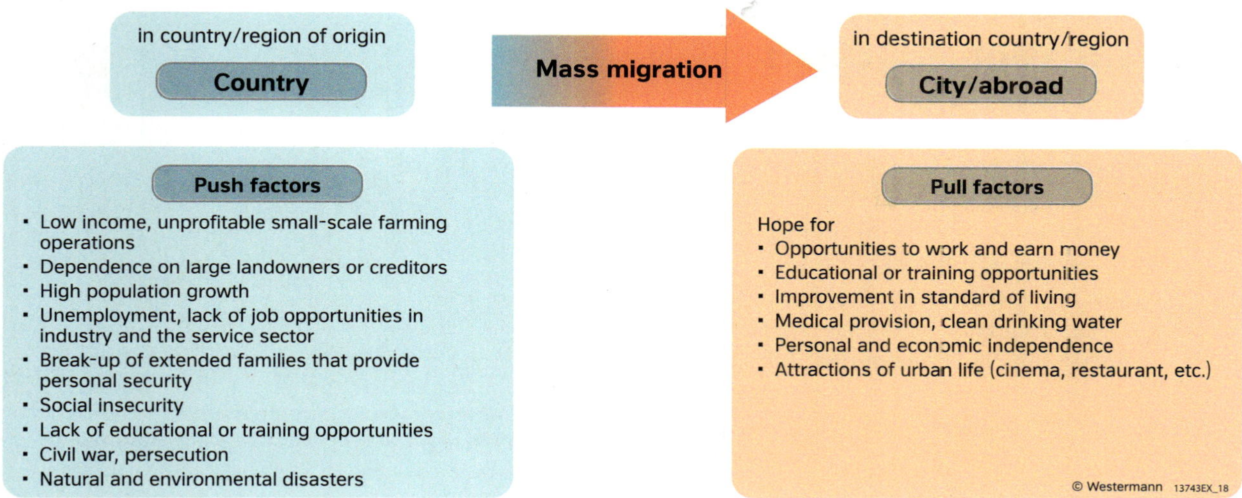

M4 **The model of push and pull factors**

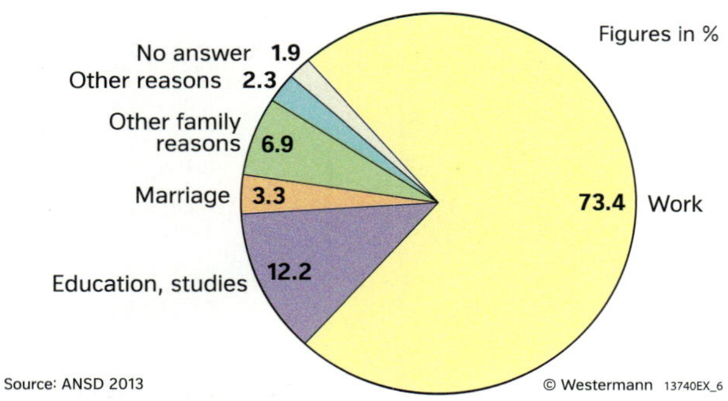

M5 **Reasons for emigration from Senegal**

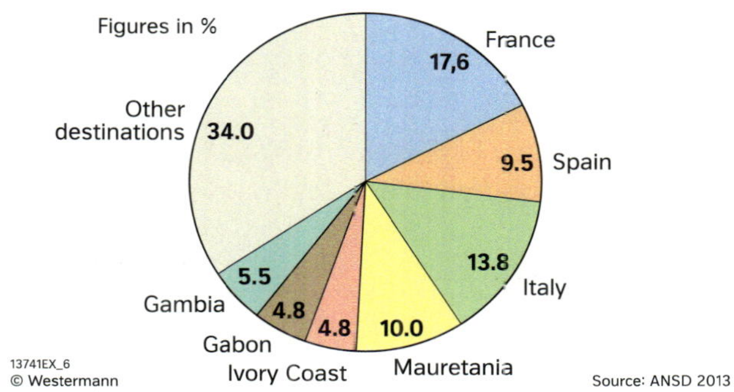

M7 **Destination countries of Senegalese migrants**

Quite a few recipients of remittances in their countries of origin use them to set up small businesses, enable their children to attend school and pay for medical care.

The remittances secure the livelihood of the families and raise the standard of living. However, they also contribute to increases in prices and thus the cost of living. All those who do not receive remittances from abroad suffer from this.

Associations and networks of migrants often get involved in their home countries with charitable projects or they become economically active. In this way, they strengthen the local economy as a whole and the infrastructure.

Often, migrants who are in employment return because they can and want to build a livelihood in their home country. With the knowledge and experience, the capital and contacts that they bring with them, they can contribute to the development of their home countries.

The emigrants have been trained in their home countries. Considerable resources were spent there for this. With emigration, the countries lose their investment in the education of the people.

The majority of financial transfers disappear into consumption, education, health care. They are rarely invested in a way that creates new income. In crises, dependency becomes apparent because less money is sent.

The migrants' remittances accustom the recipients to the fact that others pay for their lives.

It is above all the educated who emigrate. It is them that are especially important for the development and functioning of these countries.

M6 **The effects of migration on the countries of origin are controversial. Here are various arguments.**

Wenn du diese Aufgaben erfolgreich bearbeitet hast, kannst du ...

- *... Ursachen für Migration nach Push- und Pull-Faktoren gliedern und erläutern.*
- *... Auswirkungen von Migration in den Herkunftsländern und -regionen insbesondere durch Rücküberweisungen und Braindrain erläutern.*
- *... die Fachbegriffe **push & pull factors** (**Push-/Pull-Faktoren**), **brain drain** (**Braindrain**) und **remittance** (**Rücküberweisung**) erklären.*

Causes and consequences in the destination countries

In many regions of the world, immigration is limited by entry regulations. Attempts are also made to secure the borders with fences, walls and other measures. On the other hand, there is competition among industrialised countries for highly skilled migrants. What are the effects of immigration in the countries of destination of migration? How are consequences of migration to be assessed?

1. Compare the age pyramid (M2) with the age pyramid of Germany (p. 33, M3). 217
2. Explain the reasons why many people assume that Germany and many other industrialised countries are dependent on immigration (M1, M2, p. 33, M3).
3. Explain why it is possible to speak of a "gain" from immigration in the destination countries M1 – M3).
4. Work with the atlas (map: El Ejido, Almería): The province of Almería went from being an emigration area to one of the richest areas in Spain due to the construction of greenhouses. In the "mar de plástico" (plastic sea) around El Ejido, about 1.5 million tonnes of vegetables and fruit are produced annually. 201
Explain the natural preconditions, cultivation techniques and living and working conditions.
5. "In the 'plastic sea' of Almería some people wonder, if the 'illegality' of immigrants really is unwanted." Discuss this statement. 205

agricultural workers' union	–	Landarbeitergewerkschaft
construction sector	–	Bausektor
entry regulations	–	Einreisebestimmungen
exposure	–	Belastung
fence	–	Zaun
fence in	–	einzäunen
in descending order	–	in absteigender Reihenfolge
lung cancer	–	Lungenkrebs
nationality	–	Staatsangehörigkeit
protective gloves	–	Schutzhandschuhe
restricted	–	begrenzt
skin cancer	–	Hautkrebs
wages	–	Löhne
working conditions	–	Arbeitsbedingungen

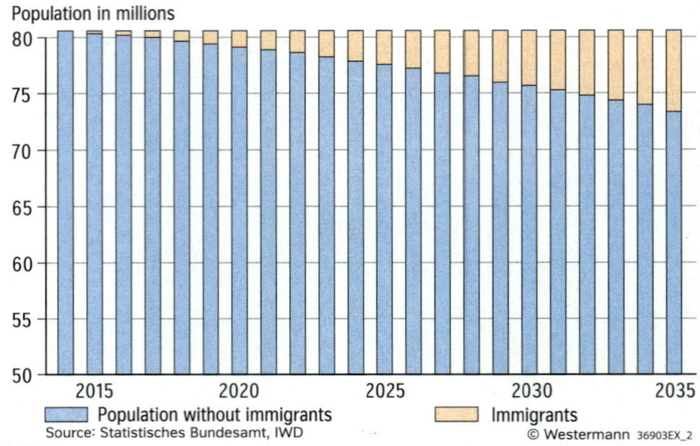

M1 **The amount of immigration that Germany would need to maintain the size of its population**

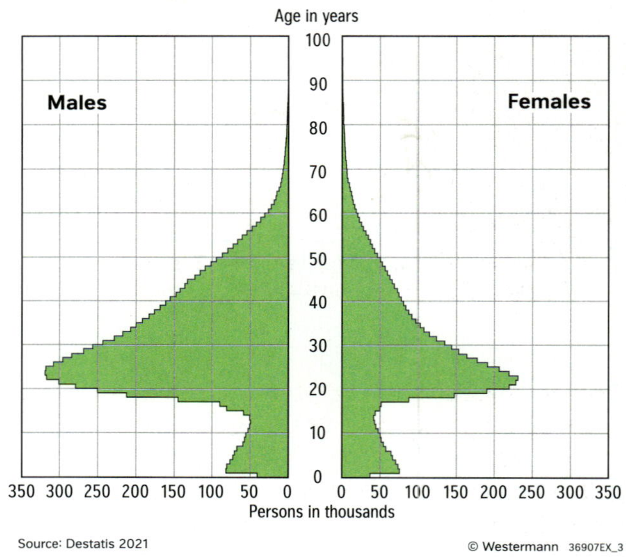

M2 **Age pyramid of immigrants coming to Germany from 2010 to 2019**

Is it mainly the poor who come to us?

No. Except for the former colonial powers, the majority of migrants in Europe come from other countries on this continent. And the majority of trans-regional migrants worldwide come from middle-income countries. [...] Also, migrants from the South do not usually come from poor backgrounds in their home countries. [...] It is those who have the necessary means – money and a certain education – who go to the North. It is also important to have a network of compatriots or relatives in the destination country to help them get started.

Source: Ludermann, Bernd: Migration. Fakten gegen die Panikmache. In: Weltsichten, 3/2016. p. 14.

M3 **Which groups migrate?**

INFO

M4 **Foreign national** in Germany means a person who has a nationality other than German.

M5 An estimated 80,000 Africans and Eastern Europeans work in Almería, half of them without entry or work permits. They mostly live in shack settlements on the outskirts of the villages.

Where exactly do the workers come from who are employed in the 13,000 vegetable farms in Almería?

Chacha: Most are from Morocco, many are also from sub-Saharan countries, then, in descending order, from Ecuador and Romania. Very few are Spanish. Among the immigrants, there are both legal and illegal immigrants without papers.

There have long been reports about the miserable working conditions in the greenhouses. Has anything improved in recent years?

On the contrary. The economic crisis has made the situation much worse. Because there is not much going on in the construction sector, many additional workers have come here. So there is strong competition. The motto now is: "You work and keep quiet." The companies can do whatever they like.

How much money do the workers earn?

The collective agreement in the province is the lowest in the whole of Spain. This is despite the fact that the work in the greenhouses is hard. Nevertheless, most companies do not comply with it. The contract provides for a wage of 46.08 euros per day, with eight hours of work and a 20-minute break. But many work more than eight hours a day, and often the pay is only between 35 and 40 euros. [...] Most of the time, the workers only want to enforce their rights after they have been fired. Before that, they are simply afraid.

The farmers use less strong pesticides today than in the past. Is the workers' exposure lower in this respect?

I will only be able to speak of an improvement when the companies abide by the rules. But there are workers who still work without masks, without protective gloves, without shoes.

Often the poison is still dripping from the plants when they start harvesting again. We have seen cases of skin cancer, lung cancer, many suffer from headaches. [...]

There have been several reports in the German media about the conditions – for example about the huts made of wood, cardboard and plastic foil in which some live.

These huts still exist. They are mostly built near the greenhouses where people work. Because these shack settlements have given the companies bad press – journalists often go there especially to see them – the companies have started to fence in their properties for about two years now. [...]

Source: "Das hat was von Guantanamo" von Katharina Meyer, www.Badische-Zeitung.de, 13.03.2015, accessed: 06.11.2017

***M6** Interview with Abdelkader Chacha (53), who works as an advisor to a farm workers' union in Almería

Wenn du diese Aufgaben erfolgreich bearbeitet hast, kannst du ...

... erläutern und begründen, welche Gruppen im Allgemeinen migrieren.

... Auswirkungen von Einwanderung erläutern.

... den Fachbegriff **foreign national (Ausländer/in)** *erklären.*

Living together

In the years 2015 and 2016, around 1.2 million refugees came to Germany. Due to this influx, cities and municipalities in particular faced great challenges: The people needed a roof over their heads, food, language courses and schooling for the children. Very many people in Germany were willing to help. Even if not all refugees want to stay for a long time, the integration of immigrants is a permanent issue.

Where did the people come from and why did they come to Germany? And what does it mean for living together when new people come?

1. a) Present in-migration, out-migration and net migration in a line diagram (M2). 207
b) Explain the reasons for migration of those seeking protection. (M4, M7).

2. a) In M1, assign the drawings 1 to 4 to the descriptions a) to d).
b) Explain the advantages and disadvantages for the different ideas of integration.

3. a) Describe what interpretation of integration is apparent in the interview M6. 210
b) Sketch how the drawing "initial situation" in M1 would have to be changed, taking into account the considerations in the interview?

adapt – anpassen
adaptation – Anpassung
asylum – Asyl
forced recruitment – Zwangsrekrutierung
human rights – Menschenrechte
living conditions – Lebensbedingungen
melting pot – Schmelztiegel
merging – Verschmelzung
net-migration – Wanderungssaldo
persecution – Verfolgung
protection – Schutz
state welfare/ social welfare – staatliche/ soziale Wohlfahrt

Helpful words and phrases for task 2b:

The models show different ideas of integration ...
Mutual adaptation / assimilation / salad bowl / melting pot has the advantage / disadvantage that ...
If people completely adapt to / merge with the new country, they ...
To give up the own culture can result in ...
In fact, there is not one culture in the new country but different cultures ...
It can be a big enrichment for a society, if immigrants ...

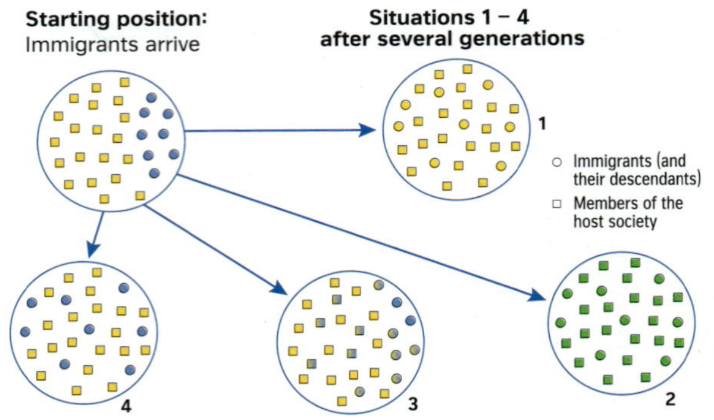

a) **Mutual adaptation:** Immigrants and host society gradually adapt to each other.
b) **Assimilation:** Adaptation of immigrants. This is understood to mean different stages of adaptation, from assimilation to merging. The term is usually understood negatively because it is associated with being forced to give up one's identity.
c) **Salad Bowl:** The ingredients of a salad all belong together, but they are also all a little different. Everyone involved retains culture and identity, cultural characteristics are protected. Diversity in the "multicultural society" is considered desirable.
d) **Melting Pot:** Different cultures and ethnic groups merge into a new, common society or nation. The original understanding in the USA. (For Latin scholars: "e pluribus unum", the motto on the dollar bill).

Source: based on Pallaske, Christoph: Was ist Integration? In: Geschichte betrifft uns. 9/2009, chart M 1.6a, amended © Westermann 18190EX_3

M1 What is meant by integration?

	Foreign nationals in total (in 1000)			Citizens from Non-EU states (in 1000)		
	Immi-gration	Emi-gration	Net migration	Immi-gration	Emi-gration	Net migration
2010	684	530	154	277	217	60
2011	842	539	303	300	204	96
2012	966	579	387	332	209	123
2013	1108	658	450	403	237	166
2014	1343	766	577	536	256	280
2015	2016	859	1157	1173	343	830
2016	1719	1084	635	925	501	424
2017	1384	885	499	609	349	260
2018	1384	924	460	594	335	259
2019	1345	961	384	600	327	273

Source: Destatis 2021 36869EX_2

M2 Arrivals and departures of foreign nationals in Germany (2010 – 2019)

INFO

M3 **Integration** means including people in a society from which they differ.
For integration to succeed, all sides must contribute. Immigrants must want to come closer. This includes, for example, learning the language. Only then can they socialise, shop, work and participate in public life.
Society must show that immigrants are welcome. It must guarantee them rights and open up job opportunities for them. Both sides must be open-minded and respect diversity.

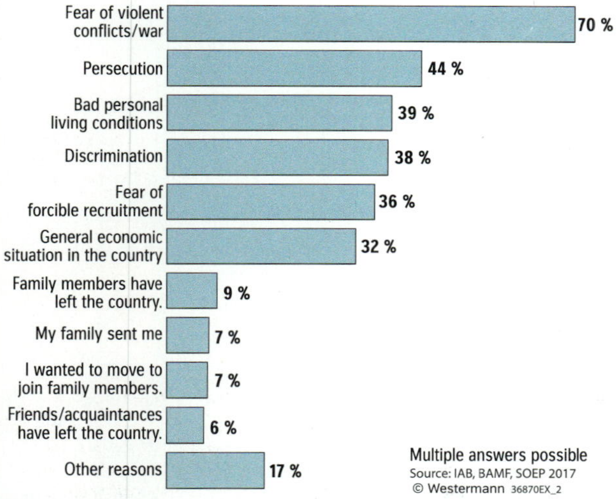

M4 Reasons for leaving the country of origin

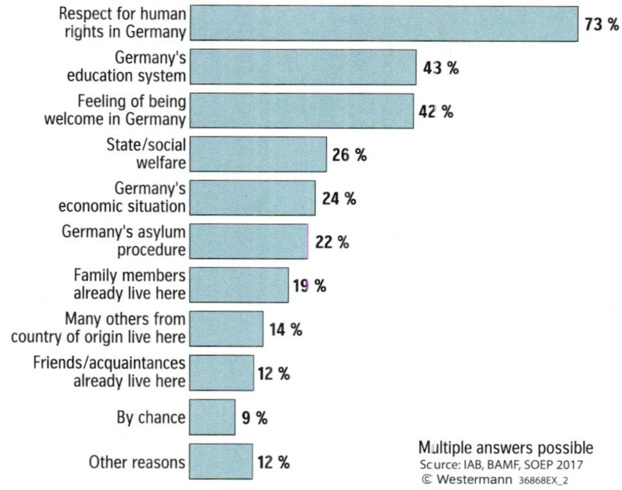

M7 Reasons for choosing Germany as destination country

M5 Arriving in Germany: learning the language, working and making friends are the most important factors

Integration – do we all understand the same thing by it?
I don't think so. There is also no scientific definition that would be viable. In Germany, it still often means adaptation, or worse: incorporation in a military sense. In Prussia, when recruits were taken in, they were lined up and put into uniforms. This image of uniformity is still present today, as is the idea of homogeneity. It is supposed that a society must be homogeneous in order to function. [...] But diversity and heterogeneity surround us, and we have many strategies for dealing with it every day.

What exactly is the society into which one is supposed to integrate?
Actually, there is no such thing. There should be much more emphasis on the fact that people participate in different areas of society and that this does not always happen at the same speed. For example, some people are wonderfully integrated at work, but have no contact with locals in their free time. Such a person is integrated into one part of society, but not into another.

There is often talk of a parallel society in which the migrants live.
This term could also be used to describe millionaires and their lifestyle, which is shared by only a few. There is a misconception that German society is very uniform and that everything that comes from outside is a problem.

Can integration succeed in a short time?
Arriving and being accepted is a very long process that takes years, often even decades. The first generation of immigrants usually has great difficulties in arriving. People have to do worse jobs than in the countries they come from. Their educational qualifications are not recognised [...]. It's different with their children, if only because they go through the educational institutions.

Source: Gehrs, Oliver: Neues Deutschland. In: fluter., 58/2016, Bonn, p.7f.

M6 From an interview with migration researcher Jochen Oltmer

Wenn du diese Aufgaben erfolgreich bearbeitet hast, kannst du ...
... die Zuwanderung von Schutzsuchenden nach Deutschland erläutern.
... unterschiedliche Auffassungen von Integration und Zusammenleben erläutern.
... den Fachbegriff ***integration*** *(****Integration****) erklären.*

What attracts people to the big cities? – Karachi

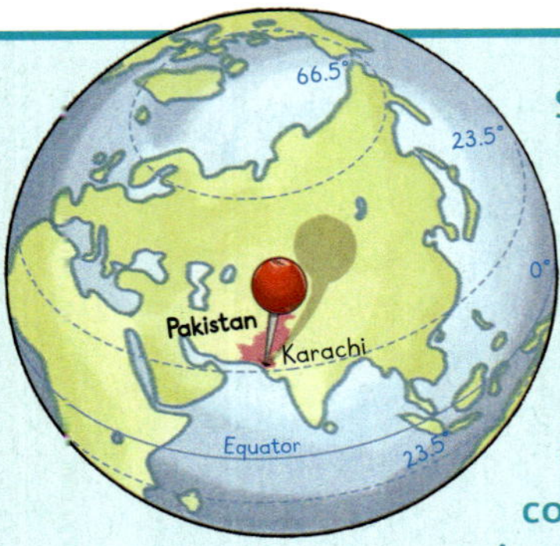

Since the beginning of this millennium, for the first time, more people live in cities than in the countryside. In 2050, around two thirds of humanity will live in cities. Cities are growing especially in developing countries and emerging economies. One example is the city of Karachi. Why do cities have such a great attraction? How is the growth of these cities proceeding?

1. Describe the growth of Karachi (M7). 201
2. Compare the population development of megacities in industrialised and developing countries (p. 38 M2).
3. Compare the development of urban and rural populations:
   a) Present the urbanisation of the world in a suitable diagram (M2). 207
   b) Compare the proportions in industrialised and developing countries and by continent (M2, M3).
   c) Examine the development in Pakistan (M4, M5, M7).
4. a) Assess whether Karachi is a metropolis (M5, M7, atlas).
   b) Explain why metropolises are particularly attractive to the rural population.
   c) Apply the model of migration (p. 43, M4) to in-migration into Karachi. Name push and pull factors (M5, M8). 210

agglomeration	– Ballungsraum
foreign trade	– Außenhandel
gross domestic product (GDP)	– Bruttoinlandsprodukt (BIP)
income tax	– Einkommenssteuer
megacity	– Megastadt
metropolis	– Metropole
rural-urban migration	– Landflucht
significance	– Bedeutung
urban dweller	– Stadtbewohner/in
urban population	– städtische Bevölkerung
urbanisation	– Verstädterung

M1 ... in the **megacity** of Karachi ...

	1950	2020	2050*
World	29.6	56.2	68.4
Industr. counties	58.5	81.9	88.4
Developing countries	17.7	51.7	65.6
Africa	14.3	43.5	58.9
Asia	17.5	51.1	66.2
Europe	51.7	74.9	83.7
Latin America	41.3	81.2	87.8
North America	63.9	82.6	89.0
Oceania	62.5	68.2	72.1

*Forecast Data: UN DESA 2018

M2 Shares of the population in cities in percent

Urbanisation refers to the increase in the urban population compared to the rural population. In the industrialised countries, urbanisation began in the 19th century. Population growth, advances in technology, increases in living standards and jobs in the new industries caused cities to grow. Housing and the necessary infrastructure were created there. After the Second World War, **rural-urban migration** began in developing countries, and cities grew into huge **agglomerations**. It should be noted that in the rich industrialised countries the proportion of the population living in cities is particularly high. Conversely, the proportion of urban dwellers is low in the poor developing countries.

M3 Encyclopaedia entry on "urbanisation"

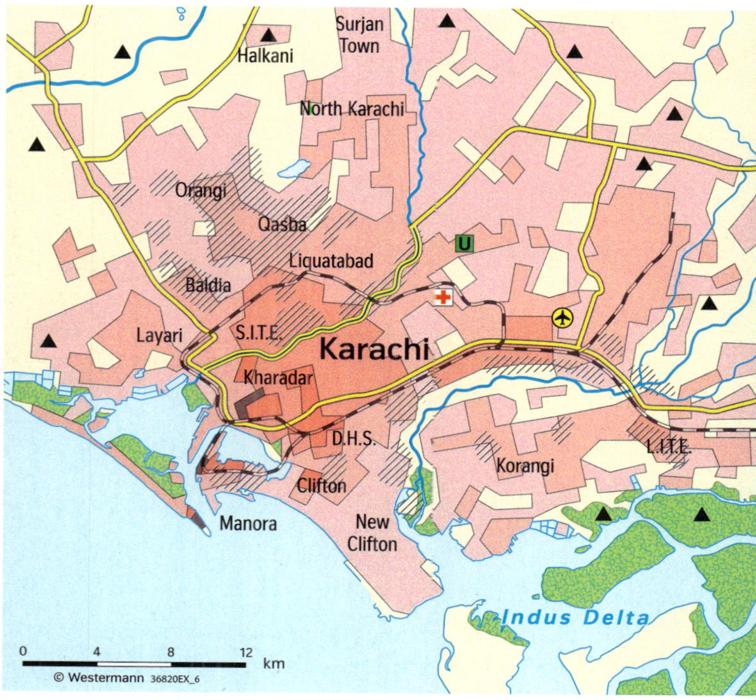

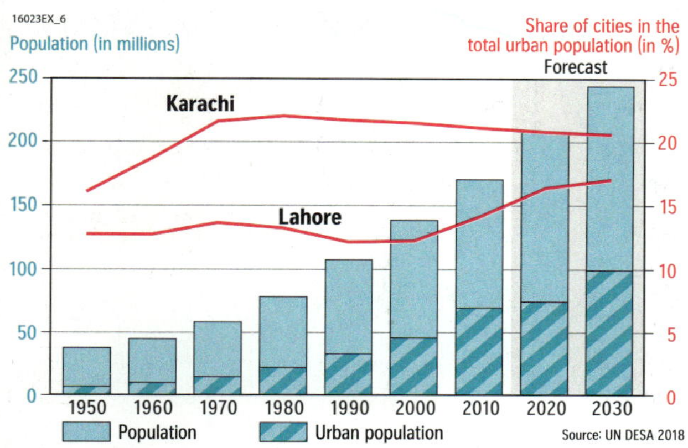

M4 Development of the population of Pakistan

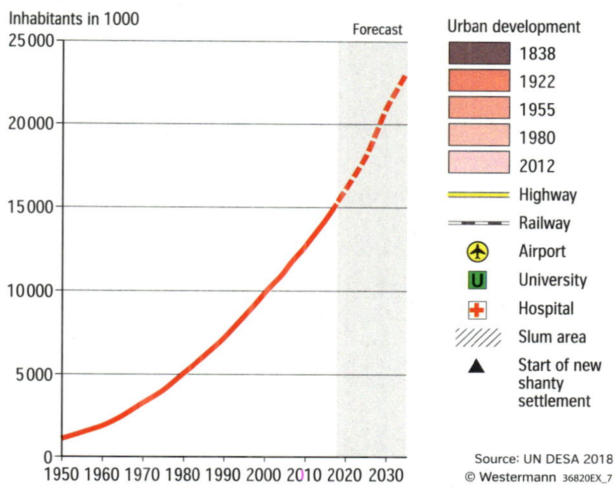

M7 The growth of Karachi

Karachi accounts for:
9 % of the population
19 % of GDP
25 % of government revenues
30 % of industrial production
40 % of employment in large-scale industry
62 % of income tax
90 % of international companies
95 % of foreign trade

M5 The importance of Karachi in Pakistan

Pakistan is a divided country, marked by poverty and violence. In the countryside, large landowners often rule arbitrarily, and in the north people must flee war and violence. Many people are also fleeing Afghanistan for the same reasons. In search of work, many people from the backward rural regions migrate to Karachi.

M8 Push factors

INFO

M6 **Metropolis** – a city with outstanding economic, political and cultural importance.

Wenn du diese Aufgaben erfolgreich bearbeitet hast, kannst du ...
- ... Unterschiede der Verstädterung vergleichen.
- ... Push- und Pull-Faktoren von Migration in Megastädte erläutern und das Stadtwachstum beschreiben.
- ... Bedeutung von Metropolen erklären.
- ... die Fachbegriffe **megacity** (**Megastadt**), **urbanisation** (**Verstädterung**), **rural-urban migration** (**Landflucht**), **agglomeration** (**Agglomerationsraum**) und **metropolis** (**Metropole**) erklären.

What are the living conditions in megacities?

With their growth, megacities have also changed rapidly. What impact does the rapid influx of people have on life in megacities?

1. a) Describe the structure of Karachi (M2). 201
b) Where in the city could the pictures M1, M4 and M6 have been taken? Give reasons for your assumption.

2. Explain the terms "no-go areas" and "no-entry areas" (M2, M5).

3. Point out the characteristics and importance of the informal sector (M3).

C **4.** In Karachi, rich and poor live as if in different worlds (M2, M6 – M8). Describe the respective "world":
A The security guard lives with his family in Korangi. Describe life in the neighbourhood from the point of view of his son Nasir.
B Describe life in Karachi from the point of view of a girl from Germany, who lives with her family in Clifton Beach for two years.

A **5.** Explain the characteristics of a metropolis and megacity like Karachi? Present the characteristics in a mind map. 199

cook-shop	–	Garküche
densely populated	–	dicht besiedelt
housing	–	hier: Wohnen
informal sector	–	informeller Sektor
kidnapping	–	Entführung
layout	–	hier: Gliederung
living conditions	–	Lebensbedingungen
nanny	–	Kindermädchen
odd job	–	Gelegenheitsarbeit
power cut	–	Stromausfall
residential area	–	Wohngebiet
robbery	–	Überfall
security guard	–	Wache
squatter settlement	–	Hüttensiedlung
street vendor	–	Straßenhändler/-in
theft	–	Diebstahl
urban poor	–	städtische Arme
value added	–	Wertschöpfung
wealthy	–	wohlhabend

M1 Heavily guarded: a gated community in Karachi

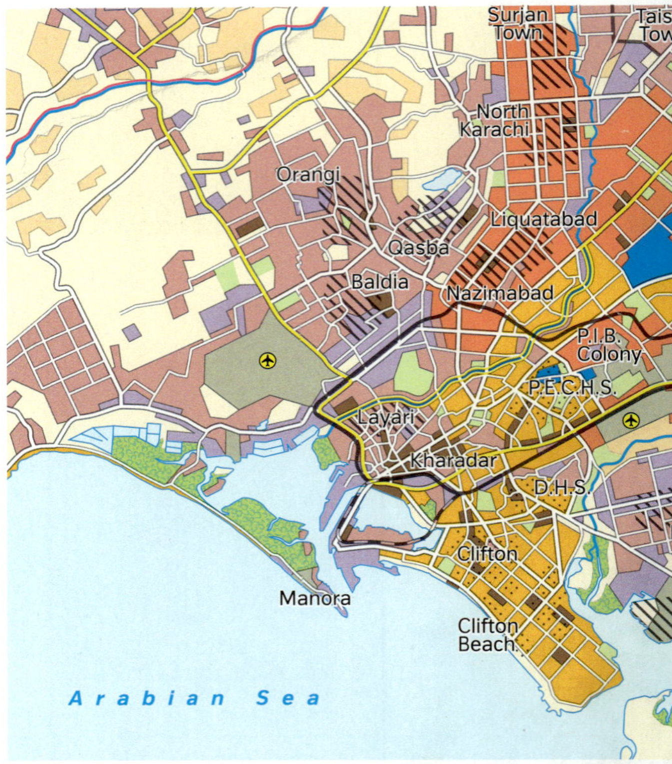

M2 The structure of Karachi

The hopes of most immigrants are not fulfilled in the big cities. Due to the rapid growth, there is a lack of housing, infrastructural facilities such as schools and hospitals, as well as jobs. People live from jobs in which they have no employment contract or social security: They engage in street trading and cookshops, work in the transport sector and do odd jobs, collect rubbish or beg. Because these activities are not registered or regulated by the state, they are counted as part of the **informal sector**. In many developing countries, far more than half of the people work in this way, in Karachi around 75 percent.

M3 From an encyclopaedia: informal sector

M4 Living in Karachi

M6 Traffic in Karachi

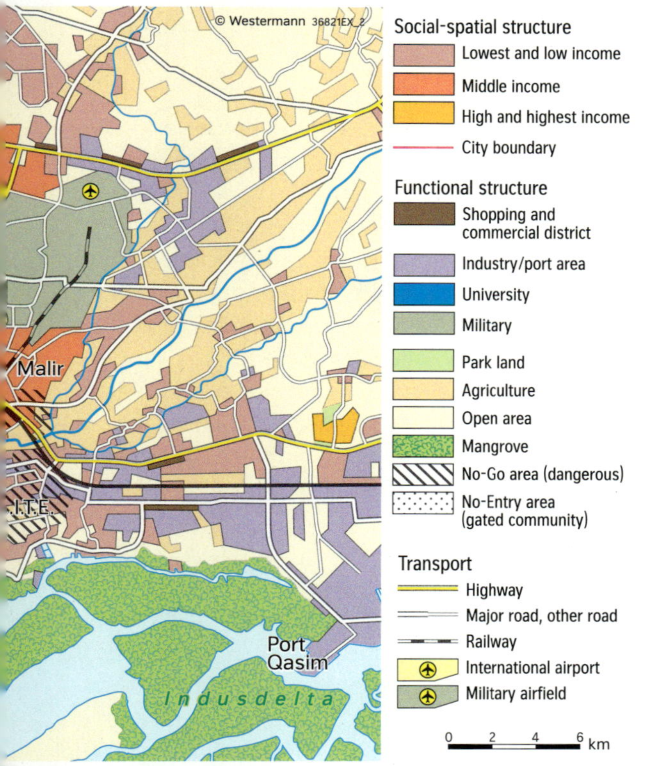

INFO

M5 **Slums** – densely populated residential areas of the poor. They consist of run-down residential buildings or sprawling **squatter settlements**. Often there are no drinking water connections and sewage systems, no facilities such as schools, refuse collection or hospitals. There is often violence and crime. Therefore, they are avoided by outsiders (no-go areas) and are also dangerous for the people who live there.

Gated communities – residential areas of the wealthy that are sealed off and guarded with walls and fences. Access is only possible for explicitly desired persons (no-entry areas). Sometimes private settlements are created, for example with a school, shopping mall and golf course.

Here in Karachi there are constant robberies, kidnappings, car thefts, etc. Many wealthy people are afraid to move freely on the streets. Strangers only come into our gated community if the residents have allowed it. We are armed! Rich Pakistanis and foreigners live here. Inside the walls, residents can go for walks, meet neighbours and have barbecues. They have drivers, cooks, cleaners and nannies.

I could never afford to eat in the restaurants around here. The children are driven and go to good schools or study in England. Everyone tries to spend as little time as possible outside the walls.

M7 Report of a security guard in New Clifton

[About half of the citizens] of Karachi live in slums. What does that mean for city life?

Hasan: [...] The biggest problem is that these settlements are moving further and further away from places where there are jobs, recreation, entertainment and better health and education facilities.

This increasing distance means an increasing division between rich and poor districts, it makes it difficult for women to work, it brings increasing costs and stress due to the miserable transport system. Crime has also increased, and the rich have created ghettos for themselves [...].

Source: Ernst, Sonja: Die Investitionen gehen an den armen Leuten vorbei. Interview for bpb, 08.06.2007. www.bpb.de, Bonn, accessed: 09.10.2017.

M8 From an interview with the architect Arif Hasan

Wenn du diese Aufgaben erfolgreich bearbeitet hast, kannst du ...

- ... *mit Karte und den anderen Materialien räumliche und soziale Gegensätze in der Stadt Karachi erläutern.*
- ... *Auswirkungen des Wachstums von großen Städten erläutern.*
- ... *die Bedeutung des informellen Sektors erläutern.*
- ... *die Fachbegriffe* **slum** *(***Slum***),* **squatter settlement** *(***Hüttensiedlung***),* **gated community** *(***Gated Community***) und* **informal sector** *(***informeller Sektor***) erklären.*

M1 Disparities in megacities, here in Mumbai, India

WHAT WE HAVE LEARNED

To go or to stay? – the significance of migration for regions of origin and destination

Migration is normal

40 One can distinguish between different forms of **migration**: Whether someone emigrates to the USA or seeks refuge from war as a **refugee** in a neighbouring country or is sent by his company to China for a few years – the reasons, the distance or whether the emigration is temporary or permanent, for example, differ considerably.

Causes and consequences

42 The causes of migration can be described in terms of **push** and **pull factors**. Poverty and a lack of future prospects, for example, are reasons that make people leave. Hope for better living conditions draws them to a certain destination. Most of the time, it is not the poorest who leave their homes, because they lack the money to pay for travel or smugglers, for example. It is also mainly younger, already educated people who set out. Therefore, the countries and regions of origin suffer a **brain drain**. On the other hand, **remittances** from migrants can support the economy and development of the countries of origin. Through returning migrants, countries can also benefit from their knowledge and experience. To what extent the "losses" of the countries of origin are compensated by the "gains" is controversial.

Consequences in the destination countries and regions

44 In many European countries, the immigration of professionals is desired. Because their populations are shrinking and ageing, many skilled workers will be needed in the long term. In addition, young people support pension insurance, health insurance and, as taxpayers, the national budget. Especially as "illegals", many migrants work for very low wages. Since their stay is not legal, they have nowhere to report about abuses or to file a complaint. Therefore, they can easily become victims of exploitation.
Immigrants trigger fears in some people. They may be seen as a threat to prosperity and jobs. On the other hand, immigrants are seen as enrichment and diversity as an opportunity for something new to emerge.

Living together

Arriving in and becoming part of a society is called 46 **integration**. What exactly is understood by this varies greatly. Success depends on all sides. For example, immigrants should want to learn the language and come closer, accept the legal system and respect cultural and social rules. Society must show that immigrants are needed, be open-minded and respect diversity, and promote education and training.

Growing cities

Rural-urban migration, but also natural growth, mean 48 that **urbanisation** is increasing, especially in developing countries. Often, it is only one city that far surpasses the other cities in a country in importance: it is the location of the most important companies, the seat of government and administration and the most important educational and cultural institutions. So it is hardly surprising that it is precisely these **metropolises** that exert a great attraction on the rural population. Some of these cities are growing into huge **agglomerations**, i.e. **megacities**.

Contrasts in big cities

The expectations of most immigrants are not fulfilled. 50 There are hardly enough jobs, large parts of the urban population work in the **informal sector**, and cheap housing is lacking. Accordingly, **squatter settlements** grow on the outskirts of the cities, inner-city neighbourhoods with dilapidated buildings develop into **slums**. There is a lack of public facilities and there is violence and crime. This often makes them no-go areas. Close to areas of extreme poverty are often neighbourhoods with no-entry areas: islands of affluence, office building districts, shopping centres, upscale residential areas and villa districts with **gated communities**.

An immigrant that has integrated in an exemplary manner! Although the doner kebab comes from Turkey, it is mainly known there as a plate dish. The mass-market, hand-held version in pita bread is said to have been invented in Germany around 1970 and is now our most popular fast food. That's what you can call a successful naturalisation. More than two million portions are said to be sold in Germany every day, generating an annual turnover of 3.5 billion euros for manufacturers and snack bars. Because there are more kebab stands in Berlin than there are even in Istanbul, it is also known abroad as "Berlin Doner". Unlike in Turkey, we add salad and all kinds of sauces to the meat. Afiyet olsun! (That's Turkish for "Bon appétit".)

Source: Gehrs, Oliver: Dönerwetter. In: fluter., 58/2016, p. 11.

M2 The doner kebab - German or Turkish?

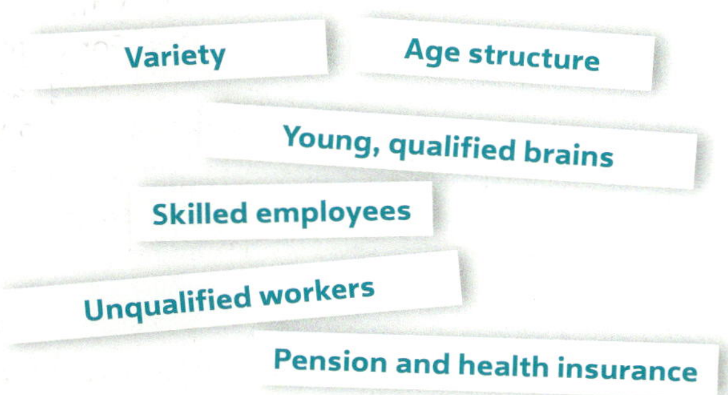

M3 Effects of migration in the destination country

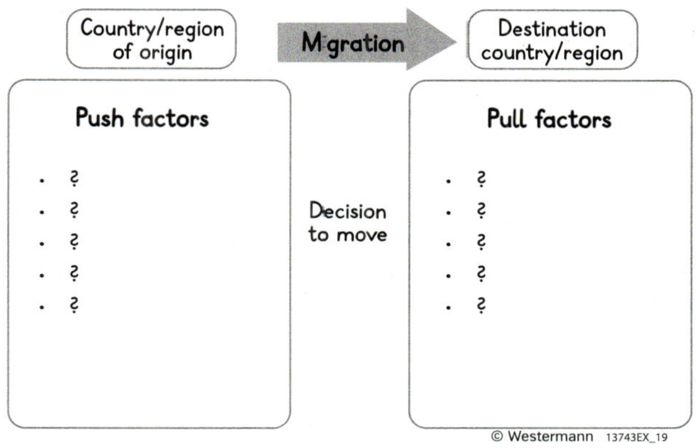

M4 Diagram: Motives and reasons for migration

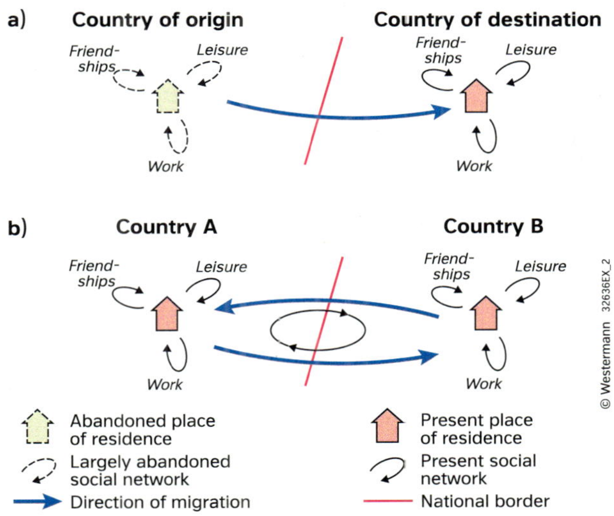

M5 Migration across national borders

1 Analyse the picture M1. 205

2 a) Show possible consequences of migration in a presentation. Use the terms in M3. 218
b) List effects of migration in the country of origin like in M3.

3 Copy the diagram in M4 into your notebook. Enter push and pull factors
a) from the perspective of an individual person,
b) for a migration movement, such as that from Senegal to Europe.

4 In the case of the doner kebab, explain wether one can speak of "exemplary integration" (M2).

5 a) Compare the representations of different forms of migration (M5).
b) Which examples presented on pages 40 and 41 match which diagram in M5? Name reasons.

Apply and practise

PROJEKT VOR ORT

Die Bevölkerung in meiner Stadt

Die Weltbevölkerung wächst, Geburten- und Sterberaten sinken in vielen Ländern, die Bevölkerungspyramiden verändern ihre Form.

Wie verhält es sich nun mit der Bevölkerung in unserer Stadt? Wie spiegeln sich hier die allgemeinen Bevölkerungsstrukturen? Stimmen sie auch überein mit denen in unserer Familie oder in unserer Klasse?

1. Recherchiert dazu in verschiedenen Projekten alleine oder in der Gruppe. Entscheidet euch für eine Fragestellung und plant die Untersuchung. Recherchiert vor Ort und mithilfe des Internets, befragt Personen aus der Familie, im Stadtteil oder bei Ämtern der Stadt. Dokumentiert eure Recherchen, etwa mit Kartierungen oder auch mit Fotos. Präsentiert und erörtert die Ergebnisse in der Klasse. 211, 218, 222

- Wie ist die Generationsabfolge über mehrere Generationen, wie der Abstand zwischen den Generationen?
- Wie viele Kinder gibt es pro Generation?
- Wie viele sind gestorben und in welchem Alter?
- In welchem Alter bekommen die Frauen ihr erstes Kind?
- Wie ist die Altersstruktur einer Familie und wie spiegelt sie sich in der Bevölkerungspyramide der Stadt?
- Wie hoch ist der Anteil der über 80-Jährigen, der unter 3-Jährigen?
- Wer ist aus der Stadt fort- und wer zugezogen?
- ...

M1 Projektvorschlag 1: Befragung im persönlichen Umfeld: in der Familie, Nachbarschaft, Klasse oder im Sportverein, Jugendtreff, Freundeskreis 222

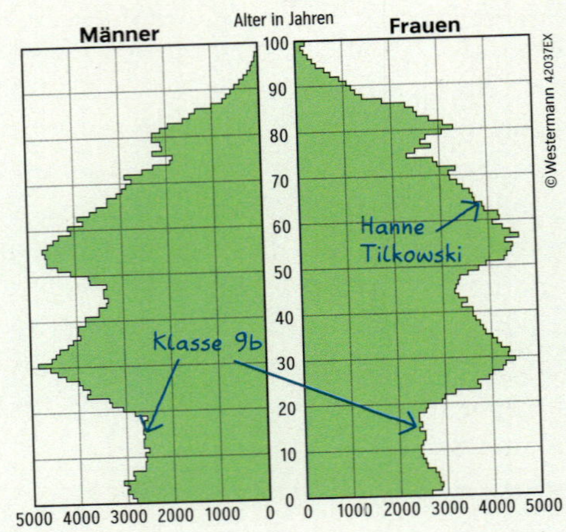

Essen Einw.	2010	2015	2020
	571 392	584 782	591 807

Bevölkerung:

unter 18 Jahre	96 589
18 – 44 Jahre	204 454
45 – 64 Jahre	163 044
über 65 Jahre	127 720

Zahl der Geburten/Sterbefälle:	5845/7017
natürliches Bevölkerungswachstum:	-2 ‰
Geburtenrate:	9,9 ‰
Sterberate	11,9 ‰
Alter der Mütter bei der Geburt:	31,2 Jahre
Durchschnittsalter:	44,1 Jahre (mit Migrationshintergrund 33,4 Jahre)
Lebenserwartung:	Frauen: 82,6 Jahre, Männer: 77,3 Jahre
Doppelstaatler und nichtdeutsche Staatsangehörige:	164 858 (27,9 %)
Wanderungsgewinn:	2,8 ‰ (31054 Zuzüge, 29385 Fortzüge)

Quelle: Amt für Statistik, Stadtforschung und Wahlen der Stadt Essen, Rau, Roland et Carl P. Schmertmann 2020

M3 Projektvorschlag 2: Recherche der Bevölkerungsstrukturen über die Webseite oder Informationsmaterial eurer Gemeinde oder Stadt. Zum Beispiel: Bevölkerungspyramide und -steckbrief zu Essen

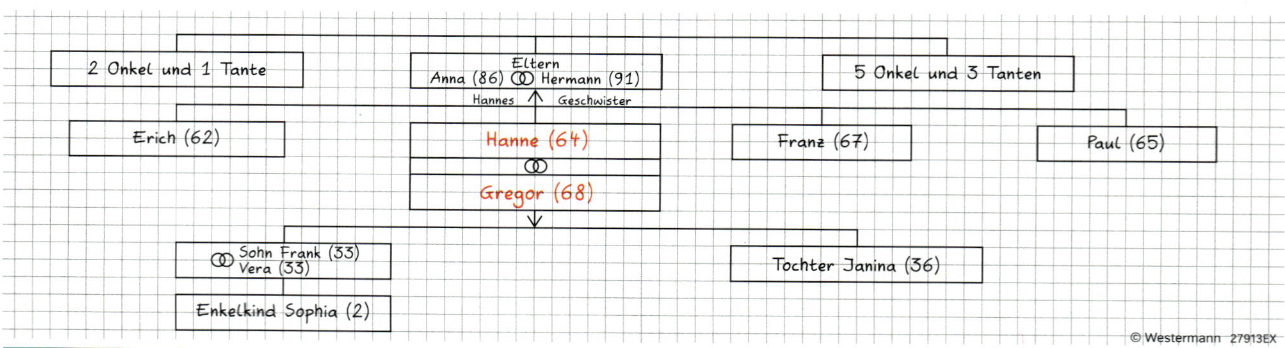

M2 Material zu Projekt 1: Schema eines Familienstammbaums (Beispiel) für die Befragung von Familie Hanne Tilkowski aus Essen-Steele (Tipp: Eintragung der Altersstruktur der Familie in die Bevölkerungspyramide)

Untersucht, wie sich die Einwohnerzahl in 15 Jahren entwickelt hat. Schrumpft oder wächst meine Stadt?

- Welchen Anteil an der Einwohnerentwicklung hat die natürliche Bevölkerungsentwicklung? Welche Rolle spielt das Wanderungssaldo? Recherchiert nach Gründen, was Menschen in deine Stadt zieht oder warum sie fortziehen.
- Vergleicht mit anderen Städten oder Kreisen in Nordrhein-Westfalen.
- Stellt die Ergebnisse als Diagramme dar. Formuliert sie abschließend als Beitrag für einen Stadtflyer.

M4 Projektvorschlag 3: Analyse der Bevölkerungsentwicklung (M5)

		2005	2010	2015	2019
Bevölkerung	E	585 430	574 635	582 624	582 760
	HSK	277 219	267 601	263 762	259 777
	D*	82,44	81,75	82,18	83,17
Geburtenrate in ‰	E	8,1	8,1	9,7	10,2
	HSK	8,9	7,5	7,8	8,5
	D	8,3	8,3	9,0	9,4
Sterbefälle in ‰	E	12,6	12,8	12,9	12,0
	HSK	10,5	10,7	12,6	12,3
	D	10,1	10,5	11,3	11,3
Wanderungssaldo in ‰	E	0,0	1,9	18,6	0,4
	HSK	-0,1	-5,5	12,9	1,1
	D	1,0	1,6	14,0	3,9

*in Mio. Daten: Destatis, IT.NRW 2021

M5 Bevölkerung in Essen (E) 2000 bis 2019, zum Vergleich Hochsauerlandkreis (HSK) und Deutschland (D)

M6 Jannis aus der 9b bei einer Befragung seiner Nachbarn.

- Erfrage Familienstrukturen, z. B. die Generationenabfolge, den Stammbaum (M1, M2).
- Welche Lebensstationen und Fortzüge aus einer Stadt oder einem Land gab es im Leben?
- Erkunde Aufenthaltsorte oder (öffentliche) Einrichtungen im Alltagsleben und dokumentiere sie (z. B. Kita, Sportplatz, Seniorencafé):
  - Durch Kartieren (Kartenvorlage: Stadtplankopie, digitale Karten)
  - Durch Fotografieren und dabei Verdeutlichung des Typischen dieses Ortes (Tipps in M11).
  - …

M7 Recherche in einem Stadtteil mit Befragung ausgewählter Bevölkerungsteile (z. B. Altersstufen). 222

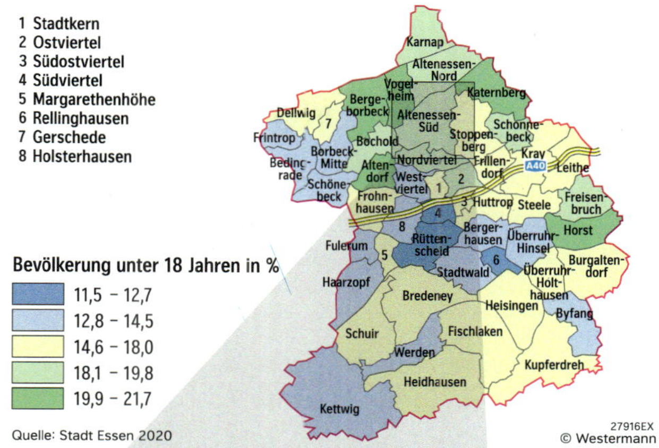

M8 Bevölkerungsanteil der unter 18-Jährigen in allen 50 Essener Stadtteilen

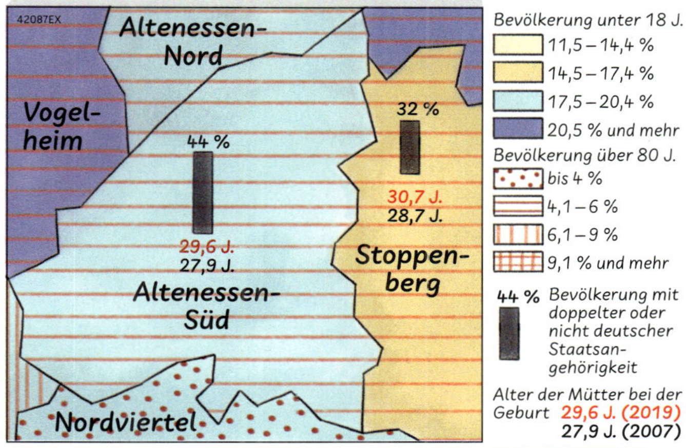

M9 Kartierung von Sara aus der 9b zu ausgewählten Bevölkerungsstrukturen in und um Altenessen – Süd

- Einwohnerzahl, Bevölkerungssaldo
- Altersstruktur (z. B. Anteil der über 65-Jährigen, der unter 18-Jährigen)
- Lebenserwartung, Geburtenrate
- Einwohner mit Migrationshintergrund
- Alter der Mütter bei Geburten
- …

Wichtig: Überlege dir eine sinnvolle Legende für jeden Indikator, überlege z. B., welchen Wert du durch Farbe, Schraffur, Zahlenwerte oder Beschriftung einträgst.

***M10 Projektvorschlag 5: einen Stadtteil unter die Lupe nehmen, dazu ausgewählte Bevölkerungsstrukturen aus digitalen und analogen Informationsquellen (amtliche Statistik, Bevölkerungsatlas, Stadtplan, persönliche Anfragen) in einer Kartierung zusammenstellen**

- das Wichtigste nicht zu klein und ins Zentrum setzen, den Betrachterblick dorthin lenken
- das Umfeld von Personen miteinbeziehen
- in Vorder-, Mittel-, Hintergrund aufgliedern
- Gegenlicht vermeiden

M11 Dokumentarisches Fotografieren

IN FOCUS

Latin America – population and migration

The countries of Central and South America are characterised by differing developments. Among them are examples of massive urbanisation – here are some of the largest conurbations in the world – as well as regions that are affected by heavy outward migration.

OPTIONAL TOPIC 1

Latin America – population

For decades, Latin America was characterised by strong population growth. Yet the development varied from country to country.

How is the population distributed in Latin America and how has it developed?

1. Prepare a short presentation that introduces the topic of this double page. 210, 211, 218

Tips for the structure:

- Describe the population distribution in Latin America (M1).
- Name factors that have an influence on distribution.
- Work with the atlas (map: Erde – Sprachen): Strictly speaking, Spanish and Portuguese are not the only languages spoken in Latin America (M7). 201
  - List the other languages.
  - What conclusions can be drawn from this?
- Calculate the growth rate of Guatemala's population and describe its development (M5, M6).
- Using M6 as a model, draw a diagram with birth, death and growth rates for Uruguay (M8). Also, draw a diagram of population growth according to M3. 207
- Classify Guatemala and Uruguay and allocate them to the corresponding diagram in terms of
  a) population growth (M2),
  b) birth rate, death rate and growth rate (M3).

conurbation	–	Ballungsgebiet
forecast	–	Vorhersage/Prognose
population distribution	–	Bevölkerungsverteilung

Helpful words and phrases:

The population is unevenly distributed ...
There are densely / sparsely populated areas e.g. ...
Examples for conurbations are...
... are relatively densely populated ...
The population development in the different countries varies ...
There are similarities and differences e.g. ...

M1 The distribution of the population

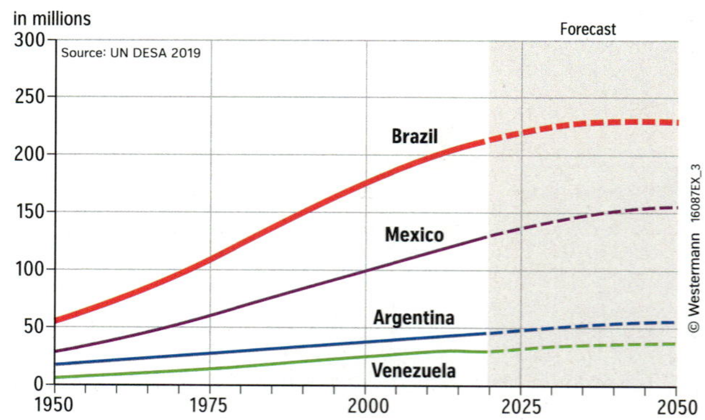

M2 Population growth in Latin America

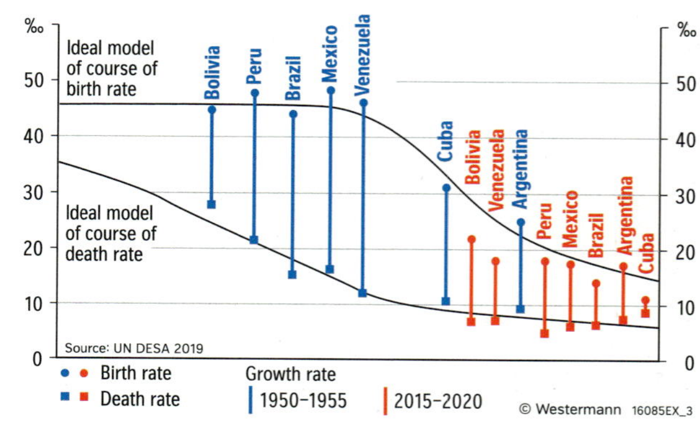

M3 Birth and death rates, 1950 – 1955 and 2015 – 2020

M4 Schoolchildren at Lake Titicaca come home by boat. Especially in the Andean countries, the number of children per family is still very high.

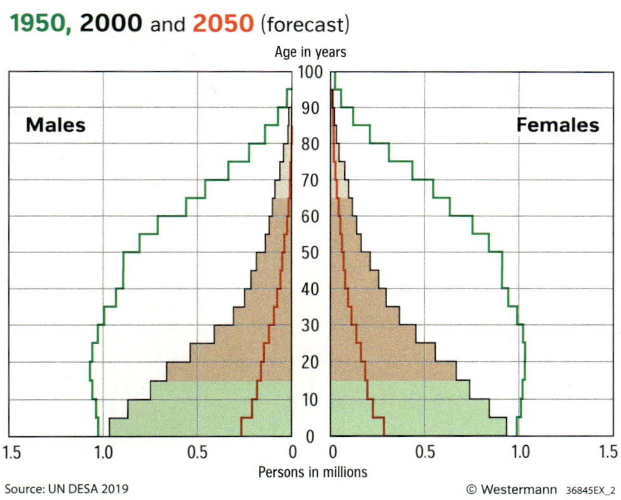

M5 The development of the age structure in Guatemala

INFO

M7 **Latin America**
refers to the states of the North and South American continent south of the USA.
The name is derived from the languages Spanish and Portuguese, which belong to the Latin language family and are spoken here with a few exceptions.
In the USA, the Spanish-speaking population group is often referred to as Hispanics.

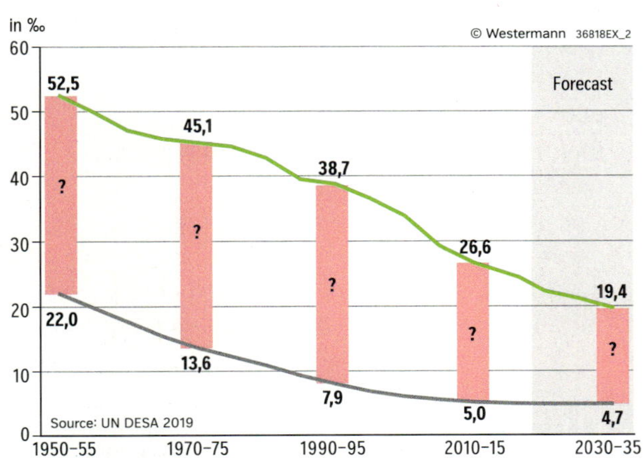

M6 The development of Guatemala's birth, death and growth rates

Uruguay			Population (in 1,000)		
Year	BR	DR	Year	Uruguay	Guatemala
1950 – 1955	21.2	10.5	1950	2239	3115
1960 – 1965	21.9	9.5	1960	2539	4211
1970 – 1975	21.1	10.0	1970	2810	5622
1980 – 1985	18.3	9.8	1980	2915	7283
1990 – 1995	18.2	9.7	1990	3110	9264
2000 – 2005	15.8	9.4	2000	3320	11651
2010 – 2015	14.3	9.4	2010	3359	14630
2020 – 2025	13.3	9.5	2020	3474	17916

BR = Birth Rate in ‰, DR = Death Rate in ‰, 2020 – 2025/2020 = Forecast
Data: United Nations 2019

M8 Population data for Uruguay and Guatemala

IN FOCUS Latin America – population and migration

The countries of Central and South America are characterised by differing developments. Among them are examples of massive urbanisation – here are some of the largest conurbations in the world – as well as regions that are affected by heavy outward migration.

OPTIONAL TOPIC 2 Migration – from Latin America to the USA

Latin America is strongly affected by emigration. The destination of many migrants are the USA, a classic immigration country. What are the reasons for the migration movement to the north? What are its effects in the USA?

1. Prepare a short presentation that introduces the topic of this double page. 210, 211, 218

Tips for the structure:

- Classify the reasons for migration according to push and pull factors (M1, M2).
- Indicate the reasons why Misael left his home and how he imagined his life in the USA (M3). 205
- Name reasons why Misael could be called either a migrant or a refugee.
- Describe how immigration to the USA has changed (M4).
- Illustrate the distribution of so-called Hispanics in the Southwest and South of the USA (M5).
- Explain the statement in the graffito in Los Angeles (M6).

aridity	–	Trockenheit
civil war	–	Bürgerkrieg
estimate	–	Schätzung
florish	–	florieren / blühen
land of unlimited opportunity	–	Land der unbegrenzten Möglichkeiten
overexploitation	–	Übernutzung / Ausbeutung
smuggler	–	Schleuser / Schmuggler
supression	–	Unterdrückung

Helpful words and phrases:

Migrant is a collective term for people who...
Refugees are people who flee because of political / religious ... reasons ...
For Misael, the term migrant / refugee / both terms could be applied because ...
The most important reason for his decision is ...
Misael is in danger because ...

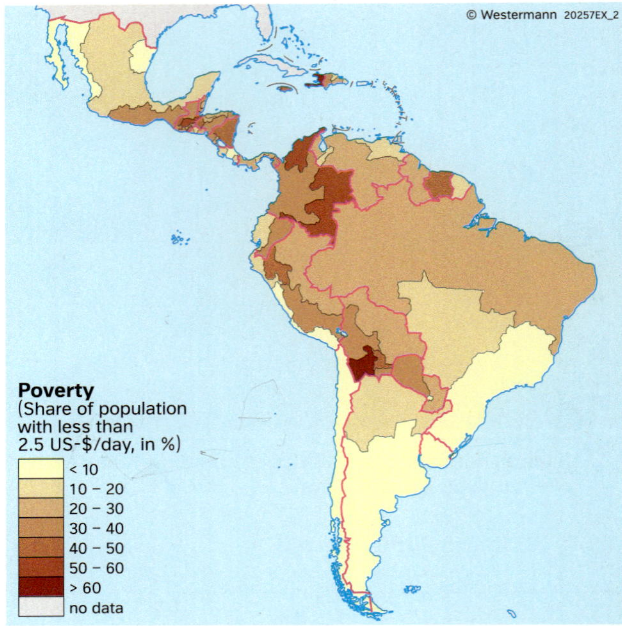

M1 Poverty in Latin America

- Unemployment, e.g. in central Mexico
- Limited educational opportunities
- War and violence, such as civil wars in Guatemala and Peru, or the drug war in Colombia
- Lack of alternatives to agriculture
- Attraction of leisure and consumption offerings
- Suppression of individual and political freedom
- Trade and industry in the hands of few rich owners
- Modernisation of agriculture, e.g. in southern Brazil, job and land loss for many people
- Overexploitation of the high mountain ranges of the Andes
- Inadequate infrastructure (drinking water, electricity, ...)
- Opportunities to earn money in trade or in the informal sector
- Increasing aridity in north-eastern Brazil through deforestation and climate change

M2 Reasons why Latin Americans leave their home country

At the age of barely 18, his family sent Misael Contreras north. The risks on the way – robberies, kidnappings, death – did not matter in view of the danger he faced at home in San Salvador: being forced to join a mara, one of those dangerous youth gangs that control entire neighbourhoods in Central America and threaten anyone with death who does not want to join. [...] His family paid 6,500 US dollars to the "coyote", the smuggler who brought him across the border. He was on the road for a month. Today, the "coyotes" collect 12,000 US dollars for such services; they have been doing their job for decades. The mass emigration of Salvadorans to the USA began in the 1980s, a consequence of the twelve-year civil war that lasted until 1992. More than two million Salvadorans - about a third of the population - now live in the United States, according to government estimates. The war is long over, but violence has steadily increased in recent years. Today, just as many people die violent deaths in El Salvador as during the civil war. [...] He saw no other option but to follow his mother. She had already emigrated illegally to the USA three years earlier. Early one morning in 2008, Misael got off the bus at a petrol station in the US state of Maryland. His mother was waiting for him there. "I was in the land of unlimited opportunity," he says. [...] In his first job, he worked as a gardener. He earned 9.50 US dollars an hour. Later he worked in construction and got between 600 and 700 dollars a week. A lot of money, because in El Salvador workers earn the minimum wage of just 242 dollars a month. [...] Contreras then teamed up with a Salvadoran friend who was a self-employed mechanic who repaired cars. He learned by watching and copying his partner, and soon the business flourished. [...]

Source: Romero, Cecibel: Auf vermintem Gelände. In: Weltsichten, 3/2016, Berlin, pp. 18/19.

***M3 Misael Contreras lives as an „illegal" in the USA**

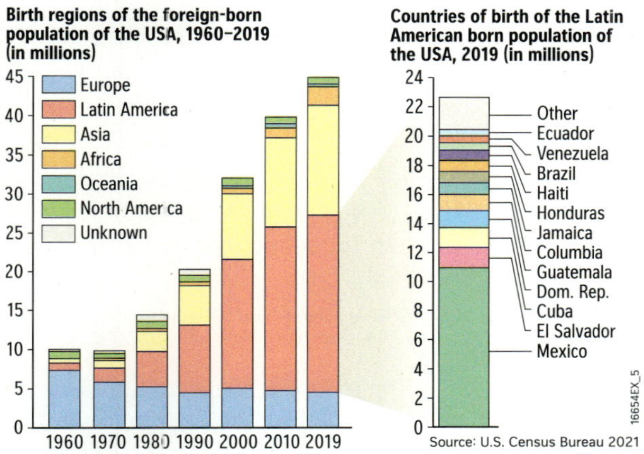

M4 Countries and regions of birth of the foreign-born population in the USA

M6 Graffito in Los Angeles

M5 Hispanics in the Southwest and South of the USA

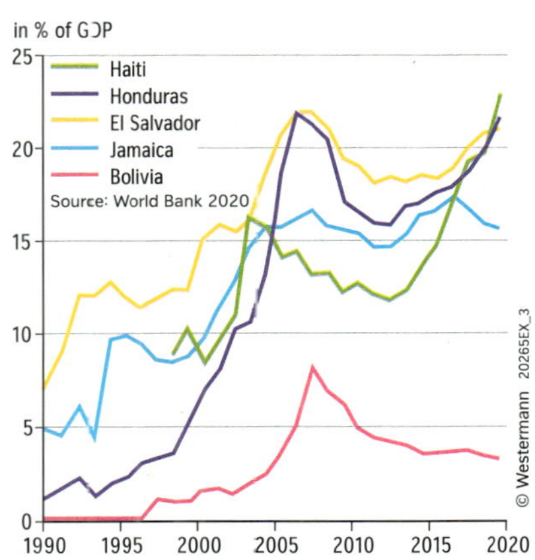

M7 Remittances

IN FOCUS

Latin America – population and migration

The countries of Central and South America are characterised by differing developments. Among them are examples of massive urbanisation – here are some of the largest conurbations in the world – as well as regions that are affected by heavy outward migration.

OPTIONAL TOPIC 3

São Paulo – a Latin American megacity

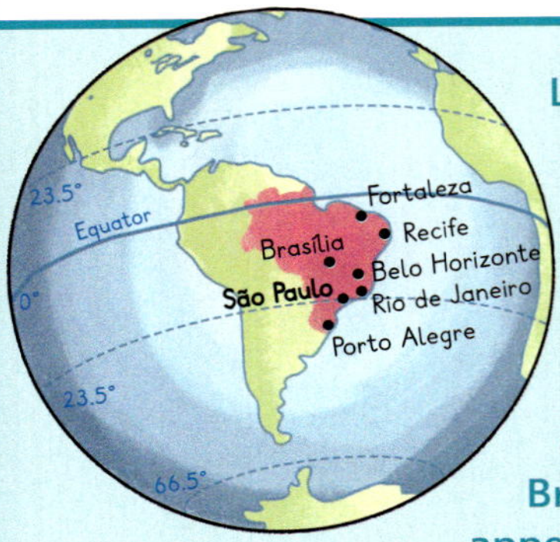

Latin Americans live in cities to a large extent. This distinguishes the continent from Africa and Asia. São Paulo is one of the largest cities in the world, one of two megacities in Brazil, and it has a great appeal. How has São Paulo grown? What are the characteristics of megacities like São Paulo?

1. Prepare a short presentation that introduces the topic of this double page. 210, 211, 218

Tips for the structure:

- Describe the growth of São Paulo (M1, M7).
- Illustrate the importance of Sao Paulo for Brazil (M6).
- Analyse the picture M4 205. Draw a sketch. Use as many technical terms from the chapter as possible.
- Analyse the materials M2 – M5: Present the causes, extent and consequences of the different problem areas in a diagram. 207.

appeal	–	Attraktivität
bypass motorway	–	Umgehungsstraße
carbon monoxide	–	Kohlenmonoxid
favela	–	Favela / Marginalsiedlung in Brasilien
fortress	–	Festung
lurk	–	lauern
nitrogen oxides	–	Stickoxide
predominant	–	überwiegend

Helpful words and phrases:

Sao Paolo has been growing rapidly since ... / prognoses show that ...
The implications of the growth are ...
For the inhabitants that means ...
Chances, the city offers are ...
On the other hand ...
The city has been expanding into different directions ...
The extension of the city is now ...
Transport is a major problem because ...

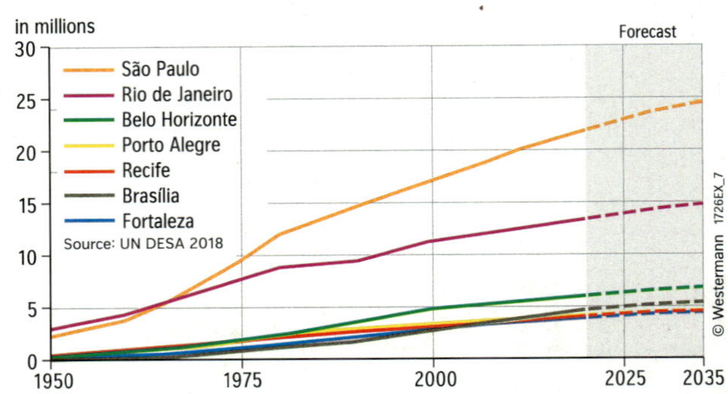

M1 Brazil – population growth of the seven largest agglomerations

The crime rate and the risk of becoming a victim of robbery or other violent crime are high in Brazil, especially in the big cities like [...] Porto Alegre, Recife, Rio de Janeiro [...] and São Paulo. There, in turn, poor settlements (favelas) are particularly hard hit. The favelas of Rio de Janeiro have recently been repeatedly affected by acts of violence, including deaths. Favelas are partly controlled by criminals and drug gangs. Armed confrontations, also with the police, often result in the death of bystanders.

Source: Auswärtiges Amt: Brasilien. Reise- und Sicherheitshinweise. As at 14.10.2020, www.auswaertiges-amt.de, accessed: 14.10.2020

M2 Travel and safety advice from the German Federal Foreign Office

Last year, more than 4,000 people were murdered. Robberies, car thefts and house break-ins are as much a part of everyday life as brushing your teeth.[...] However, the greatest danger in São Paulo lurks at red lights. Anyone carrying an expensive watch, a smartphone or cash must constantly expect to look down the barrel of a gun when the light turns red. [...] Last year, 40,000 armoured cars were sold in São Paulo. What is true for the cars is even more true for the home. High-rise buildings in better locations are secured like fortresses because of the latent danger of burglaries.

Source: Rybak, Andrzej: Metropole São Paulo. Leben in der Vorhölle. 24.06.2014, www.manager-magazin.de, accessed: 06.11.2017.

M3 Locational disadvantage: crime

M4 **Paraisópolis (Paradise City), on the left in the picture, is the largest favela in São Paulo with an estimated 60,000 inhabitants. It is located adjacent to Morumbi, one of the richest districts of the metropolis.**

Because of the massive emissions from traffic, air pollution has developed into a central environmental problem that São Paulo's residents are confronted with on a daily basis is predominantly due to emissions from traffic. Today, there are about eight million motor vehicles in São Paulo, which are responsible for the high concentration of carbon monoxide and nitrogen oxides. [...] Even the driving ban that switches every day according to even and odd numbers on the number plates has not significantly reduced the pollution. The diesel-powered trucks – due to the lack of a bypass motorway, even the through traffic must cross the city – and buses, which handle the main part of the public transport, provide an additional burden [...]. The fact that industry [...] is not the main cause of air pollution is primarily due to the fact that since the 1950s many large-scale heavy and chemical industry plants have been located outside the metropolitan area, mainly in Santos-Cubatao.

Source: Bähr, Jürgen et Ulrich, Jürgens: Stadtgeographie II. Braunschweig 2009, p. 297.

M5 **Locational disadvantage: air pollution and traffic**

São Paulo accounts for:
- 11 % of the population
- 20 % of GDP
- 21 % of the vehicles
- 30 % of employment in industry
- 36 % of services
- 55 % of industrial production
- 60 % of energy consumption
- 30 of the 50 largest companies
- 900 of the 1,200 branches of German companies

M6 **Shares of São Paulo metropolitan area in the state of Brazil**

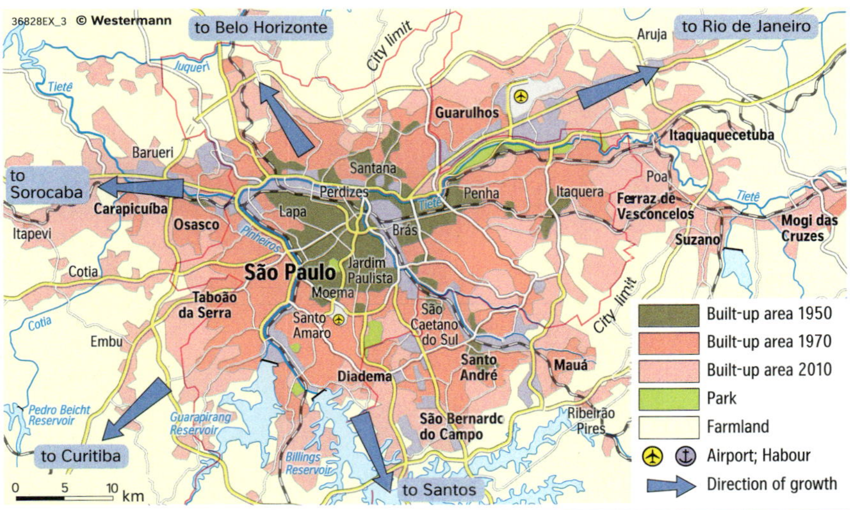

M7 **Growth of São Paulo**

2. The countries of the world – differing developments

Mali: The Natomo family on the roof of their mud-brick house in Kouakourou. They cook on a wood fire. The food is preserved by drying it in the air. Weekly expenditure on food: 21.30 euros.

USA: The Revis family in the kitchen of their house in suburban Raleigh, North Carolina. For cooking, baking and grilling they have an electric cooker, a mini oven, a microwave, and a BBQ grill outside. There is a freezer for food storage. Weekly expenditure on food: 275.60 euros.

Philippines: The Cabana family in the living room of their 19 square metre flat in Manila. They cook on a gas cooker. They have no way to preserve the food. Weekly expenditure on food: 39.80 euros.

Ecuador: The Ayme family in the kitchen of their house in Tingo. They cook on a wood fire. The food is preserved by drying it in the air. Weekly expenditure on food: 25.40 euros.

The social situation – social indicators

"Industrialised countries", "developing countries" – these are common terms for "developed" and "less developed" countries. But what does "development" actually mean? How can we measure the level of development of a country?

1. Compare the photos on pages 62 / 63. 205 What do they say about living conditions (e.g. family, housing, type of food)?
2. Illustrate to what extent the home countries of the girls in M1 are more likely to be developing or industrialised countries based on their living conditions (M1, M2, M4).
3. Describe the complex consequences of malnutrition and undernourishment (M3).
4. Record a two-minute statement on the topic: "Should the access to means of communication be defined as a basic human need?"
5. (C) Compare the social situation in Germany, Ethiopia and Peru 217
   A using the atlas (world maps on Lebenserwartung, Bildung, Gesundheit).
   B using the data on pages 240 / 241.
6. (A) "People in developing countries were even more badly affected by the COVID pandemic than those in developed countries." Explain the statement (M3, M4).

affected	–	betroffen
bottle feeding	–	Flaschennahrung
breast milk	–	Muttermilch
civil servant	–	Beamter/Beamtin
deficiency disease	–	Mangelkrankheit
exploitative	–	ausbeuterisch
graduation	–	Studienabschluss
indicator of development	–	Entwicklungsstand
level of development	–	Entwicklungsstand
malnutrition	–	Mangelernährung
means of communication	–	Kommunikationsmittel
misery	–	Elend
prosperous	–	wohlhabend
social indicators	–	soziale Indikatoren
undernourishment	–	Unterernährung
vicious circle	–	Teufelskreis

Eden (Ethiopia)*

Parents: Farmers; six siblings

at 8: Dropped out of school, help in the household, caring for siblings

at 11: Father dies

at 16: Marriage to a man chosen by the mother, 1st child

at 19: 2nd child

at 29: Fifth pregnancy, severe complications, child dies at birth

at 35: Sex education and access to contraceptives

at 38: Birth of fourth grandchild

at 64: Death

Julia (Germany)*

Father: Employee, **Mother**: Civil servant; one brother

at 8: 3rd grade pupil, ballet and riding lessons

at 11: Sex education at school

at 16: Secondary school, first boyfriend

at 19: A-Levels, work and travel in Australia, then studies at university

at 26: After graduation, work as a biologist

at 29: Marriage

at 31: 1st child, parental leave

at 34: 2nd child, parental leave

at 35: Part-time job

at 37: Full-time job

at 62: First grandchild born

at 83: Death

*Names changed for didactic reasons. Source: Deutsche Stiftung Weltbevölkerung: Eden und Julia (Poster), 25.10.2016, access: 05.03.2018.

M1 **Life in two worlds.**

The International Labour Organization (ILO) estimates that 168 million 5-17 year-olds are currently engaged in prohibited child labour. [...] This includes work for which children are too young, which is dangerous or exploitative, which damages physical or mental development, or which prevents children from attending school. Worldwide, eleven percent of all children suffer from such harmful working conditions. South of the Sahara, the proportion of child labourers is highest at around 21 percent. Around 60 million minors must toil there. The highest number of child labourers in absolute terms, almost 80 million, are in Asia and the Pacific region. [...] In India [...] it is taken for granted that children contribute to subsistence. According to estimates, working children generate 15 to 30 percent of the household income.

Source: Brot für die Welt: Arbeit statt Spielen: Wenn Kinder keine Kindheit haben. www.brot-fuer-die-welt.de, access: 16.01.2018

M2 **Work instead of play**

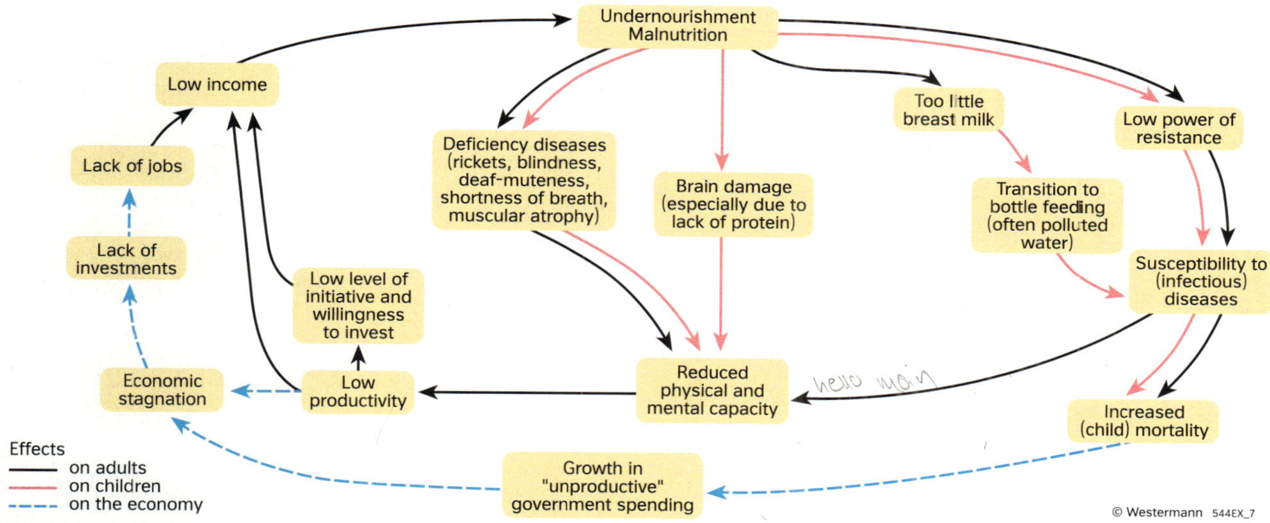

***M3** **Vicious circle of undernourishment and malnutrition: About 20 percent of the world's population is affected by malnutrition. This means that people do not get enough food to meet their calorie needs. In addition, millions suffer from malnutrition. Their diet is very one-sided, so that vital substances (e.g. protein, vitamins, trace elements) are missing. For example, many farming families in South and East Asia can hardly afford any other food than rice. Therefore, they eat rice at all meals, rarely vegetables and meat once a month at most.**

INFO

M4 The living conditions of the people – a measure of development

The satisfaction of **basic needs** is the decisive indicator to characterise the level of development of a country. Basic needs are seen as not only food, work and education, but increasingly also access to means of communication, the possibility of political participation and equal rights for men and women. The generic term **"developing countries"** or **"countries of the global south"** covers a wide variety of states, including some that appear to us to be developed. Although they are very modern in many respects and appear to be prosperous, the basic needs of many people there are not satisfied. Even in the so-called **industrialised countries** there is misery and poverty. Nevertheless, the basic needs of most people are met here (see also p. 68).

M5 Basic needs

M6 Child labour in Nepal

INTERNET

M7  For years, the photographer *Peter Menzel* has documented the living conditions of people all over the world in photo series (p. 62/63) and illustrated books.

Wenn du diese Aufgaben erfolgreich bearbeitet hast, kannst du …

… soziale Indikatoren erläutern, an denen man den Entwicklungsstand eines Landes erkennen kann.

… die komplexen Zusammenhänge von mangelnder Grundbedürfnisbefriedigung darstellen.

*… die Fachbegriffe **basic need** (**Grundbedürfnis**), **developing country** (**Entwicklungsland**), **countries of the global south** (**Land des globalen Südens**) und **developed country** (**Industrieland**) erklären.*

The economic situation – economic indicators

An essential prerequisite for development is that the country has enough financial resources. How does one measure the prosperity of states? How meaningful are such indicators?

1. a) Describe the distribution of poor and rich countries worldwide (M3, atlas).
b) Discuss the advantages, disadvantages and the intention of the map in M3.

2. Outline the importance of the informal sector for the people and the economy in developing countries (M1, M2, M4).

3. Using two examples, explain the difference in informative value of GNI and GNI per capita (absolute versus relative numbers) (M3, M6). 215

4. Evaluate the suitability of GNI as an indicator for development (M2, M6).

5. Based on M7, create a causal network or a concept map on the effects of poverty. 199

absolute/relative number	–	absolute/relative Werte/Zahlen
average	–	Durchschnitt
economic indicators	–	ökonomische/wirtschaftliche Indikatoren
employee	–	Angestellte/r
employment contract	–	Arbeitsvertrag
financial ressources	–	finanzielle Mittel
informative value	–	Aussagekraft
meaningful	–	aussagekräftig
measure	–	messen
mislead	–	irreführen/ täuschen
poverty line	–	Armutsgrenze
restriction	–	Beschränkung
rubbish disposal	–	Müllentsorgung
social disparities	–	soziale Disparitäten

Helpful words and phrases for task 1:

The bigger the the area, the higher the ...
Poorer / richer countries are predominantly on the ... hemisphere
The intention of this kind of presentation is to ...
On the on hand, one can ...
On the other ...
The visualisation is informative / eye-catching / misleading / inadequate / ...

M1 Informal activities exist in all sectors of the economy. in Mexico City, tens of thousands of informal taxis transport hundreds of thousands of customers every day. In Cairo (Egypt), all waste disposal is in the hands of thousands of people who come daily from the rubbish dumps in the surrounding areas and take care of rubbish collection, but also recycling. In Delhi (India), it is estimated that about 70 percent of all workers are employed in the informal sector. Due to the many restrictions during the Corona pandemic, many people in the informal sector lost their jobs.

INFO

M2 Gross national income (GNI)

The **GNI** is calculated by adding the values of the goods produced in agricultural and industrial production in a year. To this is added the value of all services used in that country. However, it only includes those businesses and employees that are officially registered and pay taxes, for example. But this is not the case for hundreds of millions of employees and small businesses in developing countries.

They work in the **informal sector** – without an employment contract, without insurance, without paying taxes.

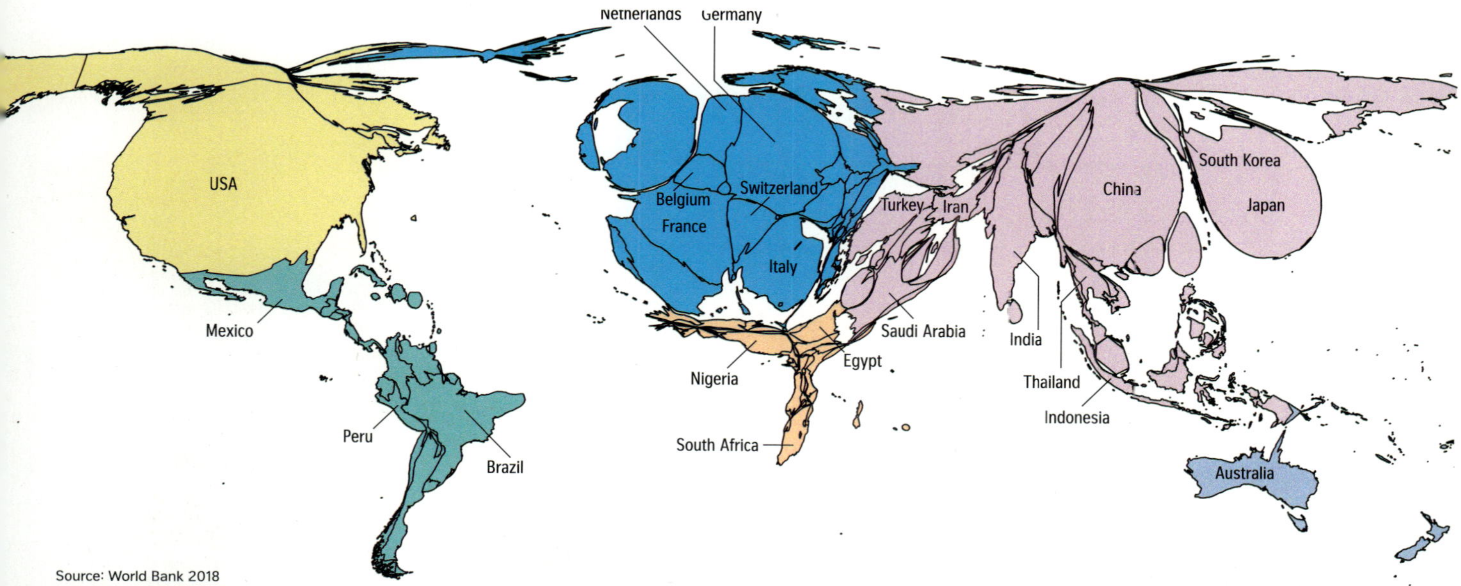

M3 **Rich and poor countries: The size of the area corresponds to the GNI of the respective country in 2016 (GNI data on p. 240).**

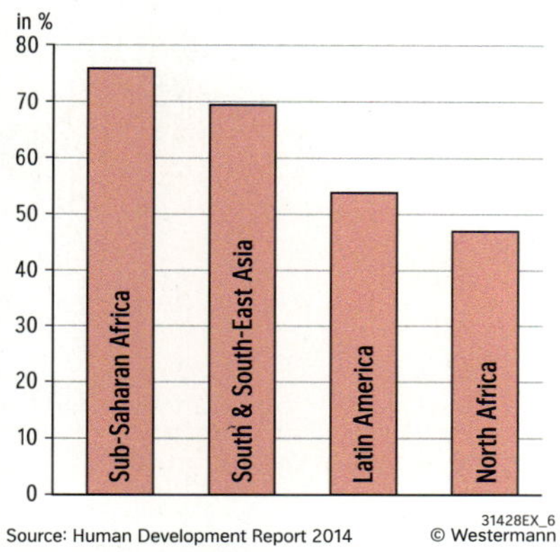

M4 **Share of employment in the informal sector outside agriculture.**

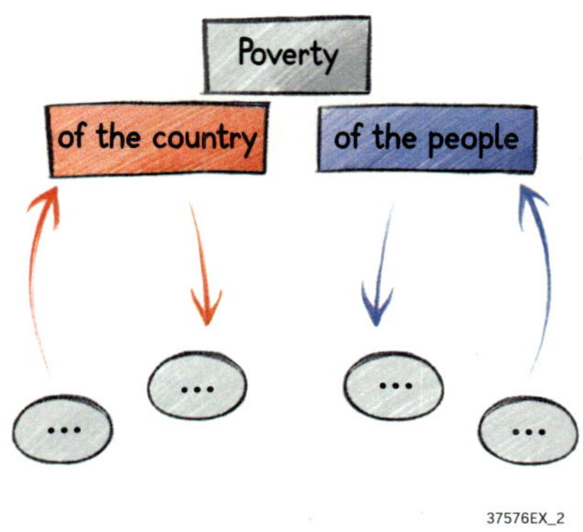

***M7** **Vicious circles ...**

Die Reichen werden reicher, die Zahl der Armen steigt.

Source: Dunz, Kristina: Klingbeil fordert Vermögensteuer als Reaktion auf Spaltung zwischen Arm und Reich. www.rp-online.de, Düsseldorf, 08.10.2019, access: 02.12.2020

Immer größere Kluft zwischen Arm und Reich in Deutschland

Source: jen/DPA: Immer größere Kluft zwischen Arm und Reich in Deutschland. www.stern.de, Hamburg, 23.02.2016, access: 02.12.2020

Globalisierung öffnet die Schere zwischen Arm und Reich

Source: Rövekamp, Marie: Globalisierung öffnet die Schere zwischen Arm und Reich. www.tagesspiegel.de, Berlin, 25.03.2014, access: 02.12.2020

Corona-Krise vergrößert Ungleichheit zwischen Arm und Reich

Source: cr/dpa: Corona-Krise vergrößert Ungleichheit zwischen Arm und Reich. www.manager-magazin.de, Hamburg, 23.09.2020, access: 02.12.2020

M8 **Press reports**

HARD TO BELIEVE!

M5

- The three richest people own more wealth than the combined GNI of the 48 poorest (mostly African) countries on Earth with 568 million inhabitants.
- 20 % of the people consume 86 % of the world's goods.
- 50 % of the world's population are so poor that they have to get by on less than 2 US dollars a day.

CAUTION

M6 **Be careful with averages!**

Economic data is often expressed as averages. For example, **per capita income** is calculated by simply dividing a country's GNI by its population.
But this average value is deceptive: especially in developing countries there are very big differences between rich and poor. In China, there are 373 dollar billionaires (2020) but also around 100 million people living below the **poverty line**.

Wenn du diese Aufgaben erfolgreich bearbeitet hast, kannst du ...

- ... ökonomische Indikatoren beschreiben, mit denen man den Entwicklungsstand eines Landes kennzeichnen kann.
- ... die Aussagekraft von ökonomischen Indikatoren diskutieren.
- ... die Fachbegriffe **gross national income – GNI** (**Bruttonationaleinkommen – BNE**), **informal sector** (**informeller Sektor**), **per capita income** (**Pro-Kopf-Einkommen**) und **poverty line** (**Armutsgrenze**) erklären.

ORIENTATION

One Earth – different worlds

Per capita income, gross national income, literacy rate, growth rate - a single indicator says little about a country's level of development. Therefore, the UN combines various indicators in a development index.
What are these indicators and how helpful are they in classifying countries?

1. Evaluate the informative value of the HDI in comparison to the GNI (M3).

2. Evaluate the informative value of the MPI (M4) in comparison with that of the HDI.

3. a) Name, if possible, two countries per continent from each of the groups shown in the map M1.
b) Which countries are classified as LDCs? Make a list by continent (M1).

(A) **4.** Compare map M1 with the HDI map in your atlas (this usually has a different survey date). Give three examples of changes that have occurred.

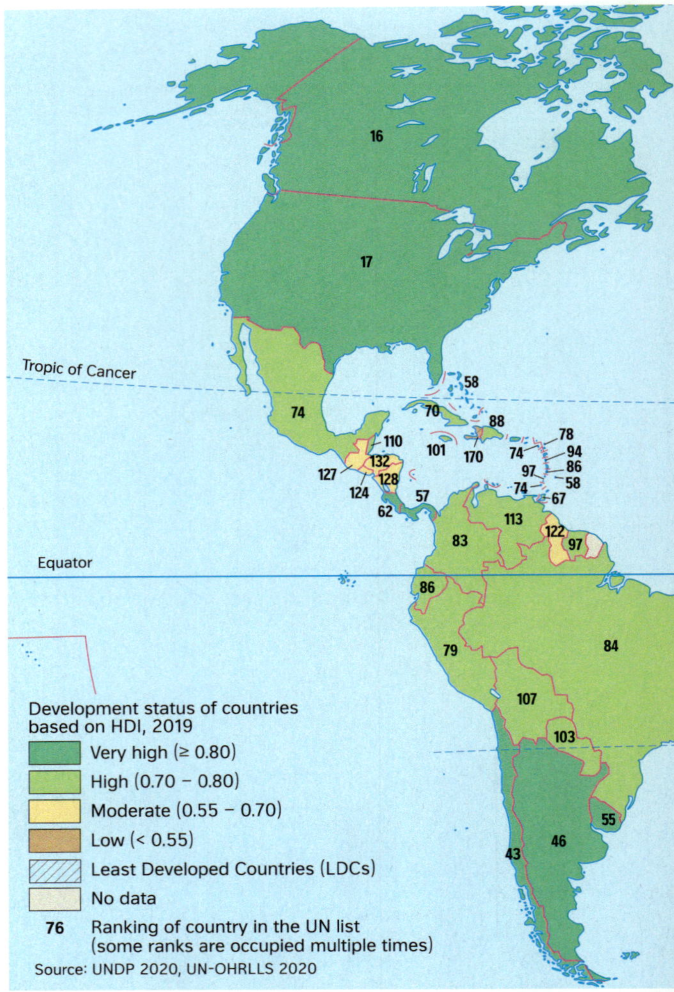

M1 **The world's development status based on HDI**

assets	– Werte/materielle Güter
child mortality	– Kindersterblichkeit
decisive role	– entscheidende Rolle
development assistance	– Entwicklungshilfe
development index	– Entwicklungsindex
expected years of schooling	– erwartete Schulbesuchsdauer in Jahren
mean years of schooling	– durchschnittliche Schulbesuchsdauer in Jahren
sanitation	– Abwassersystem
school attendance	– Schulbesuch/Anwesenheit in der Schule
significance	– Aussagekraft
tertiary sector	– tertiärer Sektor/Dienstleistungssektor

Helpful words and phrases for task 2:

Both metrics calculate ... by taking ... aspects into account
The HDI is more/less detailed, than the MPI ...
The MPI was developed because...
In my opinium MPI/HDI is the better metric, because ...
The MPI/HDI does (not) take into consideration aspects like ...
The advantage/disadvantage of ... as an indicator is that it ... / it does not

Both indicators show / hide ...

INFO

M2 **Different terms for "developing countries":**

Developing country: Poor country with low industrialisation and poor living conditions for most of the people.

Newly industrialising country – NIC: Country "on the threshold of becoming an industrialised country", e.g. Brazil.

Third World: Outdated, discriminatory term for developing countries.

Fourth World: Very poor developing countries (LDCs).

Countries of the South, Global South: Collective term for "developing" and "newly industrialising" countries. To avoid the negative term "developing countries".

One World: Countries should be considered in close interconnection and interdependence. They cannot be considered separately.

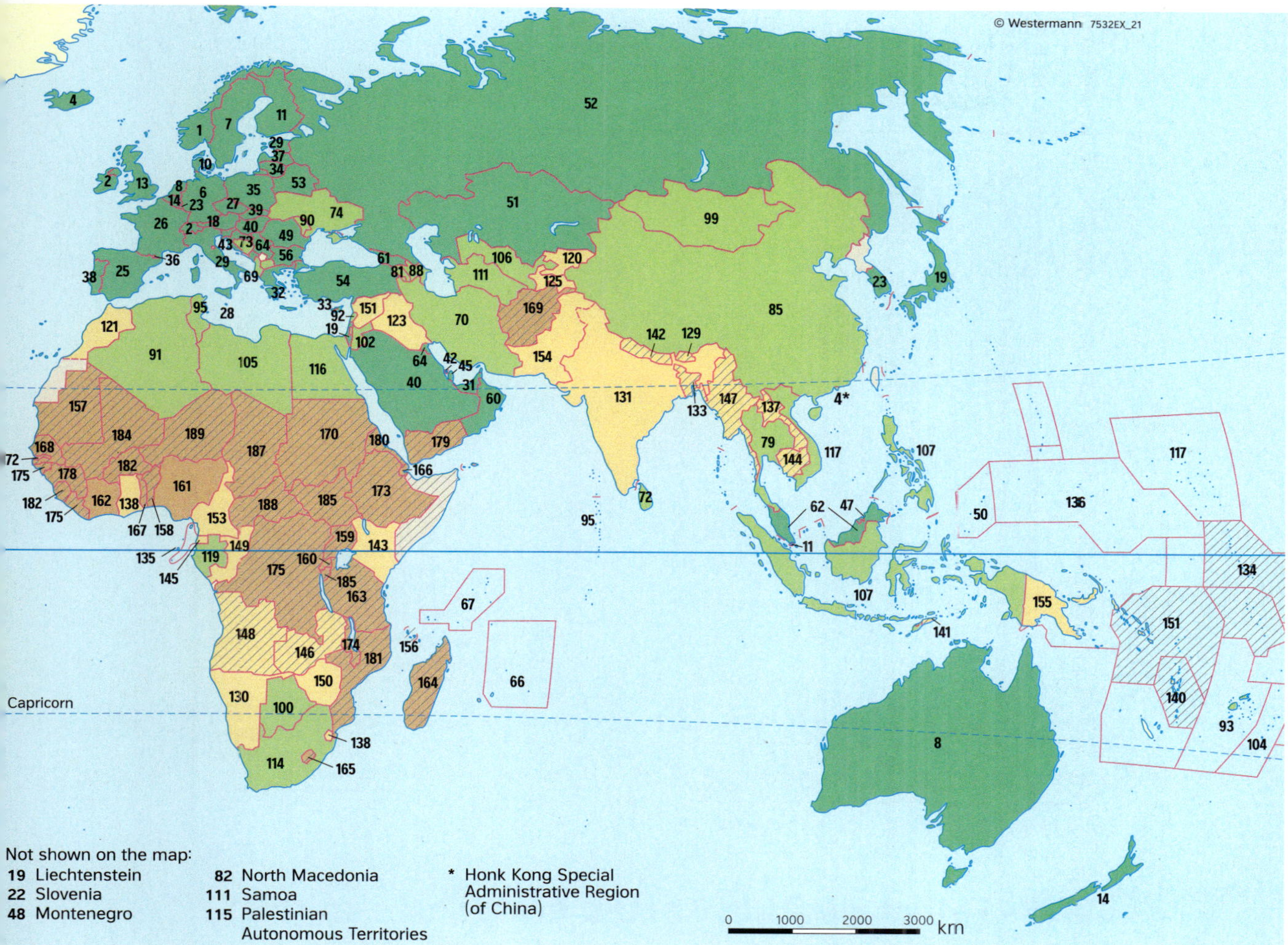

INFO

M3 **The Human Development Index (HDI)**
The UN calculates the **Human Development Index (HDI)** annually. The calculations include per capita income, life expectancy and the level of school education (years of schooling completed for 25-year-olds and expected duration for children when they start school). This results in a value that can be used to rank the countries of the world according to the „degree of human development", i.e. according to the satisfaction of basic needs. According to this classification, Germany, for example, was ranked 6th and India 131st in 2019. However, the decisive factor for the allocation of development aid remains whether a country is recognised by the UN as a **Least Developed Country** (LDC), i.e. as a particularly underdeveloped country. Economic strength (GNI, industrialisation) plays a decisive role in this classification.

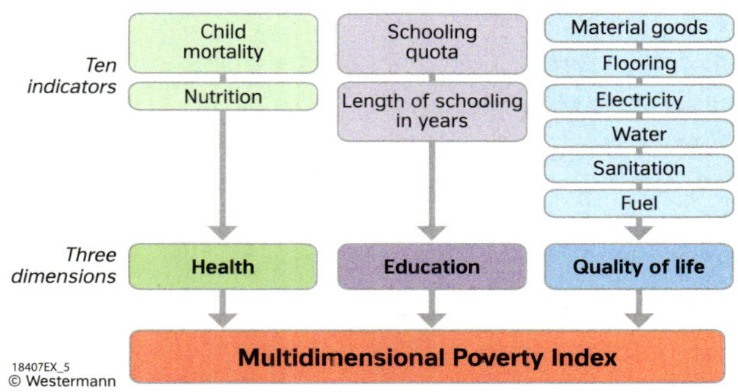

M4 The Multidimensional Poverty Index (MPI)

M5 The Gender Development Index compares the HDI of women and men in a country and can thus measure inequality of opportunity.

Wenn du diese Aufgaben erfolgreich bearbeitet hast, kannst du ...
... unterschiedliche Einteilungen des Entwicklungsstandes der Länder der Erde beschreiben.
... die Lage der Länder mit ähnlichem Entwicklungsstand auf der Erde beschreiben.
... die Fachbegriffe ***newly industrialising country – NIC (Schwellenland), Human Development Index (HDI), Least Developed Country (LDC)*** *erklären.*

industrialisation poor living condition low industrialisation

M

Mithilfe der Zoomfunktionen kann man den Kartenausschnitt verändern: Mit einem Klick auf das Plus vergrößerst du die Karte, zum Verkleinern klicke auf das Minus.

Die Checkbox vor dem Namen eines Themas steuert, ob es sichtbar geschaltet ist.

Über das Tabellensymbol öffnet man die Tabellenansicht eines Themas. Das zweite Symbol ermöglicht eine Änderung der Klassifikation in der Karte.

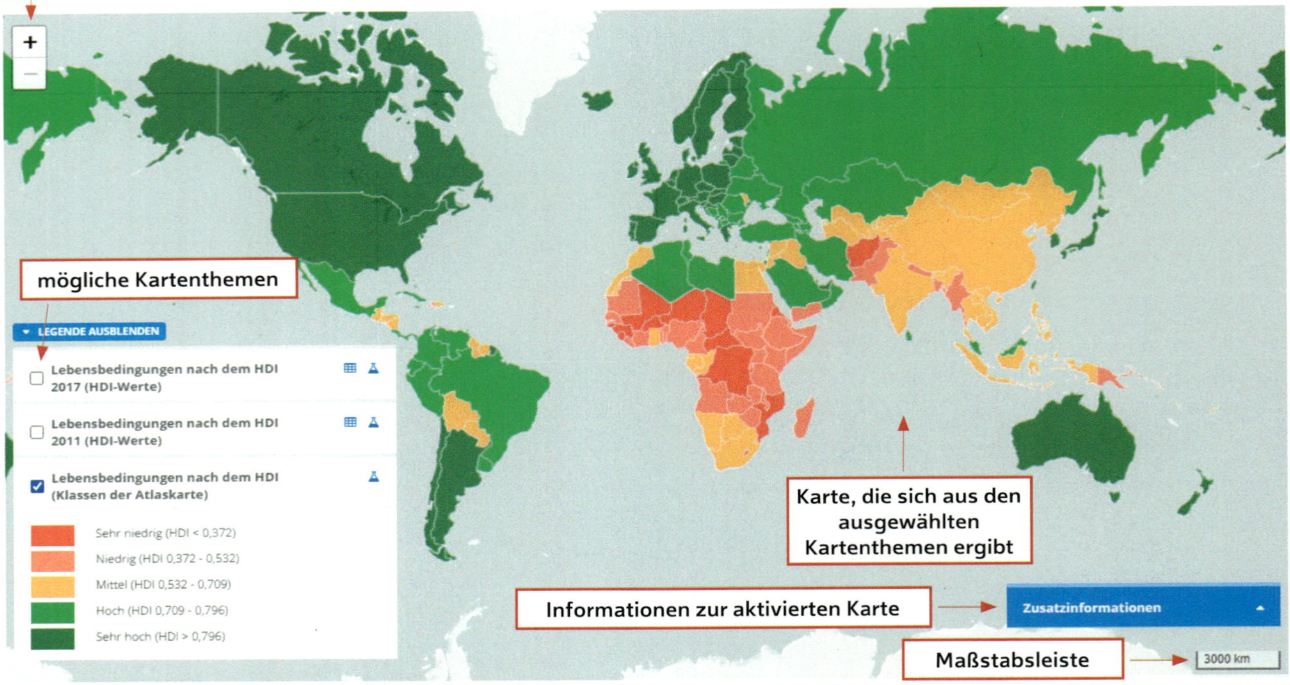

M1 Diercke WebGIS (www.diercke.de/diercke-webgis)

PROJEKT VOR ORT

Selbst digitale Karten erstellen – mit GIS

Zwei Welten, drei Welten, fünf Welten – mithilfe eines GIS kann man aktuelle Karten selbst herstellen und die Inhalte auch selbst bestimmen. Es ist interessant zu sehen, wie selbst kleine Veränderungen der Grenzwerte große Veränderungen im Kartenbild ergeben. Hier einige Anregungen:

1. a) Rufe die Karte zum Entwicklungsstand der Staaten auf,
b) rufe dann Informationen ab: eine Tabelle mit den HDI-Werten. Bestimme die fünf Länder mit dem höchsten und niedrigsten HDI.

2. Erstelle eine Karte zum Entwicklungsstand der Erde
a) mit sechs Klassen (Gruppen),
b) mit drei Klassen (Gruppen),
c) mit zwei Klassen (Gruppen).
d) Entscheide, welche Karte dir am aussagekräftigsten erscheint. Begründe.

3. Angenommen: Du bist für die Pressearbeit eines Staates zuständig und möchtest ihn für mögliche Investoren besonders entwickelt erscheinen lassen. Fertige eine entsprechende Karte. Diskutiere das Ergebnis in deiner Klasse (M2).

M2 Manipulieren leicht gemacht! 215
Ein **Geographisches Informationssystem (GIS)** 204 ist eine Software, in der Gebietsdaten (z. B. Staaten und ihre Grenzen) mit Sachdaten (z. B. Bevölkerungsdichte, Entwicklungsstand) kombiniert werden können. Wenn ihr selbst Karten erstellt, könnt ihr einmal erfahren, wie leicht sich ausdrucksstarke oder auch in ihrer Aussage manipulative Karten herstellen lassen. Dies kann zum einen über die Auswahl der Layer oder der Farben geschehen (grün hat eine positive Konnotation, rot eine eher negative). Es kann aber auch über die Abgrenzung der Klassen geschehen: Welche Grenzen zieht man bei der Zuordnung der Sachdaten? Schon eine Veränderung des Grenzwertes um wenige Hundertstel kann eine völlig andere Karte ergeben.

Diercke WebGIS bietet drei Funktionen: die Darstellung von Informationen in Karten, den Abruf von Informationen in Form von Tabellen und Diagrammen und die Erstellung von digitalen Karten nach eigenen Vorgaben.

Abruf einer Karte

1. Gehe im Browser auf die Seite ***www.diercke.de/diercke-webgis***.
2. Finde den Abschnitt **Erde – Entwicklungsstand der Staaten** und öffne den Kartendienst per Mausklick.
3. Nun kannst du links in der Legende vor den Themennamen deren Sichtbarkeit ein- und ausschalten.
4. Über die Zoomfunktion kannst du die Kartenansicht vergrößern und verkleinern.

Abruf von Informationen

Man kann zu einem Kartenthema Informationen aus der Datenbank abrufen und diese in Tabellenform darstellen lassen.

1. Stelle zunächst sicher, dass nur das gewünschte Kartenthema sichtbar ist.
2. Dann klicke in der Legende auf das Symbol **Tabellenansicht öffnen** neben dem Namen des Kartenthemas. Es öffnet sich die Tabellenansicht, in der alle verfügbaren Informationen zu einem Staat dargestellt werden.
3. Nun kannst du den Namen eines Staates, dessen genauen HDI-Wert oder dessen HDI-Rang notieren.
4. Scrolle die Tabellenansicht weiter nach unten, um die Werte bestimmter Staaten zu finden. Welcher Staat hat den höchsten Wert, welcher den niedrigsten?
5. Schließe die Tabellenansicht über das x und überprüfe anhand der Kartenansicht, wo der Staat mit dem höchsten/niedrigsten HDI liegt.

Veränderung einer Karte

Die Karten im Diercke WebGIS haben alle eine Grundeinstellung. Diese legt fest, welche Farben benutzt werden, wie viele Farbstufen es gibt und welche Farbe welchen Wert darstellen soll (z. B. HDI-Werte in sieben Stufen, grün = sehr hoch).
Man kann auch andere Einteilungen wählen: Willst du z. B. einen HDI mit drei Klassen erhalten, dann kannst du die Grundeinstellung verändern:

1. Stelle die Sichtbarkeit so ein, dass nur das Kartenthema „Lebensbedingungen nach dem HDI (HDI-Werte)“ zu sehen ist.
2. Klicke dann auf das Symbol **Ebene filtern und neu klassifizieren**.
3. Reduziere die Sichtbarkeit auf vier Klassen, indem du bei den drei mittleren Klassen auf das Augensymbol klickst.
4. Verändere die Klassengrenzen so, dass die vier Klassen entstehen (wie in der Abbildung rechts). Bitte achte darauf, dass zur Abtrennung der Dezimalen ein Punkt verwendet wird (US-amerikanische Schreibweise).

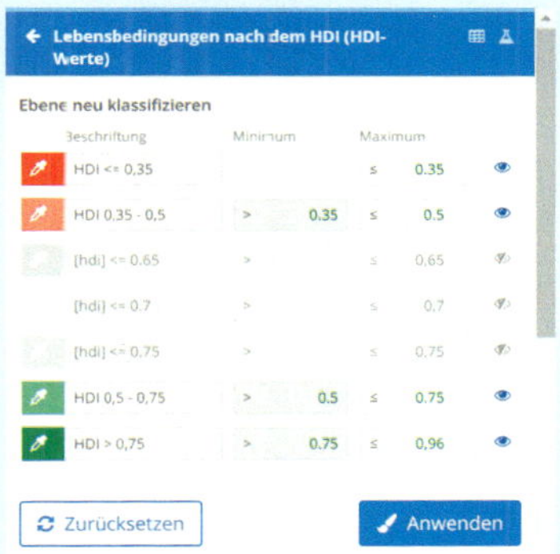

5. Ändere die Beschriftungen der Klassen, sodass sie mit der Klassendefinition übereinstimmen.
6. Wenn du zufrieden bist, klicke auf **Anwenden**. Nun erhältst du eine neue Karte mit den gewünschten Abstufungen und Farben.

M1 France: The Le Moine family in the living room of their flat in a suburb of Paris. They cook on an electric cooker with oven and with a microwave. There is a refrigerator and a freezer for storing food. Their weekly expenditure on food is 356.50 euros.

WHAT WE HAVE LEARNED

What does development mean?

The social situation – social indicators

64 What are the characteristics of development? A distinction is made between social and economic indicators which can be used to determine the level of development of a country. The social indicators include all characteristics that affect people's living conditions. These are primarily the **basic needs** of the people. Basically, it can be said that countries in which the basic needs of most people are not met can be counted as **developing countries**, the so-called **countries of the global South**, while countries in which the basic needs of most people are met can be called **industrialised countries**.

The economic situation – economic indicators

66 The economic indicators take economic development as a yardstick for the level of development of a country. This can be understood in terms of **Gross National Income (GNI)**, or **per capita income**, or the proportion of people living below the **poverty line** in a country. However, two things must be taken into account. Since especially in developing countries very many people work in the **informal sector**, GNI has only limited informative value. Moreover, one must distinguish between the financial situation of the state and that of the people. However, all values mentioned here relating to the situation of the people (e.g. per capita income) are always average values that also have only a very limited informative value.

One Earth – different worlds

One of the most meaningful and widely-used indicators of a country's level of development today is the **Human Development Index (HDI)**. It is calculated from data on education, life expectancy and average income. With the help of the HDI, the UN annually ranks the countries of the world according to their "level of human development". The countries of the world can be divided into different groups according to their level of development: industrialised countries, **newly industrialising countries**, developing countries and **countries of the global South**, **Least Developed Countries (LDCs)**. 68

Create your own digital maps – with GIS

Six, three or even only two worlds - you can try out this division yourself with a **Geographical Information System**. You can determine for yourself the limit values for defining the different groups. The basis for your new maps is a database with current data on the HDI. 70

INTERNET

M2 In its database, *Eurostat* offers data on all areas of the economy and society and thus also on many indicators of development. Most of them refer to the member states of the EU, many to all states in Europe, some also to smaller regions. 212

M

1 Interpret the cartoon M4. 209

2 Interpret the picture of the family (M1). 205
a) What does it tell us about the family's living conditions?
b) Evaluate: To what extent is it typical for an industrialised country?

3 a) Show the GNI per capita of Egypt, Nigeria and Senegal in a diagram (p. 240 / 241).
b) Evaluate the informative value of this data (also using the information in the diagram M4 on page 69).

4 Find three countries from the data on page 240 / 241 that are typical of different phases of the demographic transition because of their birth and death rates.

M4 Cartoon

TRANSFER

Disparities between the states of Europe

Even within Europe, there are large disparities in all areas. With the help of Eurostat, the European Union's database, these disparities can be easily displayed as a table, diagram or map.

C 5 Using Eurostat, first make a table and then a map
A on life expectancy in the countries of Europe.
B on another indicator of your choice.
Explain your approach and the result.

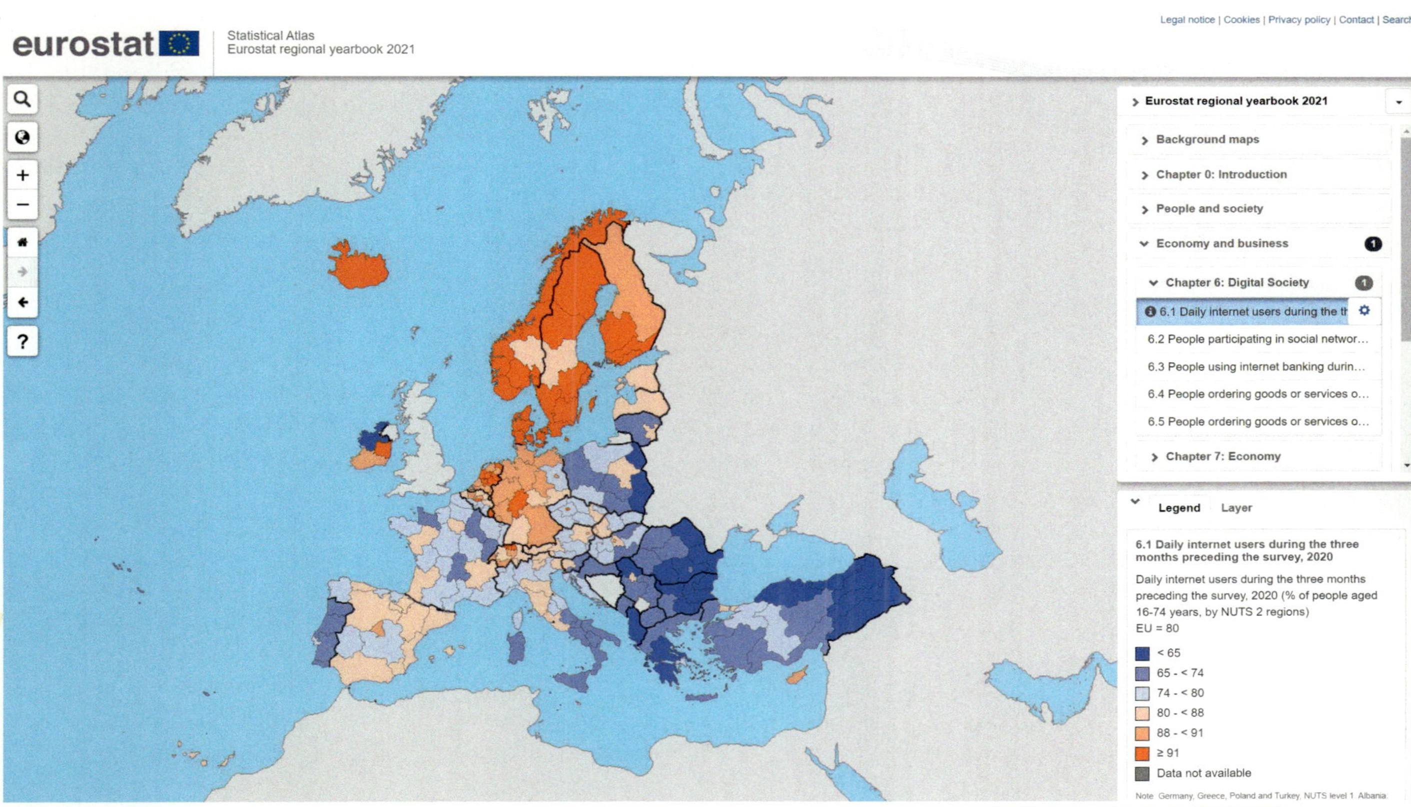

M3 **An indicator of the information society: "Digital Society – Daily internet users". This means the percentage of persons that have used the internet on a daily basis.**

What are the conditions for development? – a puzzle

1. The natural space

Why are living conditions so bad in so many countries around the world? Why are so many countries on Earth poor? Using the South American country of Peru as an example, we want to examine the most important factors that determine development. Peru shows particularly clearly how the factors for development and underdevelopment interact.
How does the natural environment influence the development of a country?

1. a) Locate Peru (atlas). 200
b) Divide Peru into different natural regions. Summarise the regions with similar climate, relief and vegetation (M1, atlas).

2. Make a chart with the three major natural regions of Peru. List (M1 – M4, atlas): 208
a) The suitability for agriculture (what is the usability, under what conditions can agriculture be practised?).
b) The suitability for livestock farming.
c) The conditions for fishing within the country and on the coast.
d) The availability of mineral resources.
e) The possibilities for developing an efficient transport infrastructure.
f) The natural conditions for tourism.

3. Compare the major regions of Peru in terms of their vulnerability (M4, M5, atlas).

4. Evaluate the natural preconditions for economic development.

Helpful words and phrases for task 4:

As per task 3 regions have vulnerabilities, like ...
Economic development depends on ...
... how to reach remote areas ...
... presence of natural resources ...
... accesability of natural resources ...
In summary, Peru has ample ...
In my opinion the economic development of Peru is ...
... could profit from ...
In conclusion, Peru's natural preconditions for economic development are ...
To develop a strong economy a country needs ...
It is problematic, if a country lacks / suffers from ...

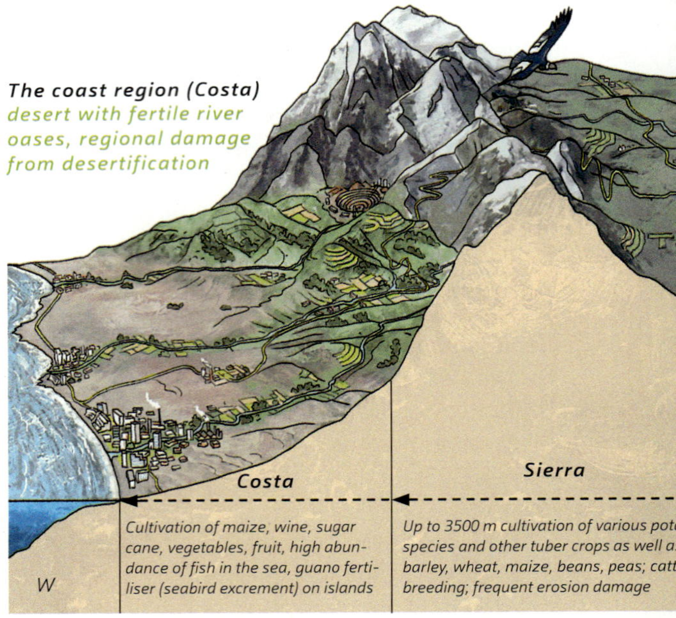

M1 Peru's natural space – three major regions

INFO

M2 Natural preconditions for the economic development of a country

Agriculture: Fertile soil, suitable temperatures, suitable vegetation period, sufficient precipitation, irrigation possibilities, cultivable land, possibilities for livestock farming.
Fisheries: Waters rich in fish.
Mining/industry: Mineral resources, e.g. oil, rare earths; possibilities for expansion of Infrastructure, e.g. navigable rivers, little relief; access to the sea (not land-locked).
Tourism: "Tourism-friendly" temperature and precipitation patterns, lakes and coasts with beaches and not too low water temperatures.

abundance of fish – Fischreichtum
agriculture limit – Anbaugrenze
barley – Gerste
breeding – Zucht
coca cultivation – Cocaanbau
economic sector – Wirtschaftsbereich
fertile – fruchtbar
fertiliser – Dünger
flooding – Überschwemmung
growth period – Wachstumsperiode
landlocked country – Binnenstaat
mineral resources – Bodenschätze
navigable river – schiffbarer Fluss
rare earths – seltene Erden
remote – abgelegen
river oasis – Flussoase
tuber – Knollenfrucht

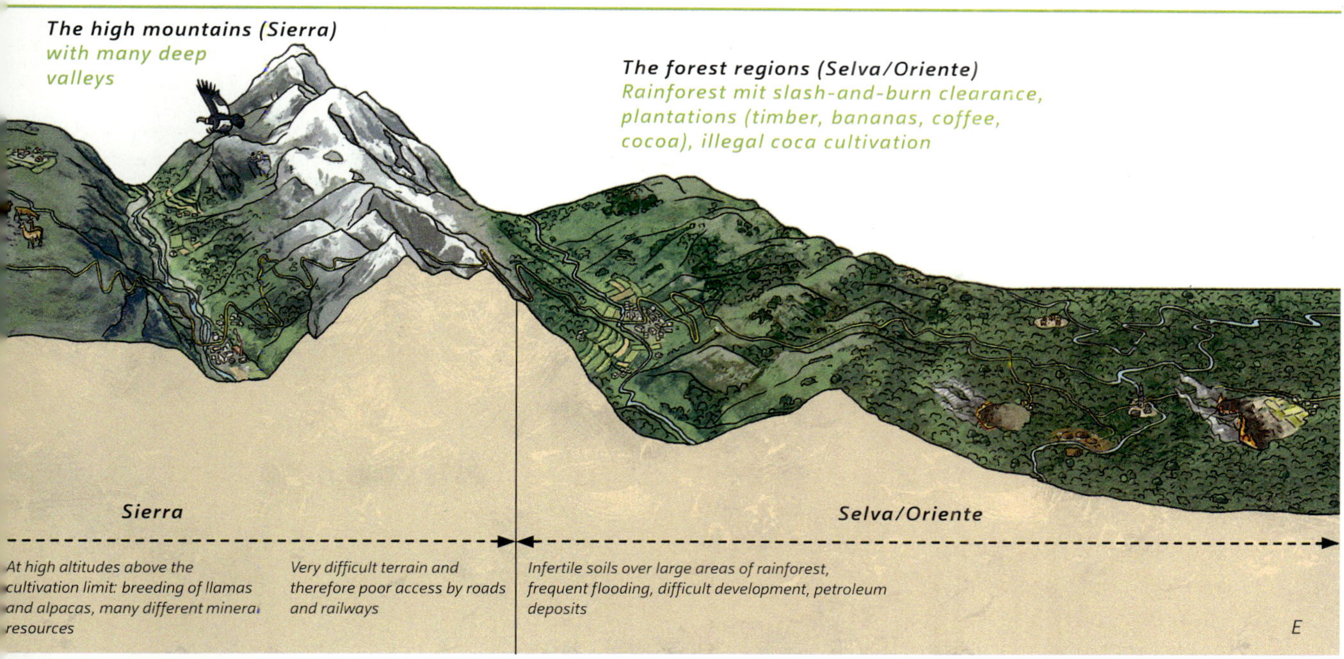

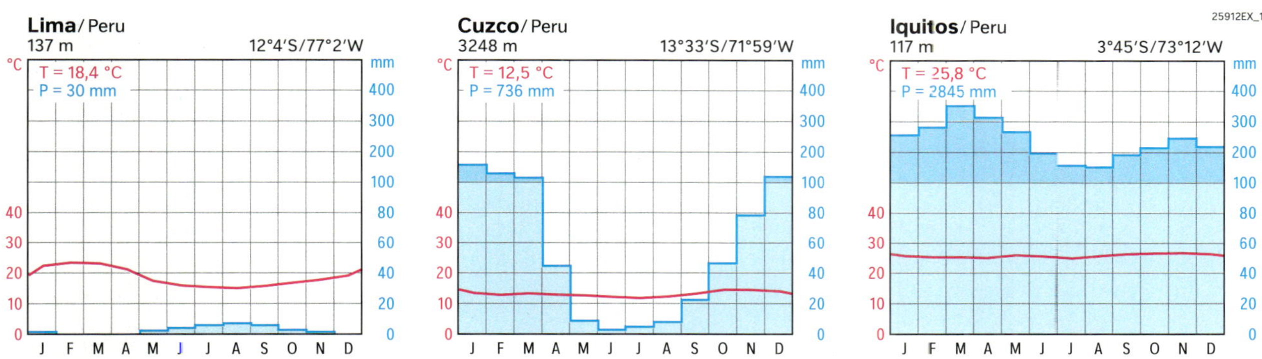

M3 The climate of Lima, Cuzco and Iquitos

An efficient **infrastructure** is an essential prerequisite for the economic development of a country. Only in this way can the remote regions and particularly **structurally weak areas** be included in development. In Peru, a very efficient road network existed even before the colonial era. This was completely neglected during the colonial period. Only the roads to mines and plantations were developed. Today, in terms of infrastructure, Peru ranks among the bottom places in worldwide comparison. There are no modern railway lines and only 13 percent of the roads are considered all-weather roads. The Andes, which are up to 6,700 metres high, divide the country like a barrier and make rail and road construction difficult. Civil engineers also face great challenges in the dune areas on the coast and in the tropical rainforest. Frequent earthquakes and floods add to the problems.

M4 Peru: Natural space and infrastructure

INFO

M5 **Vulnerability**

The probability of natural disasters (e.g. earthquakes, volcanic eruptions, droughts, cyclones, floods, landslides can pose an obstacle to development.

Wenn du diese Aufgaben erfolgreich bearbeitet hast, kannst du ...

- ... erläutern, inwieweit der Naturraum die Entwicklung eines Landes beeinflusst.
- ... die Bedeutung des Naturraums für die Entwicklung eines einzelnen Landes beurteilen
- ... die Fachbegriffe **infrastructure** (**Infrastruktur**), **economically underdeveloped region** (**strukturschwacher Raum**) und **vulnerability** (**Verletzlichkeit**/**Vulnerabilität**) erklären.

2. Population development and social and political conditions

Using the South American country of Peru as an example, we want to examine the most important factors that determine development. Peru illustrates particularly clearly how the factors for development and underdevelopment interact. What is the significance of population development and social and political conditions for the development of a country?

C **1.** Explain and evaluate:
A The composition and spatial distribution of the population (M1, M5, M6).
B The development and age structure of the population (pp. 240/241, M6, M8, atlas). 209
C The social situation of the population (M3, M8).
D The development of the population and usable land (M5, M8, pp. 13 M7).

2. In Peru, 22.3 % of arable land is concentrated among 0.4 % of landowners. Report on the unequal distribution of economic and political power within Peru (M2, M3, M8).

3. Illustrate how bad governance hinders a countries development (M3, M4, M7, M9).

governance	–	Regierungsführung
impeached by	–	abgesetzt von
indigenous	–	indigen/einheimisch
infant mortality rate	–	Säuglingssterblichkeit
investigation	–	Ermittlungsverfahren
margin	–	hier: Rand
mismanagement	–	Misswirtschaft
MP/member of parliament	–	Abgeordnete/r
nepotism	–	Vetternwirtschaft
obstacle	–	Hindernis
predecessor	–	Nachfolger/in
restore confidence	–	Vertrauen wiederherstellen
spatial distribution	–	räumliche Verteilung
sworn in	–	vereidigt

Helpful words and phrases for task 3:
... lack of transparent decisions
... lack of legal security
According to Transparency International ...
Interests of the ruling class prevent ...
Social and economic opportunities are denied ...
Obstacles for investment ...
As a consequence ...
People cannot trust the state

M1 The distribution of population in Peru

The coastal towns and economic centres of the Costa are predominantly inhabited by mestizos and Peruvians of European descent. Political and economic power is also concentrated among these population groups. The largest part of the **indigenous population** lives in the Sierra. Here, the old languages Quechua and Aymará are still spoken, and indigenous traditions are cultivated. In the Selva, the population density is very low. Several indigenous peoples live here alongside settlers, oil workers and gold prospectors.

M2 Ethnic and social disparities

Indigenous people make up the highest proportion of the poor. They live in the mountains or on the outskirts of the big cities. They have little access to services, information, transport, media, markets, education, electricity and water. They have hardly any share in political life and are neither considered nor adequately represented. Indigenous and Afro-Peruvian people are discriminated against more in Peru than in any other country in South America.

M3 Discrimination against the indigenous population

INTERNET

M4 *Transparency International* is an NGO that publishes a corruption index every year. (Peru, rank 94, Germany, rank 9). The *Länderinformations-portal (LIP)* provides reliable country portraits, also for Peru.

M5 The markets are well filled. But population pressure is also increasing in the Sierra. Less and less agricultural land is available per person. There is hardly any work to be found outside agriculture. Many young people emigrate.

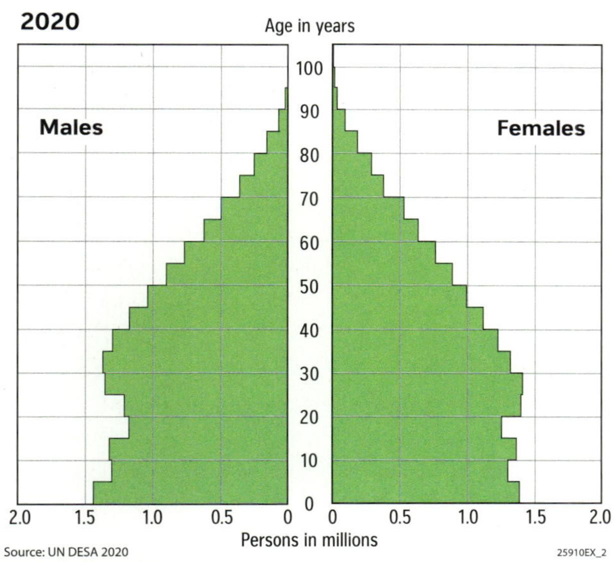

M6 The age structure of Peru's population.

Today, Francisco Sagasti was sworn in as President of Peru - the third president within a week. His predecessor had been deposed by parliament earlier in the week after allegations of corruption. His successor then resigned because of violent demonstrations. The country is in a deep crisis: it has one of the highest mortality rates worldwide in the pandemic and is dramatically losing economic strength. Investigations are underway against about half of the members of parliament ...
It is hoped that the former UN advisor Sagasti will restore confidence in the political process.

Source: Based on radio reports on 16.11.2020

M7 Hope for **good governance**

	2000	2019
HDI ranking	82	82
GNI per capita (real purchasing power)	5,310 in US-$ (2002)	12,760 in US-$
Informally employed outside agriculture	77%	59%
Land area used for agriculture per inhabitant	0.88 ha	0.75 ha
Infant mortality	3.8%	1.3%

Source: FAO, INEI, UNDP, UNICEF, World Bank, for other data see p. 389/390

M8 Developments in Peru

INFO

M9 **Bad governance as an obstacle to development**

Most developing countries are economically and politically dominated by a small, rich, educated upper class. They are not interested in reducing social disparities. Thus, enrichment, nepotism and corruption often paralyse development. This often involves international companies that are interested in government contracts or mining rights, for example.
Part of this **bad governance** is also mismanagement, which often leads to extreme inflation. Zimbabwe holds the record with an annual inflation rate of over 89 trillion (= 89 x 10^{21}) percent (2008).

Wenn du diese Aufgaben erfolgreich bearbeitet hast, kannst du ...
- *... die Auswirkungen einer ungerechten Macht- und Besitzverteilung erklären.*
- *... Bevölkerungswachstum und Bevölkerungsstruktur als mögliche Hemmschuhe für Entwicklung darstellen.*
- *... die Fachbegriffe **indigenous people** (**indigene Bevölkerung**), **bad governance** und **good governance** erklären.*

3. Integration into the world economy – now and in the past

Using the South American country of Peru as an example, we want to examine the most important factors that determine development. Peru shows particularly clearly how the factors for development and underdevelopment interact. What is the importance of integration into the world economy for the development of a country – now and in the past?

1. Describe the economic structures that have their roots in colonial times (M1, M4, M5). 210
2. a) Describe the development of raw material prices (M3).
   b) Give reasons why the use of index numbers in M3 makes sense. 217
3. Explain why it is problematic for countries to export mainly agricultural and mining raw materials (M3, M5 – M7).
4. (A) Find out which African countries used to be colonies of the German Empire and what their current level of development is (atlas).

colonial era	–	Kolonialzeit
colonial power	–	Kolonialmacht
commodity prices	–	Rohstoffpreise
conqueror	–	Eroberer/in
German Empire	–	Deutsches Kaiserreich
exploit	–	ausbeuten
fluctuate	–	schwanken
large estates	–	Großgrundbesitz
mineral resources	–	Bodenschätze
peace and order	–	Ruhe und Ordnung
pest infestation	–	Schädlingsbefall
ratio	–	Verhältnis
rubber	–	hier: Kautschuk
supply and demand	–	Angebot und Nachfrage
tariffs	–	Zölle

Helpful words and phrases for task 3:

... the price of ... fluctuates
Prices are dependent on ...
Primary products are often cheaper than manufactured goods ...
Manufacturing goods creates jobs and income ...
In the meantime, prices for commodities steadiliy ...
As a result ...
A country can become vulnerable to ..., if it is dependent on ...

The Spanish conquistadors were already attracted by the rich mineral resources in the Andes region in the 16th century. After the Viceroyalty of Peru was founded in 1542, the colonial rulers set up numerous mines, mainly to exploit gold and silver. The indigenous population was forced to work in the mines. 3,840 tons of gold were shipped to Spain during the colonial period – a cornerstone for the economy that flourished in Europe.
In the **colony**, large tracts of land were distributed to the conquerors. They had the responsibility to ensure "peace and order" on their lands. They did this with great brutality. On the large estates, the workers produced agricultural raw materials for export, for example wool in the highlands, cotton on the coast and rubber in the Amazon lowlands. In order to be able to ship raw materials to Europe quickly, the city of Lima was founded on the coast, which was also declared to be the capital. The country's entire transport network was oriented towards this. The raw materials were used to produce finished goods in Europe, which were then sold back to the colonies. To ensure that this trade continued, the colonial powers hindered or prohibited the development of industry in the colonies.

M1 The colonial period

UNBELIEVABLE

M2 Due to poor living conditions in the colonies, wars and epidemics, Peru's indigenous population was decimated from about 10 – 15 million to about one million during the colonial period.

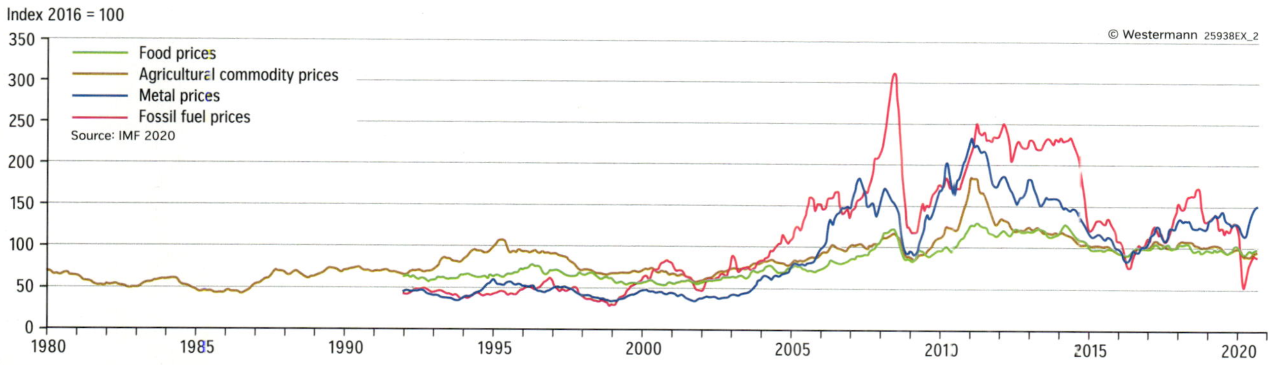

M3 The development of commodity prices on the world market

Imports: US-$ 42.4 billion; goods: 14% fuel, 13% machinery, 10% electronics, 10% food, 9% motor vehicles and vehicle parts.
Exports: US-$ 46.1 billion; goods: 66% mining products (e.g. silver[1], copper[2], zinc[2], lead[1], molybdenum[3], gold[6], petroleum, uranium[7], in future: lithium), 16% agricultural products (e.g. blueberries[1] and asparagus[2], mangoes, sugar cane).

[1-7]: respective world rank Source: Fresh Plaza, UN Comtrade, World Bank

M4 The structure of Peru's foreign trade (2019)

INFO

M6 **Terms of trade**
The relationship between the prices of export goods and import goods. The relationship deteriorates for a country, for example, if export prices fall and import prices rise or if export prices rise more slowly than import prices.

Even today, many developing countries mainly export raw materials and import finished goods. However, the prices for raw materials on the world market fluctuate very strongly and have also often fallen when there is an oversupply. Zinc and copper, for example, were cheaper at times around 2000 than 30 years earlier. In addition, in agriculture it is precisely the export-oriented monocultures that are susceptible to pest infestation. In contrast to raw material prices, the prices of high-value finished goods (e.g. machinery) have risen steadily. Many developing countries must therefore pay more and more for imported manufactured goods (even for the same quantity), while the revenues for exports do not increase to the same extent. This means that the **terms of trade** are getting worse.
In order to still be able to buy manufactured goods that are urgently needed for development, a developing country can react in two ways. The first way is to increase the production of raw materials – but this can lead to a greater supply on the world market and a fall in prices. Many states have therefore incurred debts to the rich countries. Thus, **foreign debt** has been increasing for years.

M5 The developing countries in world trade

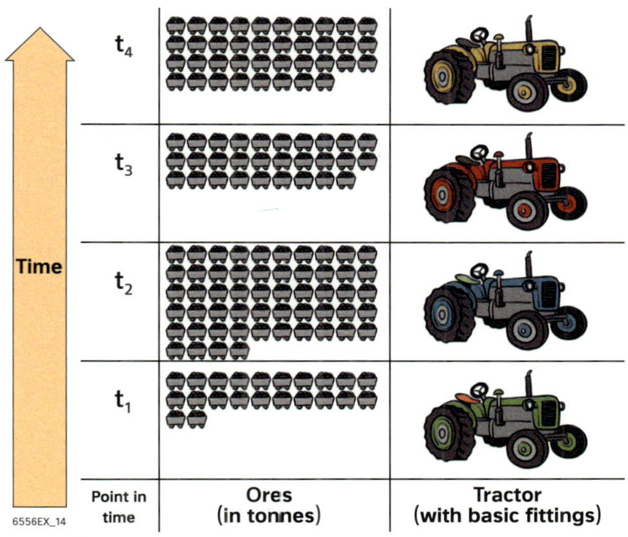

M7 Development of the terms of trade over decades (model)

INTERNET

M8 Country information portal of the *Gesellschaft für internationale Zusammenarbeit (GIZ)*

Wenn du diese Aufgaben erfolgreich bearbeitet hast, kannst du …
- *… erklären, inwieweit die Kolonialzeit die heutigen wirtschaftlichen Strukturen der Entwicklungsländer geprägt hat.*
- *… erläutern, inwieweit der überwiegende Export von Rohstoffen problematisch ist.*
- *… die Handelsstruktur eines Landes beurteilen.*
- *… die Fachbegriffe **colony** (**Kolonie**), **Terms of Trade** (**Handelsbedingungen**) und **foreign debt** (**Auslandsverschuldung**) erklären.*

4. The extent of disparities within the country

Using the South American country of Peru as an example, we want to examine the most important factors that determine development. Peru shows particularly clearly how the factors for development and underdevelopment interact. What is the significance of the extent of regional disparities within the country for the country's development?

1. a) Describe the spatial disparities within Peru (M1, M2).
b) Give reasons why one can speak of a "fragmentation" of the country (M8).

2. Describe the effects of disparities on population distribution (M1, M2, M5).
A Draw a sketch map with probable migration flows.
B Prepare a two-minute radio report on this.

3. a) Describe the importance of Lima for Peru (M4, M5).
b) Give reasons why one can speak of a "fragmentation" of the city (M7, M8).

4. Disparities hinder the development of the country. Comment on this statement.

branch	–	hier: Niederlassung
business	–	Wirtschaftsbetrieb
district	–	Stadtteil
headquarters (HQ)	–	Hauptsitz
land use	–	Flächennutzung
location factors	–	Standortfaktoren
rural areas	–	ländliche Gebiete
significant	–	nennenswert
theme park	–	Vergnügungspark
World Heritage Sites	–	Weltkulturerbe

Helpful words and phrases for task 4:
Lack of purchasing power / ... opportunities / ... investment / ... jobs / ... social services
The economic activity is concentrated ...
Investment by the government concentrates on ...
Parts of the population are (not) ...
Contribution by parts of the population is ...
Basic needs of the population are not met, which means that ...

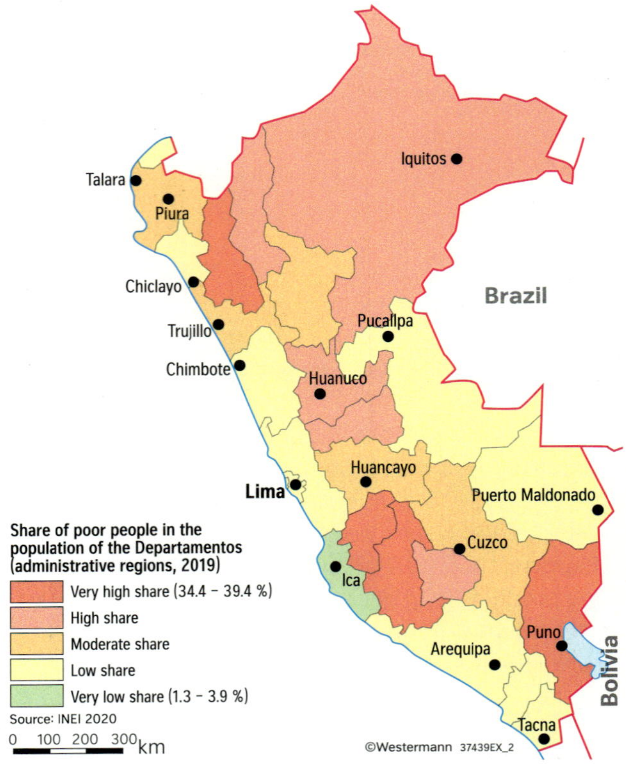

M1 **Spatial disparities** within Peru: the distribution of poverty

Economic power is concentrated on the coast: modern, export-oriented large-scale agricultural enterprises with very high productivity are found here. As in all developing countries, economic life and political power are concentrated in the cities. That is why by far the majority of businesses are located there. Because it is only there that good location factors exist, e.g. good transport connections, sufficient energy supply, telecommunication infrastructure, trained skilled workers and sufficient workers with basic education. Finally, the cities are also the sales market with the most purchasing power for industrial products. Lima, with its port city of Callao, is home to all the important state institutions, the management, sales and research departments of national companies, as well as branches of foreign corporations. The wealthy upper class, which includes the employees of international companies, isolates itself in gated communities in certain parts of the city. It is pleasant to live here: there are parks, air-conditioned flats with swimming pools and shopping malls – right next to the growing slums and shanty towns. These are the first port of call for tens of thousands of migrants who stream from the rural areas to the coastal cities and especially to Lima every year.

M2 In the Costa – the cities on the coast

M3 Urban districts of Lima: Miraflores and a shanty town on the Panamericana (North Lima)

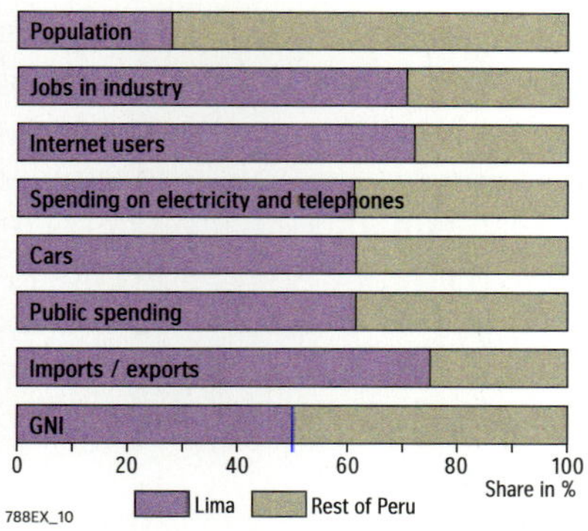

M4 The importance of Lima within Peru

Year	Population.
2020	11.0 million
2000	6.3 million
1981	4.0 million
1961	1.3 million
1940	0.5 million

For comparison, Peru's second largest agglomeration: Arequipa (2020): 1.1 million.

M5 The population development of Lima (with the Port of Callao)

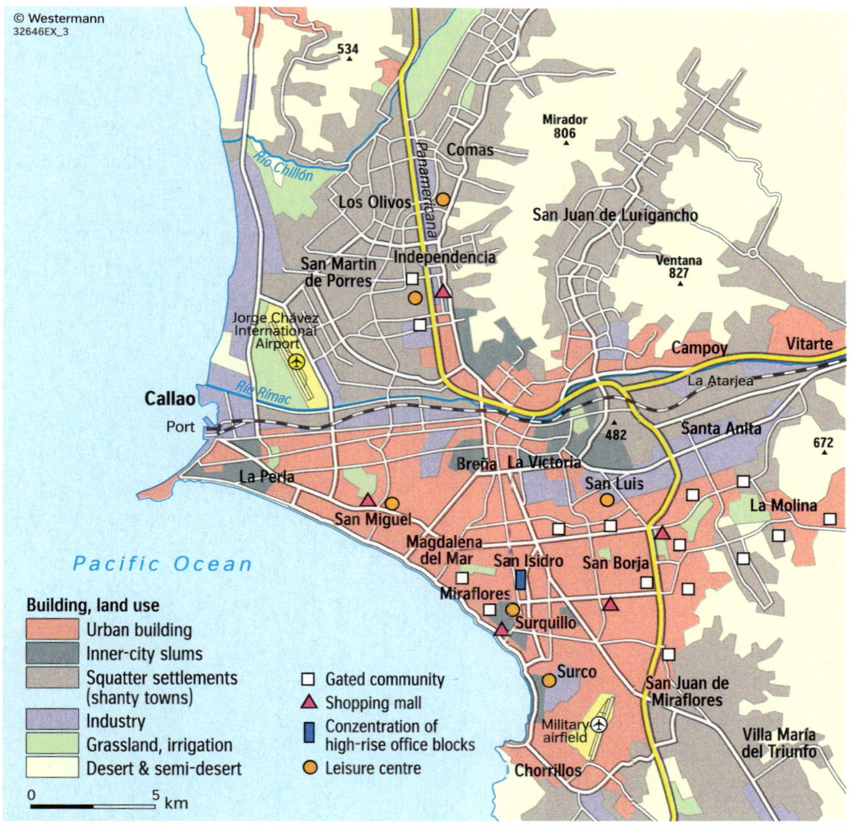

*M7 Lima with the Port of Callao – land use

The rural areas in the Sierra and Oriente are largely cut off from modern economic life. Work is almost only available in mining and agriculture - mostly around small farmers. There are hardly any modern businesses. Only at the culturally important sites, such as the pre-colonial Inca capital of Cusco and the Inca fortress of Machu Picchu, is there any tourism worth mentioning. For years, the growing population pressure, poverty, unemployment and underemployment have led to a strong rural-urban migration.

M6 Sierra and Oriente

INFO

M8 **Fragmentation** – Division of a space into different elements: Poor residential areas bordering high-rise administrative buildings of global players next to regional companies or businesses in the informal sector.

Wenn du diese Aufgaben erfolgreich bearbeitet hast, kannst du ...

... erläutern, warum Lima das wirtschaftliche und demographische Zentrum Perus ist.

... die Stadt-Land-Disparitäten in Entwicklungsländern und ihre Auswirkungen am Beispiel Perus beschreiben und begründen.

*... die Fachbegriffe **spatial disparities (räumliche Disparität)** und **fragmentation (Fragmentierung)** erklären.*

WHAT WE HAVE LEARNED

What are the conditions for development? – a puzzle

1. The natural space

74 There are different reasons why the living conditions of people in developing countries are so bad and why the countries are so poor.

One reason for some countries is the natural conditions. It is important for the development of a country that there are favourable conditions for agriculture (e.g. enough rainfall at the right time, not too steep land for the fields), for fishing, for mining and industry (mineral resources, infrastructure) and for tourism.

It is particularly important that a country is equipped with good **infrastructure** and that the peripheral, **structurally weak regions** are integrated into the economy. In this context, the **vulnerability** of the country also plays a major role: frequent natural disasters can quickly destroy any progress.

2. Population development and social and political conditions

76 The composition of the population and an unfavourable population distribution can also hinder the development of a country. The same applies to a strongly growing population. This increases population pressure, especially in rural areas.

In many countries, minorities are discriminated against. This is especially true for the **indigenous population**. In a large proportion of developing countries, economic and political power is concentrated among a small, wealthy population group. As a result, the development of the country is dependent on this small class. With **good governance**, development opportunities are offered to the broad population, but with **bad governance**, only a few can benefit. Bad governance is often accompanied by corruption and mismanagement.

3. Integration into the world economy – now and in the past

78 The economies of developing countries are often shaped by the colonial era. At that time, the main focus was on the exploitation of raw materials and their export to Europe. To this end, mines were built in the **colonies** and large farms were established that cultivated agricultural raw materials in monocultures for the European market. Ports were built on the coasts and the road network was designed to get the raw materials there as quickly as possible. In Europe, the raw materials supplied were used to produce finished goods, which were then exported back to the colonies.

Even today, this structure of foreign trade has often not changed much; developing countries mainly export raw materials and import finished goods. This brings with it various problems: raw material prices on the world market fluctuate strongly and in agriculture there are repeated crop failures (e.g. due to pest infestations).

In addition, the prices for imported finished goods are rising faster than the revenues for exported raw materials. This deterioration in the **terms of trade** increases **foreign debt**.

4. The extent of disparities within the country

Within developing countries there are often very large **spatial disparities**. Within the country, rich regions lie next to poor ones, economically active cities with international companies next to poor agricultural regions. This **fragmentation** hinders the development of the country. It leads to migration to the cities. Often, an aged population remains in the countryside. 80

The cities are the centres of the economy (Lima, for example, is by far the most important city in Peru in all sectors). They have very good location factors and many cheap and well-educated workers. For this reason, business operations are also concentrated in the cities.

Here too, there is fragmentation: rich districts border on poor ones, gated communities lie next to shanty towns and slums. Industrial areas with international companies adjoin companies active on the local market and businesses in the informal economy.

M1 Cartoon (based on a Peruvian magazine): "Education in times of pandemic"

1 Interpret the cartoon M1. 209

2 Allocate the terms to the different boxes (M2).

There are different factors that determine whether a country progresses in its development or not. In the last few pages you have learned about some important factors for development and underdevelopment. They all interact:

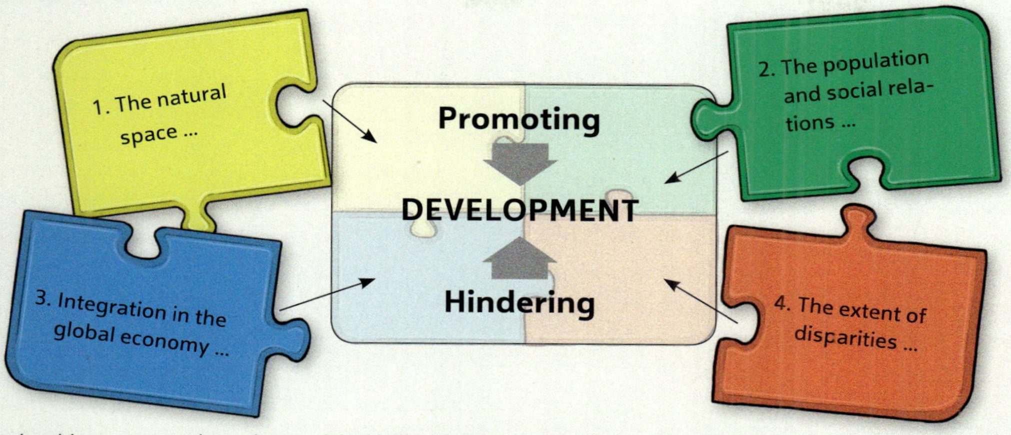

Terms: bad harvests – beaches – big landowners – climate – colony – commodity prices – corruption – droughts – earthquakes – economically active cities – export of raw materials – external debt – flooding – fluctuation in world market prices – foreign trade -bad governance – fragmentation – gated communities – good governance – import of manufactured goods – indigenous population – inflation – infrastructure – international companies – marginal settlements – mineral resources – mismanagement – monocultures – poor remote areas – population distribution – population growth – population structure – relief – rich fishing grounds – rural-urban migration – slums – social conditions – soil fertility – sufficient precipitation – terms of trade – tourist attractions – traffic network only leads to harbours – tropical cyclones – very good location factors – volcanism – vulnerability

M2 **Factors that promote or hinder the development of a country.**

TRANSFER

Disparities and structurally weak regions within the countries of Europe

There are also major spatial disparities within the European states (unemployment, population development, employment structure).
These regional disparities can be clarified and illustrated with the help of the Eurostat database or Diercke WebGIS.

3 Using Diercke WebGIS, create a map of Europe that shows the disparities within the countries of the EU. 64, 204

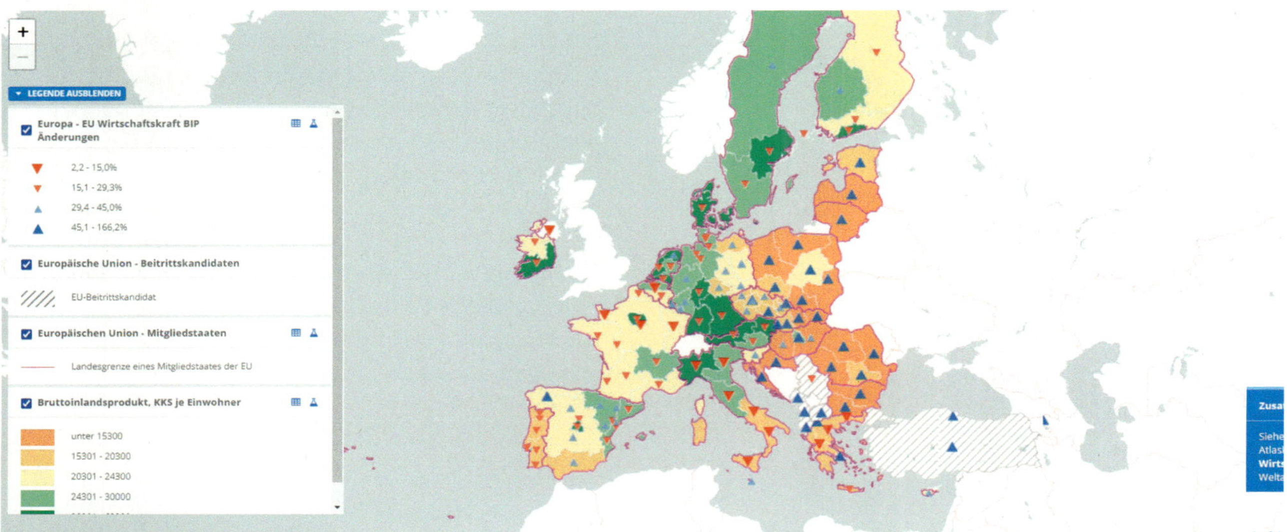

M3 **Economic power in the European Union – a digital map from Diercke WebGIS**

Improvement in sight? – strategies and measures for development

Which is the right way?

All people on the Earth should be able to live in a way that is fit for humans. What are the strategies for development? Where do measures have to start? Are changes in the developing countries enough or do the industrialised countries also need to change?

1. Give an example of an obstacle to development, a resulting strategy, a possible development project and its impact (M2).
2. Describe the guiding principles of development cooperation (M1, M3).
3. Compare the technical terms "development cooperation" and "disaster relief".
4. Development has many facets. Report on a development project of your choice (M1, M6).

development cooperation	–	Entwicklungszusammenarbeit (EZ)
development strategies	–	Entwicklungsstrategien
disaster relief	–	Katastrophenhilfe
emergency shelter	–	Notunterkunft
famine relief	–	Hungerhilfe
humanitarian relief	–	humanitäre Hilfe
joint effort	–	gemeinsame Anstrengung
refugee camp	–	Flüchtlingslager
shanty	–	Baracke
sustainable goals	–	nachhaltige Ziele
unfavourable	–	ungünstig
water treatment plant	–	Wasseraufbereitungsanlage
pursue a goal	–	ein Ziel verfolgen

Helpful words and phrases for task 3:

The term ... focuses on ...
In contrast
... long term / short term
... helps to establish economic activity / social security
... helps to survive
... sustainable development

INTERNET

M1 The *Bundesministerium für wirtschaftliche Zusammenarbeit und Entwicklung (BMZ)* is responsible for Germany's economic development cooperation.

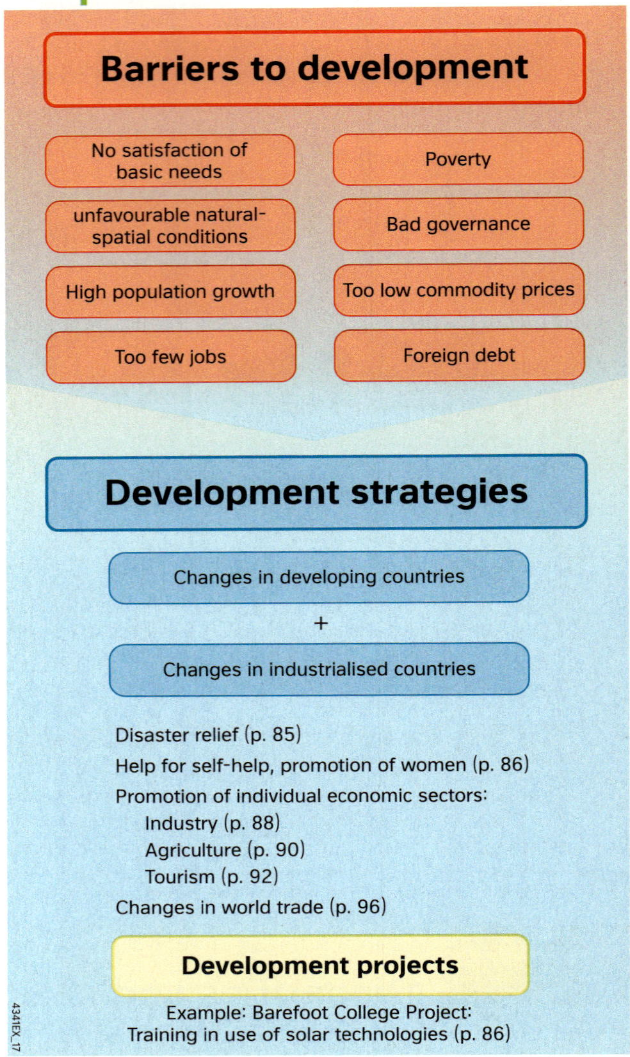

M2 Strategies

Development cooperation refers to the joint efforts of so-called industrialised countries and developing countries to reduce social and economic differences as well as differences in the general living conditions between these two groups of countries. The term development aid, on the other hand, is considered outdated because it focuses too much on the supposed need for help in less developed countries. Development cooperation, by contrast, emphasises the equal role of donor and recipient countries. It primarily pursues the goal [...] of sustainably improving living conditions and making local people independent of external aid in the long term. Development cooperation usually pursues long-term, sustainable goals and structural changes.

Source: Freiwilligenarbeit.de: Entwicklungszusammenarbeit. www.freiwilligenarbeit.de, access: 02.12.2020

M3 Short-term aid and long-term development cooperation

The Zaatari refugee camp in northern Jordan:
Established in 2012 as an emergency shelter for Syrian refugees, it developed into a permanent settlement with more than 77,000 inhabitants (2020).

The "main shopping street" of the refugee camp

M4 **Disaster relief activities aim to meet basic needs in the short term. These relief measures mainly consist of food deliveries, the installation of water treatment plants, the construction of emergency shelters and the provision of medical care. Inadequate or even destroyed infrastructure is often a major obstacle.**

M5 **In many slums, shanty settlements and villages there are no water pipes. Drinking water must be carried for kilometres or often bought by the litre. The creation of an efficient water supply, as here in India, is also part of the satisfaction of basic needs.**

M7 **In one third of all countries worldwide, there are only ten doctors for every 10,000 people – and doctors' practices are concentrated mainly in the cities. Therefore, health care in rural areas must be strengthened, as here in Niger, where – with the help of *Doctors Without Borders* – health care is made possible even far away from the cities.**

INTERNET

M6 All major **non-governmental organisations (NGOs)** active in development cooperation also present their measures and goals in the internet: e.g. *Misereor, Brot für die Welt, Welthungerhilfe, Ärzte ohne Grenzen, terre des hommes, Menschen für Menschen, Oxfam, Fian.*

Wenn du diese Aufgaben erfolgreich bearbeitet hast, kannst du ...

- *... unterschiedliche Ansatzpunkte der Entwicklungszusammenarbeit beschreiben.*
- *... die Befriedigung der Grundbedürfnisse als wichtiges Ziel von Katastrophenhilfe und langfristiger Entwicklungszusammenarbeit erkennen.*
- *... die Fachbegriffe **development cooperation** (**Entwicklungszusammenarbeit**), **disaster relief** (**Katastrophenhilfe**) und **non-governmental organisation – NGO** (**Nichtregierungsorganisation – NRO**) erklären.*

Sustainable development – help for self-help

**The goal of all development cooperation is to achieve sustainable results, to help people in a sustainable way.
But what exactly does "sustainable development" mean? What strategies and measures are there to achieve it?**

1. a) Despcribe the meaning of sustainable development using an example from the projects mentioned on this double page.
b) Present the effects of one of the projects mentioned in a causal network.

2. Report on the measures and goals of the Barefoot College project using an example of a sub-project (M1, M3), in the form of:
A a newspaper report.
B an audio document (2 – 3 minutes).

3. Which Sustainable Development Goals (SDGs) of the 2030 Agenda do you think are the three most important for the development of countries in the South? Give reasons (M4, M6, p. 15).

co-owner	–	Miteigentümer
hoe	–	Hacke
irrigation agriculture	–	Bewässerungslandwirtschaft
lend money	–	Geld verleihen
life-insurance	–	Lebensversicherung
millet	–	Hirse
Nobel Peace Prize	–	Friedensnobelpreis
official development assistance (ODA)	–	öffentliche Entwicklungszusammenarbeit
training	–	hier: Ausbildung
womens' cooperative	–	Frauen-Kooperative

Helpful words and phrases for task 2:
The Barefoot College project is...
They have different projects focussing on...
The program...
The goal is to...
So far... / The problem is / was...
Now people...
The impact is that....
Life for the people has changed because...

INTERNET

M1 
The *Barefoot College* has its own website with extensive picture material and videos.

INFO

M2

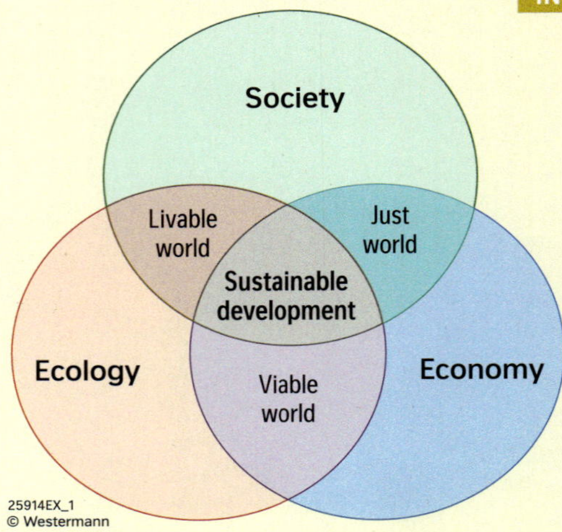

The guiding principle of **sustainable development**

Sustainability is an important principle of national and international policy. It aims to enable economic prosperity, ensure social balance and preserve the natural foundations of life for future generations. [...]
Sustainable development is defined as "development that meets the needs of the present without compromising the ability of future generations to meet their own needs".

Source: Bauer, Steffen: Leitbild der Nachhaltigen Entwicklung. www.bpb.de, 06.05.2008, access: 02.12.2020

Providing **help for self-help** is a principle for many organisations active in development cooperation.
Non-governmental organisations (NGOs) in particular support small projects that enable people to help themselves in a sustainable way. In the slums of Cairo, for example, young men are trained as carpenters; in the Bolivian Andes, there are sewing courses for peasant women and training in irrigation farming for farmers; and in the arid regions of northern India, thousands of women and children have been trained in the use of solar energy at the Barefoot College since 1989. Over 500,000 people have already been supplied with electricity in their villages as a result.

M3 **Help for self-help**

9 INDUSTRY, INNOVATION AND INFRASTRUCTURE
15 LIFE ON LAND

M4 **Little use of machinery, labour-intensive production – this is a strategy to provide a secure livelihood for as many families as possible. If other important projects can be realised at the same time (here: tarring a village road in southern India), the impact is even greater.**

- The Rural Women's Bank is not owned by individual women, but by around 12,000 small village communities (women's cooperatives).
- Each group has its own account and decides jointly on the use of the money.
- Members mutually guarantee each other's loans.
- Only when the borrowed money has been repaid is a new loan granted.
- The borrowers must explain what they are using the money for. Only sustainable investments are supported (e.g. seeds, rikshas, sewing machines, bikes, livestock)
- Life insurance for shareholders is often included.

M5 **Organisation of the Rural Women's Bank in the Indian state of Tamil Nadu**

1 NO POVERTY
5 GENDER EQUALITY
15 LIFE ON LAND

M7 **Members' meeting in the bank building on the village square (India)**

In 1983, the Bangladeshi scientist Professor Muhammad Yunus founded the Grameen Bank, the "bank for the poor". It lends money to poor people. This is in small amounts starting at one euro, usually less than 50 euros. Everyone who borrows money also becomes a co-owner of the bank – and thus feels obliged to pay it back. The clients are almost exclusively women. They use the loans to start trading in handicraft or agricultural products. This also benefits their children. Self-confident women who have their own income are also keen to educate their children. Thus, many problems are tackled directly at the root through this kind of **promotion of women**. The credit system has been adopted in over 60 countries. In 2006, Professor Yunus was awarded the Nobel Peace Prize.

M6 **Nobel Prize for the sustainable combating of poverty**

"We do not give the farmer millet, we give him a hoe. We do not give the child a handout, we give him an education so that he can feed a family later on!"

M8 **Quote from an employee of a non-governmental organisation (NGO)**

Wenn du diese Aufgaben erfolgreich bearbeitet hast, kannst du ...
- ... *die Ziele nachhaltiger Entwicklung an einem Beispiel erklären.*
- ... *einige wichtige Zielsetzungen der Agenda 2030 benennen.*
- ... *die Fachbegriffe **sustainable development** (**nachhaltige Entwicklung**), **supporting self-help** (**Hilfe zur Selbsthilfe**) und **promotion of women** (**Frauenförderung**) erklären.*

OPTIONAL TOPIC 1

Industry as a motor of development

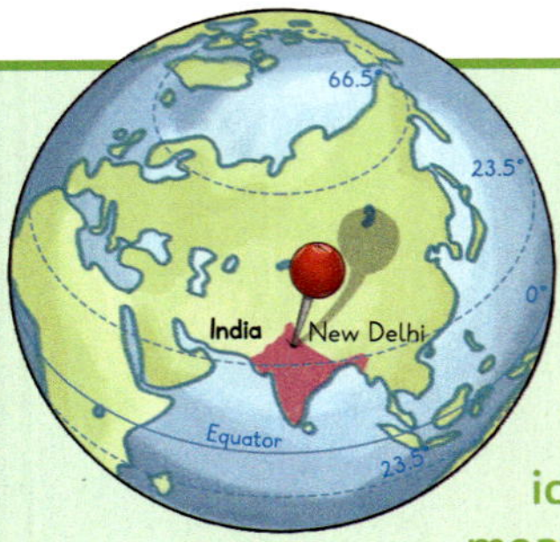

Alongside the promotion of agriculture, India's policy is built primarily on promoting industry, especially high-tech industry.
What are the goals of this policy? What are the development measures? What are the hoped-for effects?

1. Describe the strategy of the Indian government (M1, M2, M4, M5). 210
2. Contrast the interests of foreign companies in direct investment in India with those of the Indian government (M1 – M3).
3. Describe possible negative effects of the high-tech boom using the example of Bangalore (M5 – M7, atlas).
4. Assess whether the development in Bangalore is sensible and sustainable from your point of view (p. 86 M2).
5. Go on a virtual excursion (M7, internet).

air pollution	–	Luftverschmutzung
contrast	–	hier: gegenüberstellen
customised	–	angepasst
domestic	–	inländisch
driver of development	–	Entwicklungsmotor
electronic components	–	elektronische Bauteile
foreign	–	ausländisch
information technology (IT)	–	Informations-technologie
research and development (R&D)	–	Forschung und Entwicklung (FuE)
scientific publications	–	wissenschaftliche Veröffentlichungen

Helpful words and phrases for task 1:

The government promotes / supports / finances ...
They establish / found / promote ... elite universities / research centres
They offer ... quality jobs / good location factors / tax relief / qualified workforce / low production costs / excellent English speakers / round the clock service / round the clock support / business process management
They attract ... foreign direct investment / investment in research and development

Mr Murlidhar, the Indian IT industry has boomed like never before in the past decades. How did this come about?
We recognised back in the 1980s that the IT industry was a ***growth sector****. Therefore, we promoted the establishment of IT centres in some cities, for example in Bangalore. We also founded elite universities to train engineers.*
We offered foreign companies good location conditions, such as tax breaks, with the aim of creating high-quality jobs and accessing as much know-how as possible.
As one can see, with success!
Yes, hundreds of domestic and foreign companies are active in the sector today, employing more than three million people. Almost all major global corporations maintain a location here, often as a ***joint venture*** *with a local company.*
The foreign corporations appreciate the high availability and the good training of our workers, who also have a good command of English. At the same time, labour costs are often only a tenth of what is paid in the USA or Europe for the same performance.
Since the turn of the millennium, a lot of direct investment has flowed into research and development (R&D). Texas Instruments, SAP, IBM or Mercedes Benz maintain their own development centres here.
So they are no longer industrial, but service companies?
They are both! We produce electronic components and customised software, we look after computer centres all over the world and we also take over the business process management (BPM) of large international companies. That means our operations optimise the business processes in these companies.
How is that possible?
The engineer in Bangalore can connect to a computer centre in the US at any time. Because of the 13-hour time difference, India works while America sleeps – and the US customers can use their computers without restriction during their working hours. They are serviced during the night.

M1 Interview With Mr Jyotsna Murlidhar (Department of Information Technology and Electronics)

M2 In the high-tech centres, annual sales of over 100 billion US dollars are achieved. About 90 percent of this comes from exports.

M5 Over the last two decades, Bangalore, a metropolis of five million people, has developed into one of the world's leading high-tech centres with over 300,000 IT specialists. The most important companies in the Indian aerospace industry and half of all biotech companies are based here. All internationally renowned IT companies have subsidiaries or joint ventures here.

INFO

M3 Cooperation with business in the industrialised countries – India

The aim of companies from industrialised countries is to exploit the locational advantages in the countries of the South, for example low labour costs, an educated workforce and good infrastructure. The aim of **foreign direct investment (FDI)** is above all to achieve the highest possible profits. Governments in developing countries often insist that FDI should flow mainly into joint ventures. The capital and know-how come from the industrialised countries and the partner country offers the favourable production conditions.

***M6** Bangalore as a role model? "It's just like Bangalore", is what people in India now say when they think of a place where nothing works, with traffic jams, high prices, bad air, car parks instead of parks. In industrial parks like Electronic City, on the other hand, there are clean streets, fitness studios, garden restaurants - just like in a gated community.

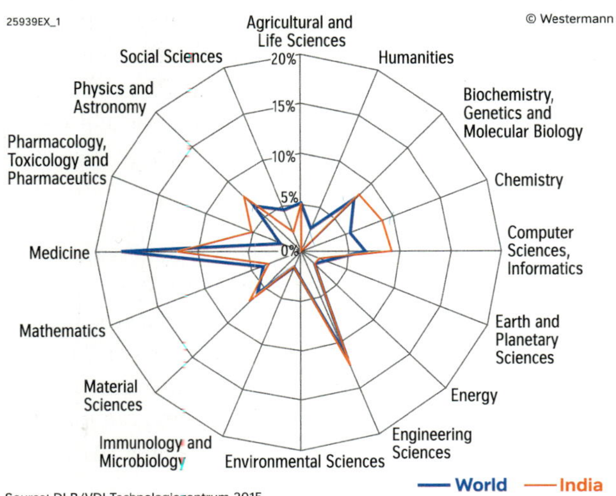

Source: DLR/VDI Technologiezentrum 2015

M4 Key topics evaluated from scientific publications. India is already one of the leading nations in information technology and biotechnology.

INTERNET

M7 Go on a virtual excursion in Bangalore. Use satellite images, photos and digital maps.
Investigate the following questions: How big is the urban area of Bangalore? What are the living conditions? Are there disparities? How is it evident that money is made here with high-tech? What development progress and problems do you see?
Start in Bangalore Electronic City.

Wenn du diese Aufgaben erfolgreich bearbeitet hast, kannst du ...
- *... Maßnahmen und Ziele der Förderung der Hightech-Industrie erläutern.*
- *... die Nachhaltigkeit von Industrieförderung beurteilen.*
- *... die Fachbegriffe **growth industry** (**Wachstumsbranche**), **Joint Venture** (**Gemeinschaftsunternehmen**) und **foreign direct investment – FDI** (**ausländische Direktinvestition – ADI**) erklären.*

OPTIONAL TOPIC 2

The promotion of agriculture

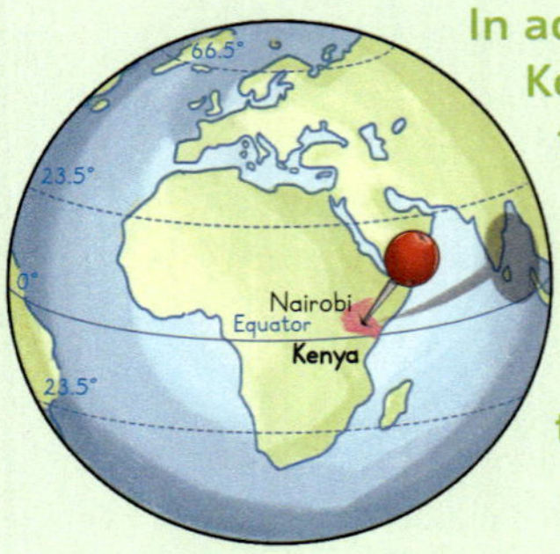

In addition to promoting tourism, Kenya's policy is also based on the promotion of rural areas and agriculture. What are the goals of this policy? What are the development measures? What are the hoped-for effects?

1. Describe the cultivation of cut flowers around Lake Naivasha (M1, M2, M4, M6, atlas).
2. Contrast the interests of foreign companies in direct investments in Kenyan agriculture with those of the Kenyan government (M1, M4).
3. Explain the support measures of Kenya's agricultural policy and their objectives.
4. Describe possible negative impacts of cut flower production (M5, M6).
5. Assess whether the projects presented here make sense to you and whether they are sustainable (p. 86 M2).
6. Go on a virtual excursion (M6, internet).

agricultural policy	–	Landwirtschaftspolitik
cut flowers	–	Schnittblumen
environmentally friendly	–	umweltfreundlich
exchange earner	–	Devisenbringer
horticulture	–	Gartenbau
irrigation techniques	–	Bewässerungstechniken
profits	–	Gewinne
reduction of poverty	–	Armutsbekämpfung
support measures	–	Fördermaßnahmen
unskilled	–	ungelernt/unqualifiziert
upstream/downstream industry	–	vor- und nachgelagerte Branche

Helpful words and phrases for task 3:

Kenya's government promotes ...
Thereby hoping to create ...
As a consequence ... is reduced
The ultimate aim is to create a sustainable ...

Mrs Mwaura, what are the goals of the Ministry of Agriculture?
On the one hand, we support our small farmers by training them in the use of new irrigation techniques and pesticides or by introducing them to our mobile phone app "M-Farm".
On the other hand, we are trying to attract foreign investors who want to invest in the agricultural sector.
Foreign investment in the agricultural sector is not as common as in industry ...
That is true, but we offer good profit opportunities. For example, our climate is better suited for growing cut flowers than in almost any other country in the world. Therefore, flower cultivation is considered a ***growth industry*** *for us. We have been promoting the establishment of flower farms at Lake Naivasha for several years. Today, every third rose bought in Germany comes from Kenya. About 70,000 people have found work on the flower farms in Kenya, most of them unskilled or low-skilled workers. With upstream and downstream industries, flower cultivation provides employment for a total of about 500,000 people.*
Where do the investors come from?
From India, from China, but also from the Netherlands and Germany. Some have set up ***joint ventures*** *with Kenyan companies. In the Flower Business Park (FBP), for example, a leading German rose breeder has expanded its business with the Kenyan company Kreative Roses Kenya Limited. This project was also supported within the framework of German development cooperation, so that FBP today has a good infrastructure (a well-developed road to Jomo Kenyatta International Airport, 100 km away, electricity, water supply and wastewater disposal).*
What does this project contribute to Kenya's development?
The main goal is to reduce poverty in the countryside and thus reduce rural-urban migration! As more and more companies produce according to fair trade standards, wages are fair and farming methods are becoming more environmentally friendly. Basically, the living conditions of half a million people have improved noticeably. In addition, horticulture is one of the biggest foreign exchange earners with 1 billion US dollars per year.

M1 Interview with Mrs Elishiba Mwaura (Bureau of Agriculture and Exports)

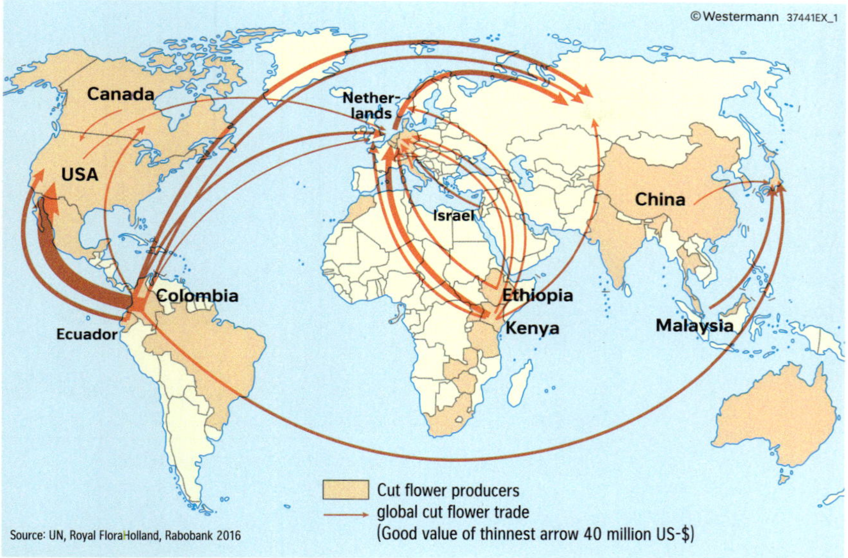

M2 The journey of the roses (not limited to Africa)

M7 With M-Farm, farmers can get the current market prices at the food exchanges for a small fee. In addition, they get tips for sowing, fertilisers or weather information. And they can offer their products for sale on an online marketplace.

INFO

M3 **Cooperation with business in the industrialised countries – Kenya**

The aim of companies from industrialised countries is to exploit the locational advantages in the countries of the South, for example low labour costs, an educated workforce and good infrastructure. The aim of **foreign direct investment (FDI)** is above all to achieve the highest possible profits. Governments in developing countries often insist that FDI should flow mainly into joint ventures. The capital and know-how come from the industrialised countries and the partner country offers the favourable production conditions.

Five years ago, the cultivation of roses at Lake Naivasha was still highly controversial. The water level of the lake had dropped dangerously, its water quality suffered from the farmers' pesticides. Environmentalists saw the unique biotope in danger. More and more hectare-sized plastic roofs stand where once not only zebras but also giraffes and rhinos roamed through untouched nature. Now, millions of transient blooms are growing under plastic tarpaulins.

Source: Palm, Christian: Die Reise der Rosen. www.faz.net, 14.02.2015, access: 15.10.2020

M5 A downside of cut flower cultivation

M4 Around one million flowers leave Lake Naivasha every day. Exports often go through the international flower exchange at Amsterdam airport. The share of fair trade flowers exported to Germany has already grown to a quarter.

INTERNET

M6 Take a virtual excursion around Lake Naivasha with the help of satellite images, photos and digital maps. Present your findings as clearly as possible.
Investigate the following questions: How big is the cultivation area at Lake Naivasha? How can you recognise the intensive flower cultivation? What are the living conditions in the settlements? Are there disparities? What development progress and problems do you see?

Wenn du diese Aufgaben erfolgreich bearbeitet hast, kannst du ...

... Maßnahmen und Ziele der Förderung der Landwirtschaft erläutern.
... die Nachhaltigkeit von Projekten in der Landwirtschaft beurteilen.
*... die Fachbegriffe **growth industry** (**Wachstumsbranche**), **Joint Venture** (**Gemeinschaftsunternehmen**) und **foreign direct investment – FDI** (**ausländische Direktinvestition – ADI**) erklären.*

Tourism in developing countries – important

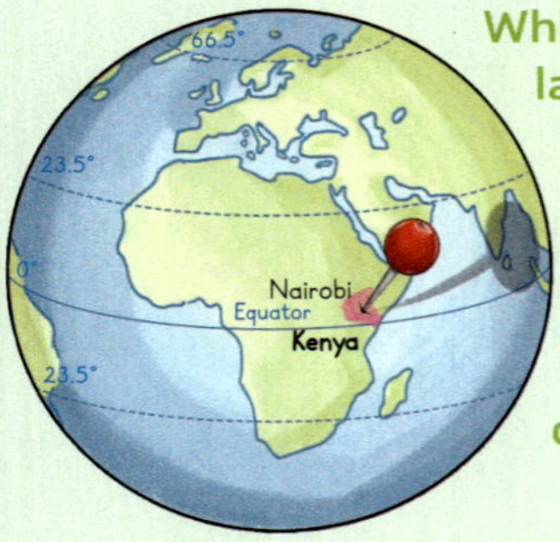

White sandy beaches, untouched landscapes, other cultures and much more have always attracted people to faraway places. What is the importance of tourism for a country's economy? And what is the tourism potential of the country based on?

1. Research Kenya's tourism potential (e.g. location, countries of origin of tourists, tourist attractions, landscape, tourism infrastructure, excursions, types of tourism) and prepare a photo presentation (M4, M5, M7, M8, internet).
2. Characterise Kenya's level of development (p. 240/241, atlas).
3. List the preconditions for tourism to be a development impulse for a country.
4. Illustrate the importance of tourism graphically (e.g. in a causal network). Supplement your presentation with concrete data on tourism in Kenya (internet, atlas).
5. (A) Explain the quotation from the World Tourism Organization in M9 (M1 – M3, M6).

accelerate – beschleunigen
beacon – Leuchtfeuer
direct/indirect employment – direkte/indirekte Arbeitsplätze
exploit – nutzen/ausschöpfen
foreign currency/ foreign exchange – Fremdwährung/ Devisen
gross national income (GNI) – Bruttonationaleinkommen (BNE)
maintain – aufrecht erhalten
protected area – Schutzgebiet
reservation – hier: Reservat
semi-desert – Halbwüste
significant – wichtig/bedeutungsvoll
SME – small/ medium-sized enterprise – KMU - Klein-/Mittelstandsunternehmen
World Heritage Site – Weltkulturerbe

*M1 The global importance of tourism, 2019

INFO

M2 Tourism brings a lot of foreign currency into a country. This **foreign exchange** is an important economic factor for many countries. One third of all developing countries earn most of their foreign exchange through tourism. The income can be used to buy products on the world market or to upgrade the country's infrastructure. Due to the strong importance of tourism, more and more developing countries are trying to exploit their tourism potential (nature, culture) and present it attractively for visitors. This often results in **mass tourism**.

The development of a tourism infrastructure is a development boost for many countries. This is because tourism creates jobs. Direct jobs in the tourism industry are tour guides, hotel employees, bus drivers or beach traders. Indirect jobs are in agriculture for the provision of food or in the construction industry for the building and maintenance of tourism infrastructure. Education is also improved because, for example, foreign language skills are an advantage. With their wages, people can feed themselves, pay taxes and afford to send their children to school. The demand for goods and services boosts the economy.
If the country takes in more money directly and indirectly through tourism, it can also invest in regions that are away from the tourist centres. In this way, tourism can also improve infrastructure and living standards throughout the country and reduce disparities.

M3 Tourism as a motor of development

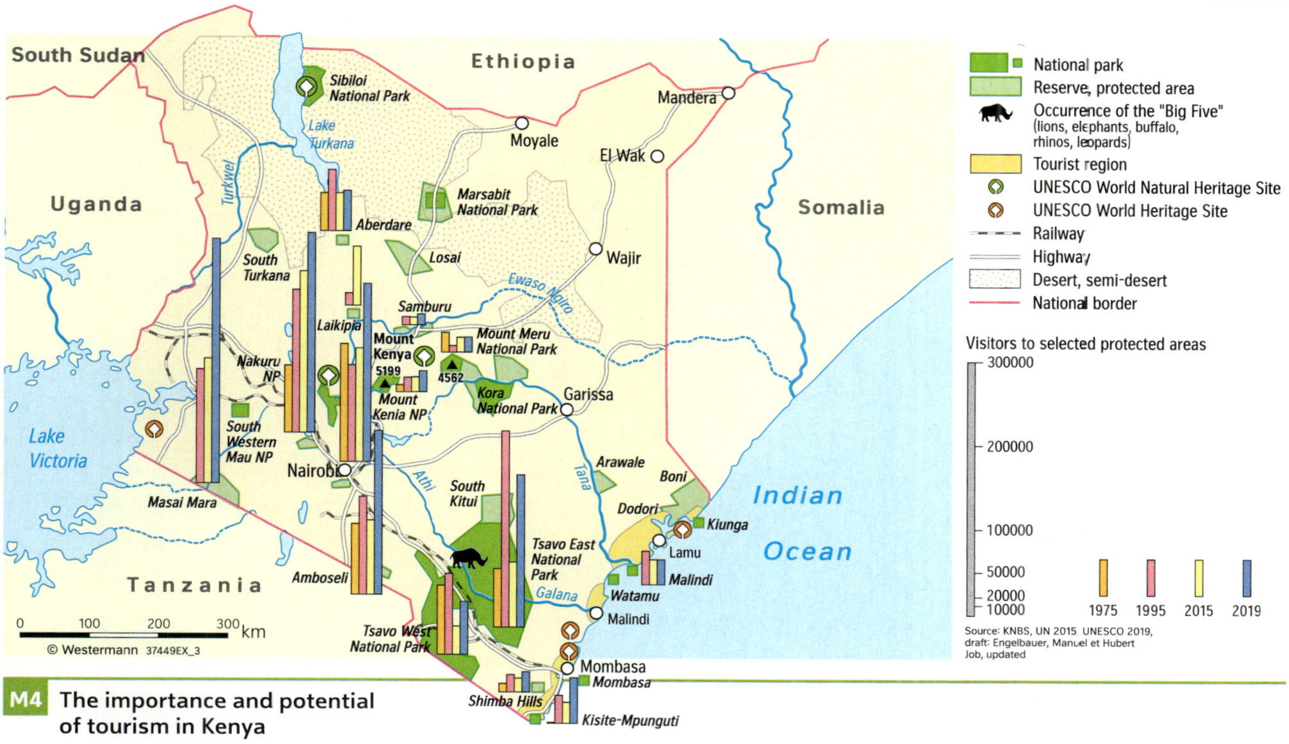

M4 The importance and potential of tourism in Kenya

M5 On safari in the Masai Mara

M8 In the Masai Mara protected area

Tourism	2000	2010	2019
International arrivals (in 1000)	899	1470	1364*
International tourism receipts (in million US-$)	500	1620	1564*
Share of GDP (in %)	9.6	11.1	8.2
Share of employment (in %)	8.9	10.4	8.5
HDI	0.446	0.533	0.579

Source: UNDP, UNWTO, WKÖ Stabsabteilung Statistik, WTTC *2017

M6 The importance of tourism in Kenya

Tourism accelerates reform. Tourism is a beacon for pro-business policies and reforms that can help SME development and stimulate foreign investment.

World Tourism Organization (2015), Affiliate Members Regional Reports, Volume 4 – Tourism in Africa: A Tool for Development, UNWTO, Madrid, p. 8. © UNWTO, 92844/04/21

M9 Quote from the World Tourism Organization

INTERNET

M7 The *Kenya Wildlife Service* website provides access to the country's different national parks.
World Travel and Tourism Council
World Tourism Organization (UNWTO)

AMAZING

M10 Kenya has over 60 national parks.

Wenn du diese Aufgaben erfolgreich bearbeitet hast, kannst du ...
- ... die Bedeutung des Tourismus für ein Entwicklungsland darstellen.
- ... das touristische Potenzial eines Landes kennzeichnen.
- ... die Fachbegriffe **foreign exchange** (**Devisen**) und **mass tourism** (**Massentourismus**) erklären.

Tourism in developing countries – important, but insecure

Tourism is seen as an important strategy for development. But does tourism only bring progress? What are the threats to development from tourism?

1. Work with an economic database. Document your results (M1, M2). 212

2. Foreign exchange earner or uncertain economic sector? Show ways in which tourism can threaten development (M3, M4).

3. Work with the chart (M3).
   a) Copy the chart into your exercise book.
   b) Fit the freestanding terms at the bottom into suitable positions in the chart.
   c) Apply the chart to Kenya (M1 – M4, pp. 92/93). Also consider possible changes (M4). Use a different colour for facts about Kenya and fill in the chart in your exercise book. 199

4. Discuss the quotation from the World Tourism Organization (p. 93 M9).

Ⓐ **5.** Tourism as a development strategy for countries of the Global South? Discuss opportunities and risks using Kenya as an example.

economic crisis	–	Wirtschaftskrise
hotel chain	–	Hotelkette
leakage effect	–	Sickerrate
run-off	–	Abfluss
seasonality	–	Saisonalität
sights	–	Sehenswürdigkeiten
uncertain	–	unsicher

Helpful words and phrases for task 4:

The ultimate aim of a development strategy is ...
Tourism enables many oppurtunities, such as ...
Kenya's assets are ...
Tourism in Kenya is dependent on ...
... the climate crisis or corona crisis show that ...
To be a motor for development, tourism must ... / needs to be ...
Jobs in tourism should be ...
The environmental costs can be / must not be ...

INTERNET

M1
- Trading Economics
- World Travel and Tourism Council
- Auswärtiges Amt – Reisehinweise

Trading Economics ist eine Datenbank. Hier kannst du Länder hinsichtlich verschiedener Indikatoren untersuchen. Auch kannst du Länder und Indikatoren miteinander vergleichen. Trading Economics ist ein kommerzieller Anbieter und greift auf die Daten von Regierungen und Organisationen wie Vereinten Nationen (UN) oder die Weltbank zurück.

Leitfragen deiner Untersuchung
Wie hat sich der Tourismus in Kenia in den letzten zehn Jahren entwickelt? Welchen Einfluss hatte die Coronapandemie auf den Tourismus und die Wirtschaft des Landes?

Schritt 1: Finden der richtigen Daten
Rufe die Website von Trading Economics auf.
Tipp: Scrollst du runter, kannst du die Sprache auf Deutsch einstellen.
Oben in der Menuleiste klickst du auf „Die Länder“. Wähle Kenia aus. Rechts sind die zu untersuchenden Indikatoren nach Themengebiet zusammengefasst. Klickst du auf „Handel“, kannst du die Ankünfte „Ausländischer Gäste“ auswählen.

Schritt 2: Erklärung der Entwicklung
Erkläre die Entwicklung der Ankünfte von Touristen in Kenia.
a) Erstelle ein Diagramm zu den Ankünften von Touristen in den letzten 10 Jahren. Tipp: in der Menüleiste oberhalb des Diagramms kannst du Einstellungen vornehmen. Speichere es ab (rechte Maustaste „Grafik speichern unter“).
b) Recherchiere zu „Corona Kenia Safari Tourismus“.
c) Erkläre mithilfe der recherchierten Ergebnisse die Entwicklung der Ankunftszahlen.

Schritt 3: Analyse der Auswirkungen der Coronapandemie auf den Tourismus und die Wirtschaft
a) Vergleiche die Ankunftszahlen mit anderen Indikatoren zu Kenia. Klicke auf „Compare“. Wähle „Kenya“ aus. Vergleiche nun mit anderen Indikatoren:
- Entwicklung des BIP (GDP-Annual Growth Rate)
- Arbeitslosenquote (Unemployment Rate)

Speichere jeweils die Diagramme ab.
b) Erläutere den Einfluss der Pandemie auf die Wirtschaft.

Schritt 4: Transfer
Finde heraus, ob sich auch in anderen Ländern die Auswirkungen der Corona-Pandemie auf den Tourismus auswirken. Vergleiche dieselben Indikatoren nur jetzt unterschiedlicher Länder (z. B. Ankünfte von Touristen von Kenia im Vergleich zu Australien, Argentinien oder Kroatien). Erläutere.

M2 Work with a database 212

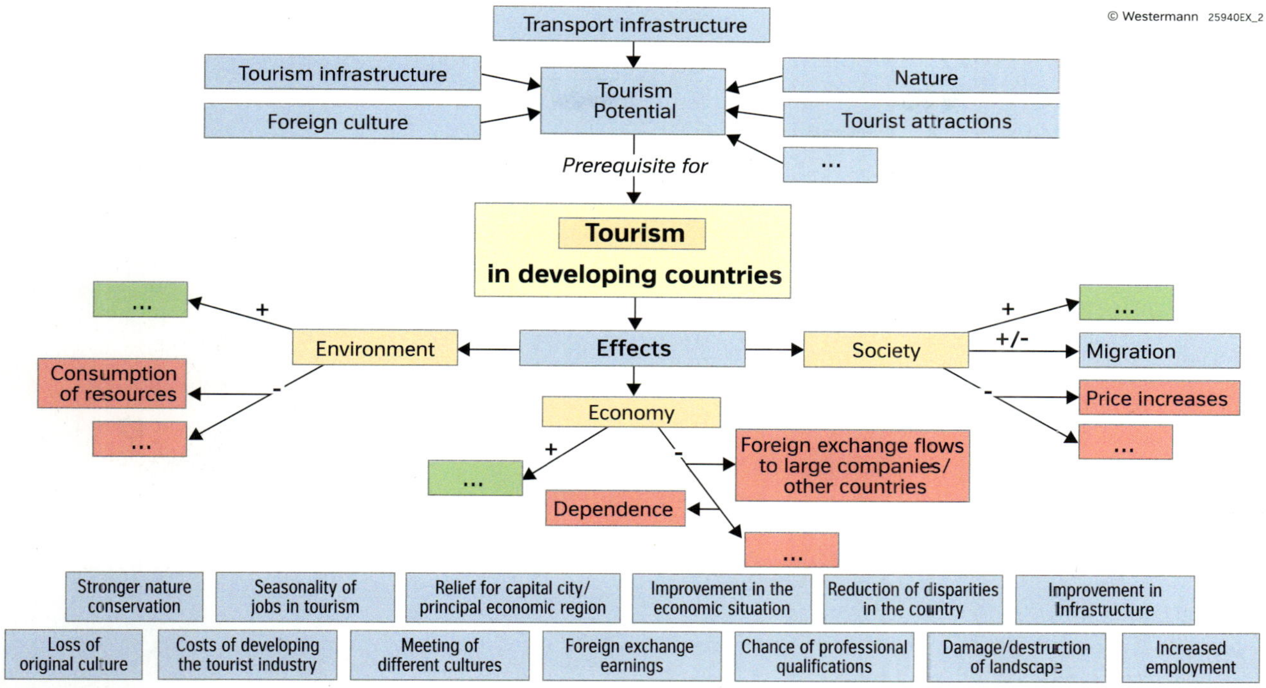

M3 Tourism in developing countries

Headlines

TOURISMUS LEIDET UNTER VULKANAUSBRUCH

Eine Aschewolke stürzt die indonesische Ferieninsel Bali in eine wirtschaftliche Krise. Der Flughafen ist nach dem Ausbruch des Vulkans Agung zwar wieder offen. Doch Tourismusmanager fürchten monatelange Einbußen.

Source: Peer, Mathias: Verheerende Wolke über Bali. www.handelsblatt.com, Düsseldorf, 29.11.2017, access: 12.11.2020

Rund 230000 Menschen waren im Jahr 2004 gestorben, nachdem ein gewaltiges Erdbeben der Stärke 9,1 vor der indonesischen Insel Sumatra einen Tsunami ausgelöst hatte. 14 Länder wurden damals von bis zu 30 Meter hohen Wellen getroffen. In Thailand waren unter den rund 5400 Todesopfern auch zahlreiche Touristen, darunter 534 Deutsche. Besonders schwer traf es neben der indonesischen Provinz Aceh und Thailand auch Indien und den Inselstaat Sri Lanka. Es war die größte Tsunami-Katastrophe seit Menschengedenken.

Source: ni/sti (dpa, afp, rtr): Gedenken an die Opfer der Tsunami-Katastrophe. www.dw.com, Bonn, 26.12.2019, access: 12.11.2020

Der Tourismus bringt also zwar viel Geld, wird aber andererseits häufig mit der Verzerrung der lokalen Entwicklung erkauft. Bauern verkaufen ihr Land an Hotelketten. Dann erleben sie, dass Produkte, die sie früher angebaut haben, für sie unerschwinglich werden. Wasser wird zum Golfplatz umgeleitet, während die Anwohner unter Wassermangel leiden. Die Straße wird bis zum Themenpark asphaltiert, nicht bis zur Schule.

Source: De Bellaigue, Christopher: Der Massentourismus zerstört seine Ziele. der Freitag, 30/2020, Übersetzung Carola Torti

„Wenn die Touristen sich jetzt abschrecken ließen und nicht mehr kämen, dann würde zusätzlich auch noch Hunderttausenden Menschen in der Dienstleistungsindustrie ihre Lebensgrundlage entzogen." Mercy Wambui (UN)

Source: Frentzen, Carola und Gold, Shabtai: Dürre im Urlaubsparadies. Kenia braucht die Touristen. www.n-tv.de, Köln, 29.07.2011, access: 12.11.2020

Nahe den Pyramiden von Gizeh explodiert ein Sprengsatz und zerstört einen Reisebus. 17 Menschen werden verletzt. Der Anschlag trifft Äygptens Tourismusindustrie, die sich gerade erholte.

Source: Ramadan, Dunja: Anschlag auf den Tourismus. www.sueddeutsche.de, München, 19.05.2019, access: 12.11.2020

Cartoon

M4 Influences on tourism (headlines)

Wenn du diese Aufgaben erfolgreich bearbeitet hast, kannst du ...

... die Chancen und Risiken des Tourismus in Entwicklungsländern abwägen.

... den Tourismus als Möglichkeit einer nachhaltigen Entwicklung in Entwicklungsländern beurteilen.

Free trade – fair trade?!

Ever since colonial times, many structures of world trade have been unjust. Are there approaches to change this? What can be done?

1. Describe some of the main structures of world trade (M1 – M4, M6, p. 79 M3 – M7)

2. Find five fair trade products in a supermarket. Note down their origin, their price and the price of a comparable product without a fair trade label. Put your results together in class.

3. a) Explain the goals and measures of fair trade using concrete examples from developing countries (M5, M6, internet).
b) Describe possible measures to further improve the system of fair trade (M7, internet).

C **4.** Illustrate the measures envisaged by the *Bundesministerium für wirtschaftliche Zusammenarbeit* (BMZ) (internet):
A the Supply Chain Law (M9).
B the initiative BMZ Aid for Trade (M7).

A **5.** Which two measures seem particularly important to you? Produce a podcast (90 seconds).

adequate payment	–	angemessene Zahlungen
certification	–	Zertifizierung
consumer	–	Verbraucher/-in
fluctuating prices	–	schwankende Preise
premium payment	–	Prämie

Helpful words and phrases for task 1:
Arbeits- und sozialer Schutz – labour- and social protection
Dumpingpreise – dumping prices
EU-Subventionen zerstören lokale Betriebe – EU-subsidies destroy local farmers
große transnationale Unternehmen – big transnational corporations
ihre Milch vermarkten – to market their milk
konkurieren – compete
Lieferkette – supply chain
lokale Milchviehhalter – local dairy farmers
Magermilchpulver – skimmed milk powder
Menschenrechte – human rights
Pflanzenfett – vegetable fat
Profite privater Unternehmen – profits of private companies / actors
Recht auf Nahrung – right to food
Staatshilfen – state aid
Subventionen – subsidies
Umweltschutz – environmental protection
ungerecht – unjust / unfair
Vollmilch – whole milk

The goals of the **WTO** are the greatest possible transparency of the trade policies of its members, the agreement on, compliance with and monitoring of common multilateral trade rules, and the ongoing liberalisation of world trade through the reduction or elimination of tariffs and other trade barriers as well as the simplification of customs procedures. Such measures should ultimately lead to the strengthening of economic power and competitiveness. The particular development interests of the poorer and poorest members are given special consideration.

Source: BMWi: Handelspolitik. Ziele, Aufgaben und Prinzipien. www.bmwi.de, access: 14.12.2020

M1 Goals of the World Trade Organization (WTO)

Beispiel Milchpulver: Bereits zwischen 2008 und 2018 sind die EU-Exporte von Magermilchpulver, das mit Pflanzenfett angereichert ist, nach Westafrika um 234 Prozent gestiegen. Konzerne [...] verpacken dieses Milchpulver in ihren westafrikanischen Werken um und vermarkten es dort – auch aufgrund von EU-Agrarsubventionen – zu Dumpingpreisen. [...] Zugleich wird dadurch lokalen Milchviehhaltern die Vermarktung ihrer Vollmilch erschwert und ihr Recht auf Nahrung gefährdet. Das geplante Wirtschaftspartnerschaftsabkommen (EPA) der EU mit dem westafrikanischen ECOWAS-Raum könnte die Problematik verschärfen [...].

Source: Schneider, Sarah et Armin Paasch: Mehr europäisches Milchpulver in Westafrika. E+Z 11/2020, S. 24

M2 Free trade – not without problems

Kritische Stimmen bewerten die globale Handelsordnung als weitgehend ungerecht und undemokratisch, da sie im Wesentlichen große transnationale Unternehmen und ökonomische Eliten bevorzuge, wettbewerbsschwächere soziale Gruppen und Länder hingegen seien dem „Diktat der Märkte" ausgeliefert. Weitere Spannungsfelder zeichnen sich hinsichtlich der Frage ab, unter welchen Bedingungen das Wachstums- und Profitstreben mehrheitlich privater Akteure mit Umwelt-, Menschenrechts-, Arbeits- und sozialem Schutz vereinbar sind.

Source: Öztürk, Asiye: Editorial. In: bpb: Welthandel. APuZ 64, Bonn 1 – 3/2014, S. 2

M3 Criticism of the structures of world trade

M4 **Global companies target markets in developing and newly industrialising countries.**

M8 **Protest against world market policies**

Goals and features:
- The products must be produced under fair conditions (reasonable wages and working hours, holidays, insurance, no child labour).
- Producers should receive adequate payments for their products, essentially independent of the (fluctuating) world market price.
- Producers (often small farmers) also receive Fairtrade premium payments, to finance social institutions.

Problems and criticisms:
- Products of fair trade are not always clearly identifiable for consumers.
- Certification as recognised partners is often too difficult and too expensive for producers.

M5 **Fair trade (with the example of the Fairtrade organisation) – change with the help of consumers?**

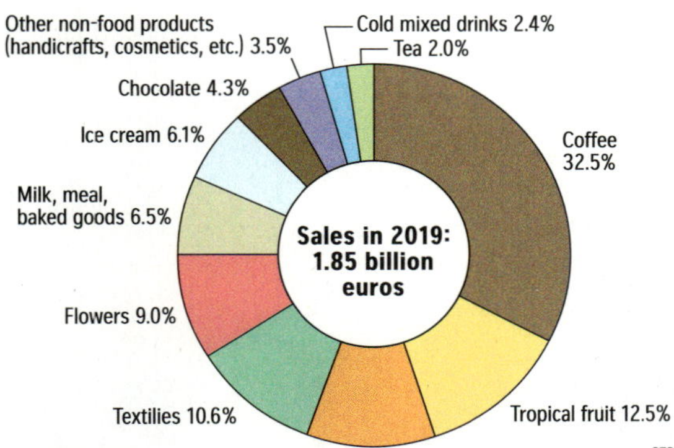

Source: Forum Fairer Handel e.V. 2020

27873EX_2 © Westermann

M6 **The range of products from fair trade (2018)**

Mr Müller, why does Germany need a Supply Chain Law?

The **supply chains** of German companies reach into all parts of the world. A normal pair of jeans, for example, [...] travels through dozens of places before it reaches our shops. In Bangladesh, they are produced for 5 euros [...]. This is often only possible because the women in the textile factories slave away for 14 hours a day and ecological and social minimum standards such as the ban on child labour are disregarded. [...]

What should change?

The new law will ensure that basic standards of human rights, such as the ban on forced and child labour, are also respected at the beginning of our supply chains. [...]

I have spoken to many companies. And many already meet the standards. For example, the 52 companies that participate in the state textile label "Grüner Knopf". Among them are sustainability pioneers [...], sports labels [...], medium-sized companies [...] and large retailers like Tchibo, Lidl, Aldi or Kaufland. They show that it is possible! But it should not be allowed that others produce without regard to human rights standards and thus gain competitive advantages. Markets need clear rules. [...]

Source: Stiftung Warentest, test.de: Firmen sollen mehr Verantwortung übernehmen. Interview mit Entwicklungshilfeminister Müller: Schluss mit freiwilligen Standards! www.test.de, 18.09.2020, access: 14.12.2020

M9 **The Supply Chain Law – change with the help of the state?**

INTERNET

M7 
- BMZ: → Lieferkettengesetz, Aid for Trade, Welthandel
- FAIRTRADE Deutschland, Gepa, Unternehmen Weltladen und Fairtrade Kritik

Wenn du diese Aufgaben erfolgreich bearbeitet hast, kannst du ...

... *wesentliche Strukturen des Welthandels beschreiben und bewerten.*
... *Schritte zu einem gerechteren Welthandel erläutern und bewerten.*
... *die Fachbegriffe **World Trade Organization – WTO (Welthandelsorganisation)**, **fair trade (fairer Handel)** und **supply chain (Lieferkette)** erklären.*

WHAT WE HAVE LEARNED

Improvement in sight? – strategies and measures for development

Which is the right way?

84 There are different ways, different strategies to develop a country. They start from the factors that inhibit development and try to mitigate or eliminate them. Both German governmental **development cooperation** and the projects of **non-governmental organisations (NGOs)** aim to improve the living conditions of people in the countries of the South and to make them independent of external support measures in the long term.

Sustainable development – help for self-help

86 Improving people's living conditions in the long term is the principle of **sustainable development**. "Sustainable" means that no irreparable damage is done to the environment, that economically long-term, viable solutions are found and that the social conditions and living conditions of the people, are permanently improved.

One way of achieving this is through **help for self-help**. People are enabled to help themselves to improve their own living conditions.

This sustainable development is also the goal of the UN's **2030 Agenda for Sustainable Development**. In addition to government organisations, many non-governmental organisations are particularly involved in the field of sustainable development and helping people to help themselves.

Promoting industry

88 One development strategy is to export more manufactured goods instead of raw materials. To this end, developing countries strive to build up industries in **growth sectors** in particular, because this creates many jobs in the long term and generates economic growth. This is why these industries are also interesting for **foreign direct investment**. Foreign companies often work together with companies from the developing country in **joint ventures**. This benefits both sides.

In India, for example, one focus of development has been on building up an IT industry. Today, India is one of the world's most important locations in this sector.

Promoting agriculture

Certain areas of agriculture are also among the **growth sectors**. In Kenya, for example, this is cut flower production. Roses and other cut flowers are sought-after goods on the markets of industrialised countries. Therefore, as in other sectors of the economy, there are **foreign direct investment** and **joint ventures**. Thousands of new jobs have been created through the promotion of cut flower production. 90

Developing tourism – opportunities and risks

Many developing countries rely on the promotion of tourism as a development strategy. This also creates many new jobs, both directly in hotels and tourist facilities and indirectly in agriculture or construction. In addition, **foreign currency** comes into the country, which can then be used to buy goods from abroad. 92

If all the framework conditions are favourable for foreign tourists, **mass tourism** can develop.

However, a prerequisite for positive development is that the country is also attractive for tourists. In a favourable case, a tourist destination can promote the development of an entire region. However, tourism is not a consistently secure source of income. Tourists can be deterred from travelling for a variety of reasons – with serious consequences for the entire economy. And tourism itself can also have negative effects. 94

M1 Products from fair trade

Free trade – fair trade?!!

96 Free world trade – yes or no? The **World Trade Organisation (WTO)** advocates free trade with as few tariffs and trade barriers as possible. But in many countries, the structures of world trade are criticised as very unfair. This is due to the predominance of raw material exports with dependence on world market prices as well as the terms of trade of the developing countries, the measures of the industrialised countries to increase the sales of their industrial products, the power of the corporations, and the poor working conditions in the developing countries at the beginning of the **supply chains**.

Many initiatives are trying to change this. Through **fair trade**, non-governmental organisations support producers by, for example, paying higher prices and thereby helping to improve living conditions. The Bundesministerium für Zusammenarbeit und Entwicklung (Germany's Federal Ministry for Cooperation and Development) is trying to eliminate unfair structures in world trade, among other things, through legal measures such as the Supply Chain Act.

M2 Nigerian-German joint venture for industrial plant

M5 Consulting hour! – medical provision in the Ethiopian highlands

M3 Well for drinking water in Malawi

M6 "Africa's Amazon" - the online trading company Jumia delivering food in Kigali, Rwanda

M4 Cartoon

1 Choose one of the development measures (M1 – M3, M5, M6) and
a) describe the strategy behind the measure,
b) discuss whether the measure is sustainable and represents help for self-help.

2 Interpret the cartoon M4. 209

Apply and practise

Global denken!
Lokal handeln!

PROJEKT VOR ORT

Auch du kannst helfen!

Wir sind alle besorgt. Wir würden auch alle gerne helfen!
Aber was können wir schon tun? Es ist wie beim Umweltschutz, wie in der Politik, wie überall: Ich allein kann nichts tun – erst recht als Jugendliche oder Jugendlicher!

Unsinn!

Jede und jeder kann etwas tun! Jede und jeder kann helfen, dass sich in unserer Welt etwas verändert!
Du kannst diskutieren und helfen, Einstellungen zu verändern. Du kannst etwas organisieren und die Menschen in den Entwicklungsländern direkt unterstützen!

Hier findest du einige Anregungen:

Du kannst zum Beispiel:
- Arbeiten verrichten,
- mit Freundinnen und Freunden oder der Klasse einen Sponsorenlauf durchführen, bei dem jeder Kilometer bzw. jede Runde auf dem Sportplatz bezahlt wird,
- einen Kuchen-, Saft- oder Waffelstand an der Schule oder in der Einkaufszone aufstellen (Achtung: Genehmigung einholen!).

Die Menschen kaufen und spenden besonders gerne, wenn sie wissen, für welchen konkreten Zweck gesammelt wird. Informationsmaterial gibt es bei den Hilfsorganisationen.

M2 **Spenden sammeln? Kein Problem! Ideen muss man haben!**

Sozialer Tag – mach doch.
Zehntausende Schüler*innen bundesweit tauschen ihre Schulbank gegen einen Arbeitsplatz und spenden ihren Lohn an Schüler Helfen Leben. Somit unterstützen sie nicht nur Jugend- und Bildungsprojekte für Gleichaltrige in Südosteuropa und Jordanien, sondern haben auch die Möglichkeit, einen Tag lang in ihren Traumberuf zu schnuppern. [...] Schüler Helfen Leben ist der beste Beweis dafür, dass Jugendliche gemeinsam etwas bewegen können – wenn sie nur wollen.

Quelle: Schüler Helfen Leben e.V., Neumünster: Sozialer Tag – mach doch. www.schueler-helfen-leben.de, Zugriff: 16.05.2018

***M1** **Bei Schüler Helfen Leben entscheiden die Schülerinnen und Schüler selbst, welche Kooperationsprojekte mit den Geldern des Sozialen Tages gefördert werden.**

Sommer 2018. Wieder einmal haben sich Los Masis angesagt. Alle drei Jahre kommt die Musikgruppe aus Bolivien nach Deutschland. Hier tritt sie in Kirchen, Schulen, bei Vereinen und auf Plätzen auf. Die Musiker sammeln Geld für ihre Schule in Bolivien, dem „Centro Cultural Masis" in der Hauptstadt Sucre.
Hier werden benachteiligte Kinder und Jugendliche unterstützt. Viele von ihnen sind Migranten vom Land oder Nachfahren von zugewanderten Landbewohnern in zweiter oder dritter Generation. Viele leben am Stadtrand von Sucre, wo Land- und Stadtleben aufeinanderstoßen.
Das Centro bietet pädagogische und psychologische Unterstützung und hat zum Ziel, die Identität der Jugendlichen durch Musik, Tanz und Erhalt der kulturellen Werte zu stärken.

M3 **Auch ihr könnt einen Auftritt von Los Masis organisieren: in eurem Ort, in eurer Schule, in eurer Nachbarschaft. Du kannst Los Masis auch auf Facebook besuchen.**

Brot für die Welt
www.brot-fuer-die-welt.de

OXFAM Deutschland
www.oxfam.de

HUMAN HELP NETWORK
Die Kinderhilfsorganisation
www.hhn.org

MISEREOR
IHR HILFSWERK
www.misereor.de

terre des hommes
Hilfe für Kinder in Not
www.tdh.de

unicef
für jedes Kind
www.unicef.de

DSW
Deutsche Stiftung Weltbevölkerung
youth can
www.dsw.org

welt hunger hilfe
www.welthungerhilfe.de

M4 **Große NGOs, die sich mit Hunderten unterschiedlichsten Projekten in den Ländern des Südens engagieren.**

- Berichte deinen Eltern und Verwandten, was du inzwischen über die Verhältnisse in Entwicklungsländern gelernt hast und mache Vorschläge zur Hilfe.
- Überprüfe dich: Achtest du darauf, wenn es möglich ist, Produkte aus fairem Handel zu kaufen?
- Verwendet ihr solche Produkte in eurem Haushalt?
- Gibt es Fair-Trade-Produkte bei euch in der Schule?
- Willst du dich nicht in einem Eine-Welt-Laden engagieren?
- Frage nach, welche (kleine) Hilfsorganisation es in deinem Heimatort oder Schulort gibt, bei der du ehrenamtlich mitarbeiten kannst.
- Engagiere dich beim „Sozialen Tag". Hilf bei der Organisation, der Auswahl der Einzelprojekte.
- Überzeuge Mitschülerinnen und Mitschüler, sich auch zu engagieren.

M5 **Man kann nicht nur Spenden sammeln ...**

INTERNET

M6 *Make me smile Kenya – Aktion gegen den Hunger – Menschen für Menschen – help2kids – Ärzte ohne Grenzen – Freiwilligenarbeit Südamerika*
Es gibt zahllose Organisationen, die sich für die Menschen in den Entwicklungsländern einsetzen.
Aber aufgepasst! Nicht alle sind seriös! Auch hier gibt es „schwarze Schafe"! Zum Beispiel fließt bei einigen mehr Geld in die Verwaltung und Organisation, als bei den Bedürftigen ankommt. Das DZI (Deutsches Institut für soziale Fragen) testet die Organisationen auf Seriosität und gibt Tipps zum Spenden.

Wenn du diese Aufgaben erfolgreich bearbeitet hast, hast du ...
... wirklich einen Beitrag geleistet, die Lebensbedingungen von Menschen in Entwicklungsländern zu verbessern!

IN FOCUS

What development strategy is China pursuing?

The People's Republic of China was founded in 1949. It was a developing country at the time. But since about 1980, the country of China has undergone profound economic development.

C Aspect 1: What development strategy has the Chinese government pursued and what stage of development has China reached?

China is no longer a poor country. There is evidence that it has caught up with the leading industrialised countries.

1. Prepare a short presentation on this key topic. 211, 218

Tips for the structure:
- Describe the state of development of China in 1978/1980.
- Describe China's development in terms of social, economic and demographic indicators. Also use the atlas and the country data on pages 240/241.
- Work out which development strategy China has pursued.
- Evaluate China's development strategy using Shenzhen as an example.

agglomeration – Ballungszentrum
foreign trade volume – Außenhandelsvolumen
isolated – abgeschottet
manufacturing – Herstellung
processing – (Weiter)-verarbeitung
prosperity – Wohlstand
services/service industry – Dienstleistungen/ Dienstleistungsbranche
settle – (sich) niederlassen
share – teilen, Anteil
special economic zone (SEZ) – Sonderwirtschaftszone

INTERNET

M1 The *Statistische Bundesamt der Bundesrepublik Deutschland (Destatis)* has compiled data on China's development in numerous country profiles.

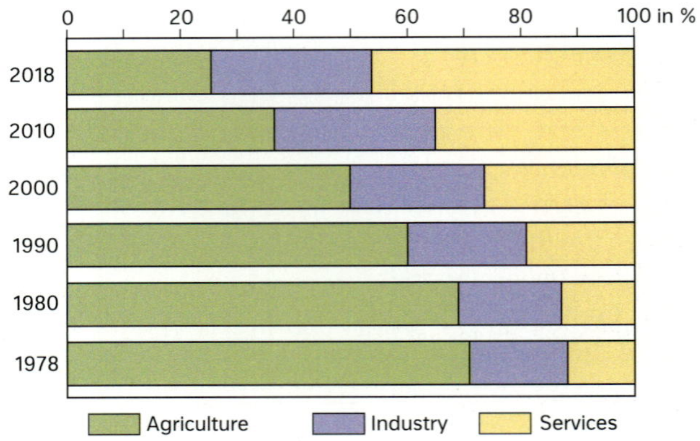

Source: Schucher 1999, Brooks Ray et Tao Ran 2003 (IMF Working Papers), China Statistical Yearbook 2019, CEIC Data 2017

25944EX_2 © Westermann

M2 China – employment by economic sector

	1978	1980	1990	2000	2010	2016	2018
	(in billion US dollars)						
Import	11	20	53	225	1396	1588	2136
Export	10	18	62	249	1578	2098	2487

Source: China Statistical Yearbook 2019

M3 China – development of foreign trade

INFO

M4 1978 was the year of change for China. Until then, the government had isolated the country from the rest of the world. In order to increase the prosperity of its own population, the government decided to push ahead with the industrialisation of the country. For the first time, it allowed foreign companies to establish themselves in China.

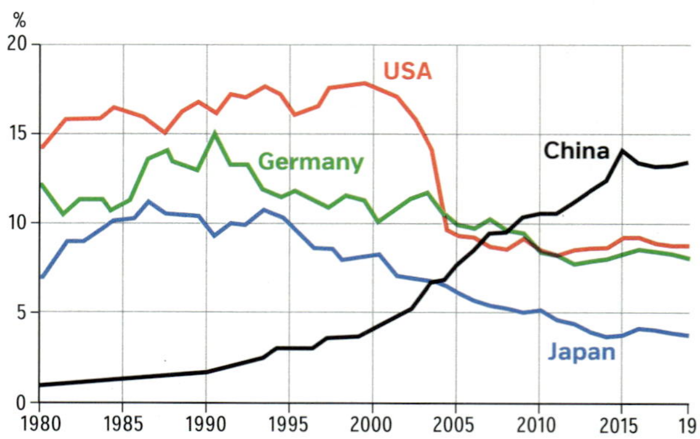

Source: UN 2020, own calculations

© Westermann 25941EX_2

***M5** Shares of total world exports

	1980	2000	2018
Population in millions	987	1267	1394
Share of urban population in %	19.4	36.2	59.6
Birth rate per 1000 inhabitants	18.2	14.0	10.9
Death rate per 1000 inhabitants	6.3	6.5	7.1
GNI per capita in US-$	220	940	9460
Human Development Index (HDI)	0.42	0.59	0.76
Direct investments in million US-$			
from abroad	57	40715	135000
from China to abroad	44*	916	179100
International tourism (arrivals in millions)	5.7	83.4	142.0

*1982

M6 China – development indicators (selected)

M9 Shenzhen is a symbol of economic development in China. Today, around 13 million people live in the city, which had only 13,000 inhabitants in 1965. Shenzhen is one of the fastest growing cities in the world.

Shenzhen [...] is today, next to Shanghai, the richest and most modern urban centre in mainland China. With exports worth US-$ 237 billion in 2016 [2019: 220 billion], Shenzhen is the largest exporter of all Chinese cities, contributing 11% [2019: 8.4%] of the national foreign trade volume. The main contributor to the city's success is the high-tech manufacturing sector, especially in electronics, IT, hardware and telecom equipment. Shenzhen is a world leader in the manufacture of electronic hardware [...]. All the important manufacturers, including Samsung, Apple or Microsoft, and of course the Chinese producers [e.g. Huawei, Lenovo], have production here.

Source: Außenwirtschaft Austria (Hg.): Außenwirtschaft – Regional-Update Guangdong und Südchina. Vienna 2017, p. 4/5.

M7 The economic importance of Shenzhen

M8 The distribution of per capita income and economic development zones in China

IN FOCUS What development strategy is China pursuing?

The People's Republic of China was founded in 1949. It was a developing country at the time. But since about 1980, the country of China has undergone profound economic development.

C Aspect 2: What has the Chinese government done to reduce spatial disparities between Western and Eastern China?

Western China and the other greater regions of China have developed at different speeds.

Export oriented industrial centres developed in the eastern, coastal areas, whereas the hinterland, the mountainous Western China stayed behind.

1. Prepare a short presentation on this key topic.
211, 218

Tips for the structure:
- Describe the development strategy of the Chinese government and its goals.
- Illustrate the importance of individual measures for the economic development of Western China. Also use the atlas.
- Finally, make an assessment of the future economic development of Western China.

dam	–	hier: Staumauer
demand	–	Nachfrage
downstream	–	flussabwärts
environmental regulations	–	Umweltschutzbestimmungen
establish	–	einrichten / ansiedeln
high-elevation-railway	–	Hochgebirgsbahn
incentive	–	Anreiz
inland port	–	Binnenhafen
ocean-going vessel	–	hochseetüchtiges Schiff
promote	–	fördern
regional disparities	–	regionale Disparitäten
seaport	–	Seehafen
sluice	–	Schleuse
tax	–	Steuer / Abgabe
tourism revenues	–	Tourismuseinnahmen
transshipment	–	Umschlag (im Hafen)
upstream	–	flussaufwärts

Tibet is to be developed into a tourist region. For this reason, the railway through the high mountains from Golmud to Lhasa was built (Lhasa Railway: completion in 2006, costs around 3.3 billion euros). In Lhasa, which was visited by 19 million tourists in 2018, for example, up to ten five-star hotels are to be built to increase the number of tourists. The main aim is to attract more affluent tourists from Europe and North America. As a result, tourism revenues are to be increased from three to seven billion euros per year.

M1 Development measures in the field of tourism

Since 2000, the Chinese government has been promoting twelve provinces and autonomous regions between the Gobi Desert and the South China Sea with the Go West Policy. [...] Beijing pumps billions into these provinces for the construction of infrastructure and offers generous incentives in terms of taxes or land use for companies willing to settle there. Yet for a long time, neither Chinese nor foreign companies were attracted there. Companies shied away from the long transport routes [...]. But when wages in the booming coastal areas began to rise a few years ago, the west suddenly became interesting. Especially since, with increasing development, it is becoming a sales market itself.

Source: Kühl, Christiane: Projekt Aufbau West. In: Außenwirtschaft. Deutscher Sparkassenverlag Stuttgart, Vol. 5 (Oct/Nov 2013), p. 10.

M2 The "Go West" programme

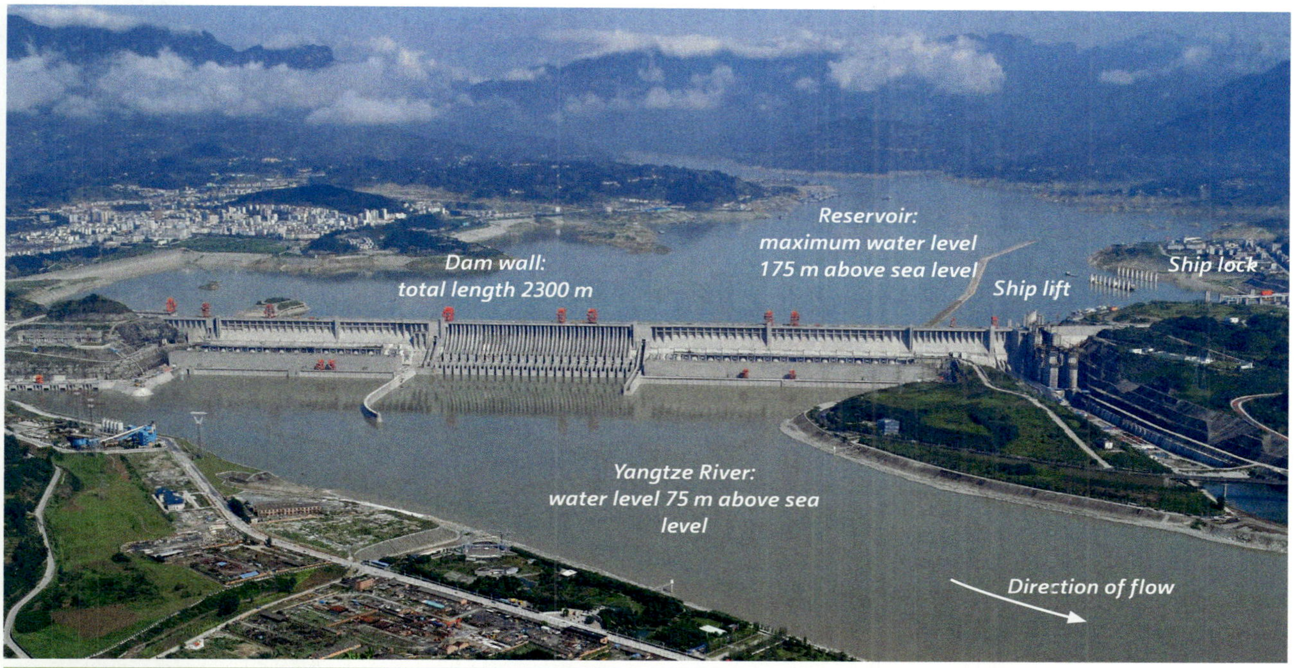

M3 The Three Gorges Dam on the Yangtze River, China's longest river, was completed in 2009. Since then, the river can be navigated by ocean-going vessels all year round. Goods can now be transported cost-effectively between Western China and the seaports around 1500 km away in Eastern China. The throughput of the port of Chongqing (about 650 km upstream of the Three Gorges Dam), for example, has doubled since 2009 to about 173 million tonnes (by comparison: Europe's largest inland port - Duisburg 133 million tonnes).

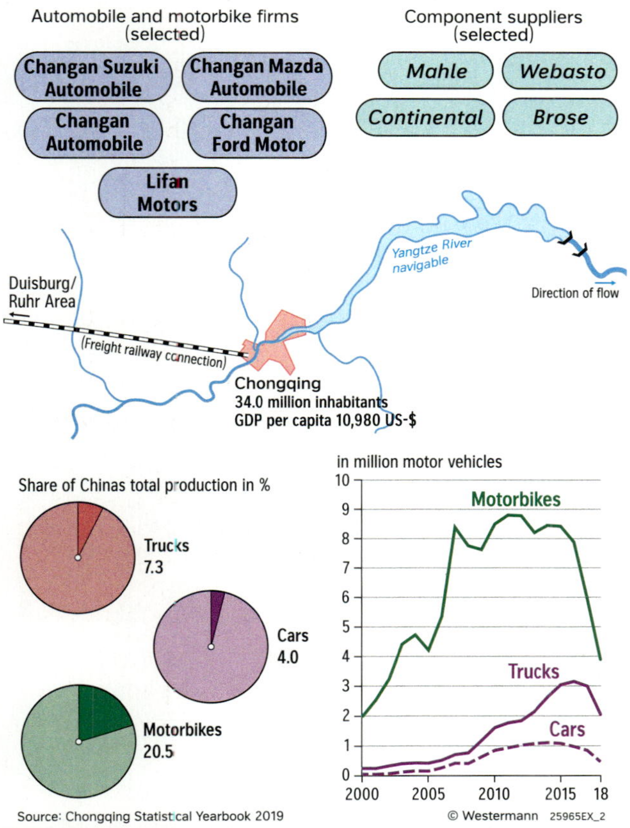

M4 Development measures in the industrial sector: In Chongqing, the Chinese government has succeeded in establishing various industrial sectors. The production of motor vehicles has declined in recent years, partly due to stricter environmental regulations (CO_2 emissions) and lower sales figures (e.g. lower economic growth in China).

	West	East	West	East
	Length in km		Share in %	
Roads				
2000	553 900	848 800	39.5	60.5
2018	1 991 700	2 854 800	41.1	58.9
Railways				
2000	22 100	36 500	37.7	62.3
2018	52 800	78 900	40.1	59.9
Waterways				
2000	21 900	97 500	18.3	81.7
2018	33 900	93 200	26.7	73.3

Note: Lengths in km rounded
Sources: China Statistical Yearbook 2001 and 2019, own calculations

M5 Development measure: building the infrastructure

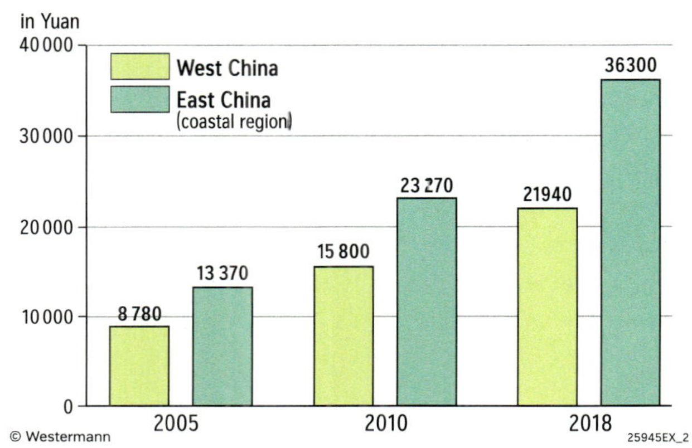

M6 Western China: disposable income per capita compared to Eastern China (coastal region)

IN FOCUS

What development strategy is China pursuing?

The People's Republic of China was founded in 1949. It was a developing country at the time. But since about 1980, the country of China has undergone profound economic development.

C Aspect 3: What is China's development strategy to become the world's leading economic power?

The Chinese government has set itself the goal of making China the world's sole leading economic power by the time the country celebrates its 100th anniversary in 2049. To achieve this goal, it has developed a global strategy and has already implemented the first development measures.

1. Prepare a short presentation on this key topic.
211, 218

Tips for the structure:
- Describe China's development strategy and its goal.
- Work out how China is driving forward the modernisation of industry.
- Illustrate the Chinese government's intentions for its expensive "New Silk Road" project.
- Finally, make an assessment of China's future economic development.

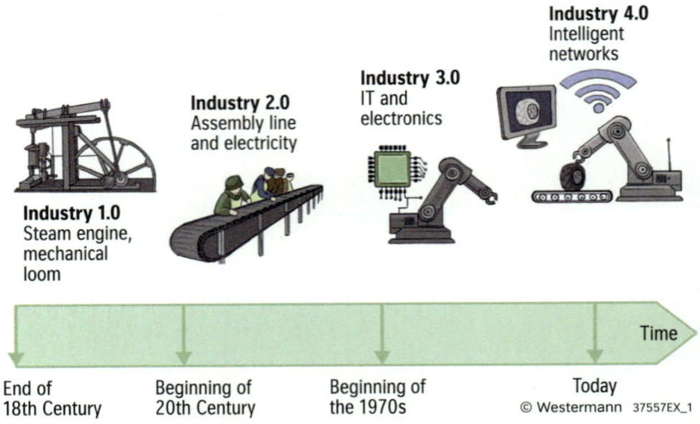

M1 **At the heart of the strategy "Made in China 2025" is the automation and digitalisation of industry. The strategy resembles the German Industry 4.0 strategy. The Chinese strategy aims more strongly at raising production to the level of Industry 3.0.**

INFO

M2 The Chinese government wants to become the world's leading industrial and trading power with the help of the "Going Out" development strategy. In the future, China will primarily produce high-quality and internationally competitive goods for export. Economic experts believe that China can only achieve this goal if its industry is modernised. In addition, global trade in goods requires a large number of trading centres that are well networked with each other.

acquisition	–	(Unternehmens-)Kauf
automotive industry	–	Automobilindustrie
dependence	–	Abhängigkeit
electric vehicles	–	Elektrofahrzeuge
industrial robots	–	Industrieroboter
key industry	–	Schlüsselindustrie
participations	–	(Unternehmens-)-Beteiligungen
revitalise	–	wiederbeleben
shipbuilding	–	Schiffbau
smart factory	–	intelligente Fabrik
trade relations	–	Handelsbeziehungen
world market share	–	Weltmarktanteil

In Sany's factory hall 18 in Changsha, central China, the factory of the future is already a reality. China's largest machine manufacturer produces asphalting machines and concrete mixers here. The factory is permeated through and through with electronics: The machines are networked with each other and continuously collect data about the production process. The position of components and delivery units can be called up at any time. Sany uses this information to optimise production. This brings the company closer to its goal of a self-organising [...] "smart factory".

Source: Wübbeke, Jost und Conrad, Björn: Industrie 4.0: Deutsche Technologie für Chinas industrielle Aufholjagd? In: China Monitor (Mercator Institute for China Studies) Berlin, Nr. 23 (11.03.2015), p. 2.

M3 **"Smart factories" (Industry 4.0) like the one at Sany in Changsha are still the exception in China.**

Key industry	Target by 2025
Automobile industry	Increase in production of electric vehicles; sales of 3 million units, i.e. world market share of 80%
Information technology	Reduce dependence on foreign countries; increase the share of Chinese IT from 75% (2020) to 90%
Shipbuilding	Increased use of high technology; increase in world market share from 40% (2020) to 50%
Industrial robots	Increase the share of robots manufactured in China to 70%

M4 Key industries (selection) in which China wants to be the world leader by 2025

Company	Industry	Investment in million € (Shareholding in %)
Daimler	Car manufacturer	7200 (10)
Kuka	Robots for the automobile industry	4660 (95)
Biotest	Pharmaceutical manufacturer	1300 (90)
KraussMaffei	Specialist for plastics machinery (e.g. car fittings)	925 (100)
Robert Bosch Starter Motors	Starter motors, electric drive components	545 (100)
Manz CIGS Technology	Engineering (e.g. solar modules)	54 (30)

M7 On a buying spree – Chinese investors are taking over or investing in German companies

	2012–2015	2016–2019
Investments in billion US-$	79.9	191.9
No. of company buyouts and participations	631	934
of which in Germany	130	196

Source: Ernst & Young (2020): Chinesische Unternehmenskäufe in Europa. Eine Analyse von M&A-Deals 2006–2019. o.O., p. 3/4/7, own calculations

M5 China's investments in Europe

"What fascinated Midea about Kuka was that it stands for German high technology, has excellent skilled workers and engineers and good management."

Source: Stahl, Stefan: Deshalb haben die Chinesen Kuka gekauft. Augsburger Allgemeine Zeitung, www.augsburger-allgemeine.de, 20.01.2017, access: 09.12.2020

M8 Till Reuter, CEO of Kuka, on the takeover by the Chinese company Midea

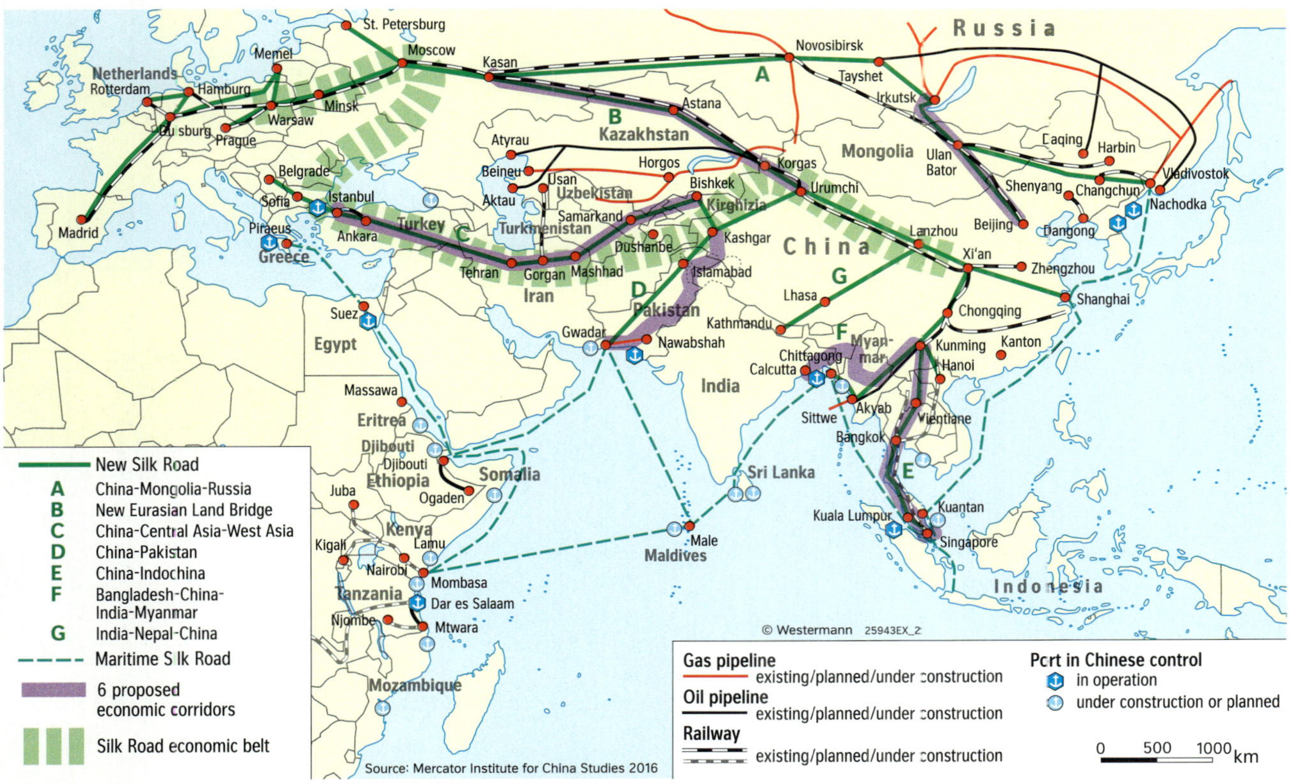

M6 The "New Silk Road" is the Chinese government's largest infrastructure project to date as part of its "Going Global" strategy. According to estimates, it will cost around 1000 billion US dollars. The project is intended to revitalise old trade routes and expand China's trade relations with a total of 56 countries in Asia, Europe and Africa.

3. Towns and cities made for people?

WESTFÄLISCHE MÜHLENSTRASSE
WASSERSTRASSENKREUZ
MITTELLANDKANA
WASSER-TURM
MINDEN
BIELEFELD
WESER
HERMANNSDENKMAL
DETMOLD
DORTMUND
GÜTERSLOH
SPARRENBURG
MARTa
HERFOR
RUHR UNIVERSITÄT
SSEN
BOCHUM
LIBORI VOLKSFEST
EIN
WAR-STEINER
MONT-GOLFIADE
PADERBORN
WELT JUGEND-HERBERGE
SCHWEBEBAHN
WUPPER-TAL
BURG ALTENA
SCHLOSS
SIEGEN
ROTHAARGEBIRGE
KAHLER ASTEN
GEZEICHNET VON JACQUES TILLY UND DAVID SALOMO

How do towns and cities develop?

The city centre – the pulse of the city

When urban dwellers say "I am just going into town!" they usually mean the central business district (CBD), the city centre. What are the main features of the city centre? How are city centres changing?

1. a) Does the illustration on pages 108/109 include your town or city, or one that you know? Describe how this town or city is presented.
b) Describe what you would illustrate from your town or city, and show what makes your town or city a good place to live in.

Ⓒ **2.** What makes a town or city a good place to live in, in your opinion? Compare with M1 and M2.
A Make your own wordle.
B Write a comment as a contribution for the citizen's platform in your town or city.

3. a) Describe the functions and amenities in the city centre area of Paderborn and outline the land use in a simplified sketch (M4).
b) Outline similarities and differences to the city centre in your town or city.

Ⓐ **4.** a) Illustrate why city centres are less and less marked by high-street retail shops (M5, M6).
b) Illustrate new things emerging in the city centre (M3).

large-scale retailer	–	Großhandelsunternehmen
retail	–	Einzelhandel
retailer	–	Einzelhändler/-in
surplus of importance	–	Bedeutungsüberschuss
urban dweller	–	Großstadtbewohner/-bewohnerin
wordle (tag cloud)	–	Wortwolke

Helpful words and phrases for task 4:
... experiences the strongest competition between ...
It has undergone a change that ...
... as a consequence of ...
This development is mainly due to ...

Source: Pfinder 2013, p. 10

***M1 What makes a city youth-friendly and good to live in? Survey of young people in Munich**

Young people from different small towns in Germany met in the Youth Forum for Urban Development. They discussed how living in a small town can be made more attractive. They presented their ideas to Germany's Federal Building Ministry.
The young people would like to see the preservation and expansion of infrastructure for education and leisure. To this end, they themselves are prepared to do voluntary work. It is important to them to be given opportunities and spaces to get together and to try things out. This could be, for example a place to ride a dirt bike or a climbing tower. The young people emphasise that that they want to be asked beforehand what is important to them, and that such places should be planned together with them.

M2 Urban development – what is important for young people. Results of the "Jugendforum Stadtentwicklung für Attraktivität kleiner Städte", 2016

Neuss: The former Hertie department store is now home to the regional theatre and the district administration.
Gelsenkirchen: The former department store houses the city library, adult education centre, apartments for senior citizens and catering and retail outlets.

Herne: A department store that has been empty since 2009 is being converted into the "Neue Höfe Herne". There are offices, the headquarters of a large company, and catering on the ground floor.

M3 Department stores are often closed. Today, the buildings find other uses

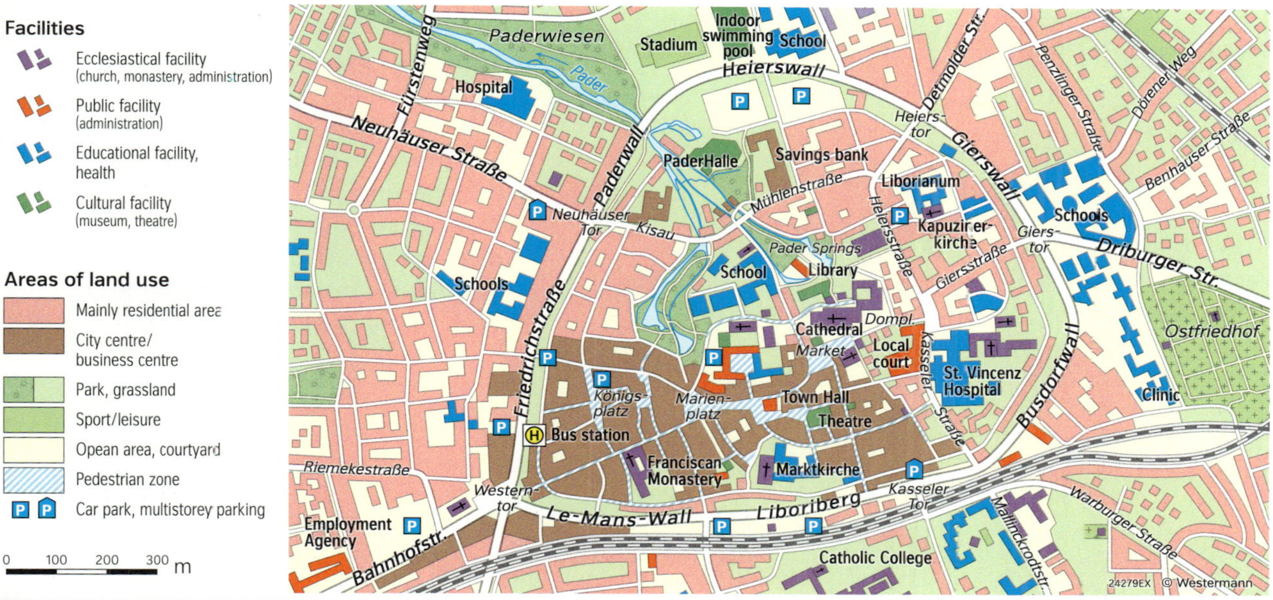

M4 The city centre of Paderborn

INFO

M5 **City centre** – the most central location in the city or town, it has a surplus of importance and fulfils a supply function that extends beyond its own boundaries.

M7 A shopping street by day and by night

Little by little, they [the department stores] are being squeezed by competitors. From the 1980s onwards, specialist do-it-yourself, consumer electronics or furniture stores sprang up like mushrooms on greenfield sites. In even larger shops, they offer even larger assortments at even lower prices. In the city centre, shopping malls and specialised retail chains like H&M compete with one another. Brands like Apple, on the other hand, manage to create a real hype around their products. When a new i-device is launched, customers are already queuing outside the Apple stores on the evening before. Department stores seem downright old-fashioned and dusty in comparison. But the biggest blow comes from online retailing. From the middle of the noughties (the decade from 2000 to 2009), it has conquered an ever-larger share of the market. While Amazon & Co. race from one sales record to the next, the profits of department stores - and of high-street retailing in general - shrink from year to year. Retailing, for centuries one of the core functions of town and city centres, is coming under pressure. Today, with the outbreak of the Corona pandemic, many are now finally threatened with extinction.

Source: Podjavorsek, Peter: Das Ende der Kaufhäuser, Malls und Einkaufsstraßen. www.deutschlandfunkkultur.de, 06.10.2020, accessed: 11.12.2020

M6 Retailing is changing: city centres under pressure

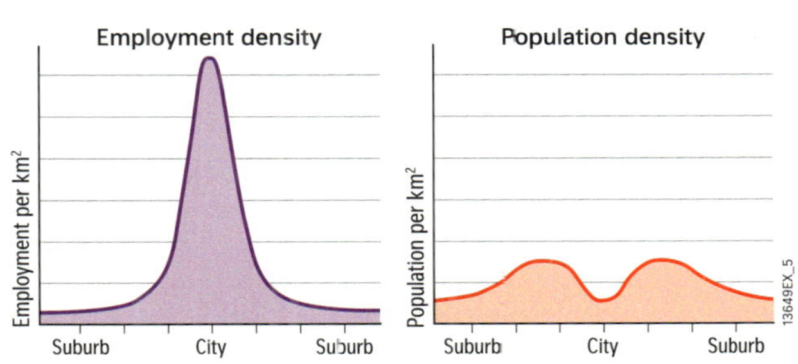

M8 Density of population and employment in the city

Wenn du diese Aufgaben erfolgreich bearbeitet hast, kannst du ...

- ... eigene Vorstellungen von einer lebenswerten Stadt erläutern.
- ... Bedeutung und Funktion der City einer Stadt erläutern.
- ... die Situation des innerstädtischen Einzelhandels und des Wandels in der City erläutern.
- ... den Fachbegriff ***city centre*** (***City***) erklären.

Centres in the city

Banks, the town hall, supermarkets, offices, museums, hotels .... towns and cities are where one finds the amenities and businesses of the tertiary sector. To a large extent, they determine the importance of the town or city. Its importance is higher or lower, depending on how well it is provided with these amenities and businesses. Why do such businesses mainly settle in towns and cities? What characterises their locations in the city?

1. a) Allocate the photos and statements (M2, M3, M5 – M13) to locations in the model M2. Give reasons for your allocation.
b) Describe the features of three different locations of the tertiary sector.

2. Name three examples of agglomeration advantages that are shown in the pictures, statements and the model, and illustrate them.

3. Which location in the model would be ideal for
a) a skateboard shop,
b) an ice cream parlour,
c) the office of the district student council,
d) the headquarters of a software company.
Give your reasons.

C **4.** **A** In many cities, office locations are built near water. Give reasons for the attraction of harbour and waterfront areas for service companies.
B Show the locations of three service companies that you know. Classify these in M2 and give reasons.

accessibility	–	Erreichbarkeit
advantages of agglomeration	–	Agglomerationsvorteile
amenity	–	Annehmlichkeit
at the seams	–	aus den Nähten
catchment area	–	Einzugsbereich
ice cream parlour	–	Eisdiele
proximity	–	Nähe
stock exchange	–	Börse
transport link	–	Verkehrsanbindung

Helpful words and phrases for task 4:

There are a set of reasons why ...
... plays a crucial role in the selection of the right location for ...
This location meets one/two ... basic requirement/s
First of all, ...
Secondly, ...

INFO

M1 **Advantages of agglomeration** are the advantages that result from the closeness of a business to other businesses. For example, a pharmacy can profit from being close to a doctor's surgery (location community), or different shoe shops in the same shopping street can benefit from the customers' knowledge that they can find a wide range of shoes there (competitive attraction).

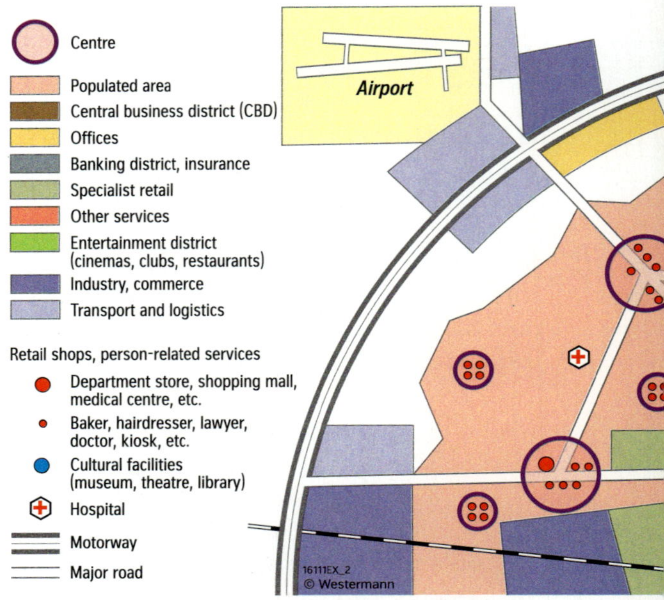

M2 **Model: Service locations in the city** 210

I have been working in this department store for nearly 20 years now. It is a shock that we are to be closed! Fewer and fewer customers come; they are voting with their feet. But in this company, there have been fewer and fewer new ideas.

M3 **Guido W., department store employee (2019)**

INFO

M4 Further important location factors for the service sector:
- Direct contact to customers in close proximity;
- Transport links/accessibility: for customers and suppliers, but also to be able to reach others easily;
- Image of a locality: a good address serves the purpose of self-promotion;
- Availability of cheap land, e.g. for large-scale retail businesses

We urgently needed office space. Our inner-city location was bursting at the seams. Now we have built multiple new office blocks and a parking garage away from the centre. The staff who move there are mainly those without direct contact to customers.

M5 Volker M., spokesman for an insurance company

The best thing about Saturday is breakfast. Then we fetch bread rolls from the baker on the corner!

M6 M6 Lukas J., school student

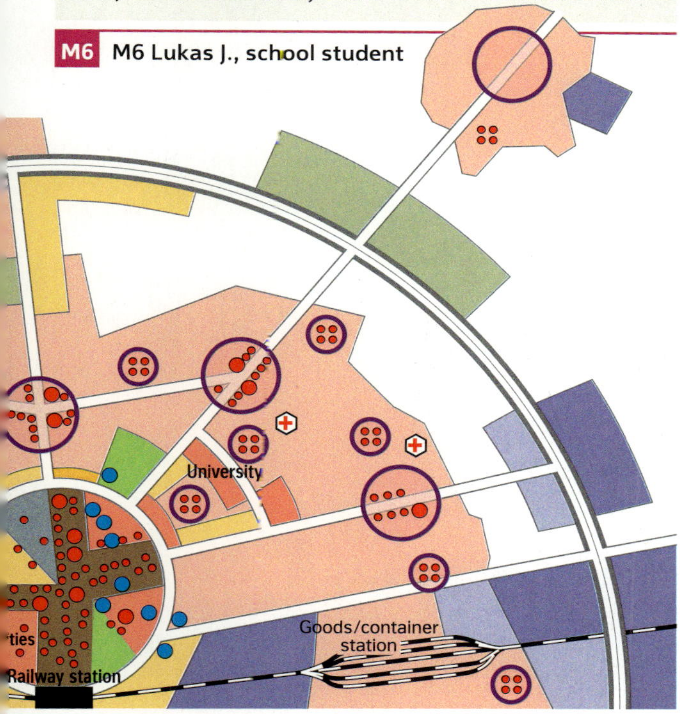

Some years ago, together with my colleague I made myself independent. Our law office is in a beautiful old house, directly on the quiet square. We can reach the city centre quickly.

M7 Georg U., family lawyer

It is important for us to be able to serve a wide catchment area. Our patients are sent to us not only from the whole city, but also from the neighbouring districts.

M8 Wiebke V., specialist physician

Almost every Saturday, wego to a discount store by car. Directly opposite, there is a supermarket with a drugstore next door. We can buy almost everything there that we need during the week.

M9 Gökce P., 15 years old

M10 Frankfurt, headquarters of the major banks, the European Central Bank, the German Federal Bank and the stock exchange. Other important facilities and business in the financial sector are also located here, many of them in one quarter of the inner city.

It is four o'clock on Saturday morning. I am now going to bed, and I will go back to work this evening. Today, with my two employees, I have roasted and sold eight large doner kebab spits. During the week there is less to do here.

M11 Jens S. operates a fast food restaurant

This household goods store was already operated by my grandparents. We are the last shop in this street that does not belong to a big chain. We are considering if it might not be more profitable to give up the business and to rent out the house and the shop.

M12 Angelika F., owner of a shop

M13 Shopping malls on greenfield sites or in inner cities have been strong competition for retailers in the CBD for decades.

Wenn du diese Aufgaben erfolgreich bearbeitet hast, kannst du ...
- ... Standortfaktoren von Betrieben und Einrichtungen des tertiären Sektors erläutern und konkrete Standorte begründen.
- ... den Fachbegriff **advantages of agglomeration** (**Agglomerationsvorteile**) erklären.

How have towns and cities developed?

Some of our towns and cities are very old. The first ones arose at the time of the Romans; the majority emerged in the medieval period. The centuries have left their marks: old buildings or structures, such as a narrow network of streets, with which we still live today. Which features of towns and cities emerged in the different epochs? What can still be found today?

1. Describe the picture M1 and explain how the town shown in the model corresponds to the medieval town (M2). 210
2. a) Is Dortmund (M4) a typical medieval town? Give reasons.
   b) Work with the atlas (map: Dortmund, Innenstadt). Describe the development of Dortmund and explain how the medieval structures are used today.
3. a) Describe the industrial town of Wuppertal at the end of the 19th century, and explain - using this example - the features of typical towns of that era (M3, M5).
   b) Explain how far the building of model villages (workers' colonies) was seen as a way to make towns and cities more people-friendly (M6 – M8).

company premise	–	Firmengelände
courtyard	–	Innenhof
kitchen garden	–	Nutzgarten
model village	–	hier: Arbeiterkolonie
network of streets	–	Straßennetz
tenement block	–	Mietskaserne
two-family dwelling	–	Zweifamilienhaus
urban sprawl	–	Zersiedelung
workers' settlement	–	Werkssiedlung

Helpful words and phrases for task 1:

In the foreground ...
In the centre ...
In the background ...
In the foreground/background on the left/right ...
Left/right of the ... you can see ...
market stalls
half-timbered house

M1 Today's presentation of a medieval town

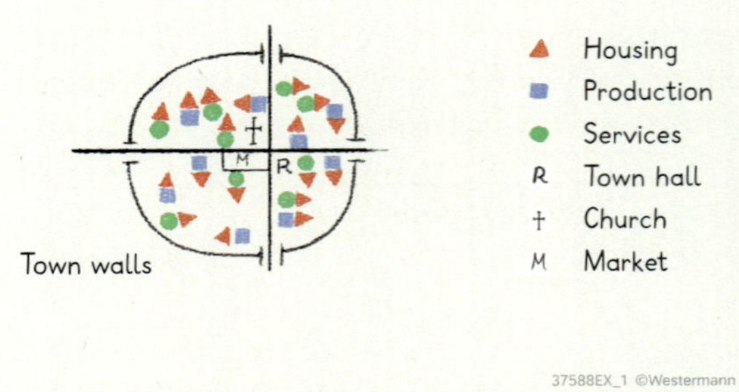

Further important characteristics:
- Above all, towns and cities are places of work and living for tradesmen, merchants and the associated transport companies.
- Town walls with gates to protect the town from attack.
- Mostly narrow, winding streets.
- Working and living quarters very closely connected to each other.

M2 Model: A medieval town 210

M3 Wuppertal Barmen, 1870

M6 The Dahlhauser Heide colony in Bochum-Horstel

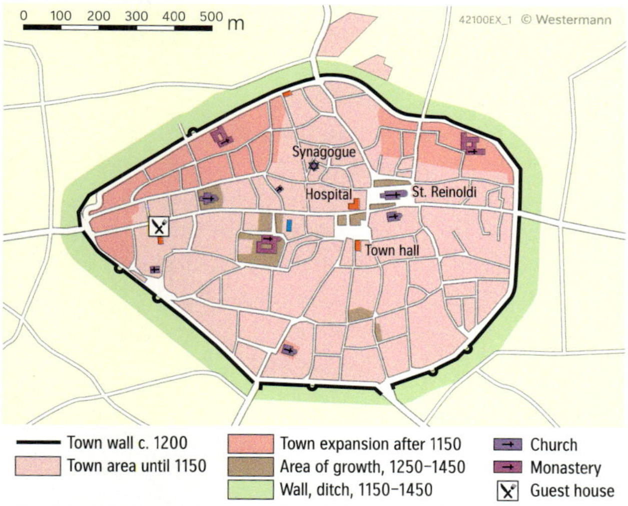

Source: Mühlhofer, Stefan et al.: Dortmund. (Deutscher Historischer Städteatlas 5), Münster 2018

M4 Dortmund in the medieval period

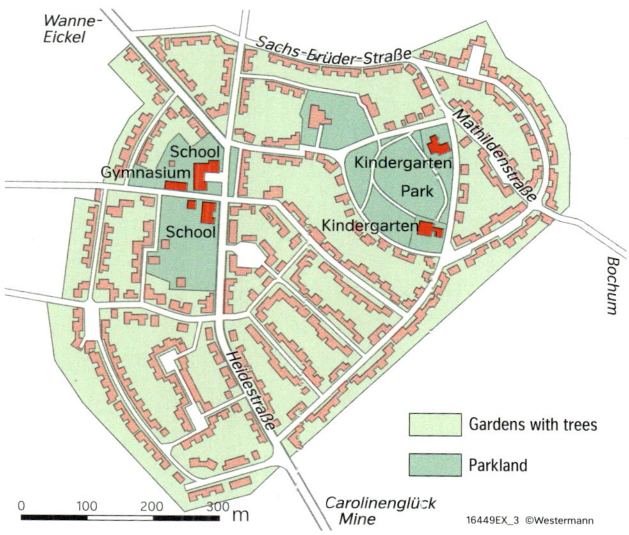

M7 The model village was built by Krupp AG between 1906 and 1915.

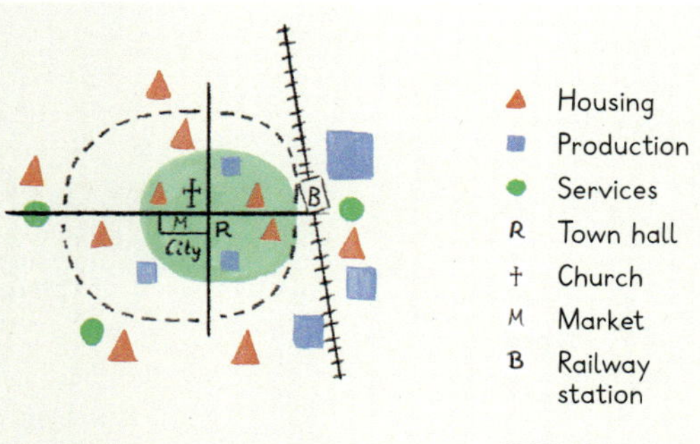

Further important characteristics:

- Railways as a motor of industrialisation.
- Urban growth due to factories and mines outside the old towns; new towns emerge; urban sprawl.
- Standardised mass building of dwellings to house as many people as possible; tenement blocks built around an inner courtyard; workers' settlements.
- The old town forms the new city centre.

M5 Model: a town in the industrial age 210

[Model villages were built] on behalf of companies. [...] The works settlements were situated close to the company premises, and therefore some distance away from existing town centres. As a result, they acquired the colloquial name of colonies. Around the turn of the century, 20 percent of the miners already lived in them. In the course of 100 years, 2000 settlements were built for two million people. [...] Alongside some houses for officials, [the Dahlhauser Heide colony] consists of 339 two-family dwellings. Each half of the house has a kitchen and a living room [...} Behind the houses, there are extensive kitchen gardens.

Source: Hoppe, Wilfried et al.: Umbau bestehender Siedlungsstrukturen. In: Ders.: Das Ruhrgebiet im Strukturwandel. (=Diercke Spezial), 2010, p. 80/82

M8 Model villages in the Ruhr Area

Wenn du diese Aufgaben erfolgreich bearbeitet hast, kannst du ...

... *Merkmale von Städten des Mittelalters und des Industriezeitalters erläutern.*
... *erläutern, dass Stadtstrukturen politische, wirtschaftliche und soziale Verhältnisse der Epoche ihrer Entstehung widerspiegeln.*

How have towns and cities spread out into their surrounding area?

In the 20th century, some cities in Europe grew into large megacities, or several cities merged into large conurbations. For a long time, Munich has led the ranking of the most attractive and successful German cities. And Munich is growing. What are the reasons for this growth and what are the effects of this development?

1. Describe the growth of Munich (M4, M6).

2. Work with the atlas: Give reasons for the attraction of Munich and its region (M3).

3. a) Describe the flows of migration in the Munich region (M3).
b) List push and pull factors for suburbanisation from the city into its surrounding area and from the surrounding area into the city (M1 – M6).

Ⓒ **4.** **A** Explain the interrelations between the city and its surrounding area (M6, M8, M9).
B Draw a meaningful diagram to illustrate this. 199

Ⓐ **5.** Assess the theory that "suburbanisation and re-urbanisation take place at the same time".

affluence	–	Wohlstand
commuter	–	Pendler/Pendlerin
conurbation	–	Städteballung
cost of living	–	Lebenshaltungskosten
increase	–	steigern/Steigerung
inexorable	–	unaufhaltsam
metropolitan region	–	Stadtregion
purchase price	–	Kaufpreis
rental price	–	Mietpreis
reurbanisation	–	Reurbanisierung
subsidies	–	Fördermittel
suburbanisation	–	Suburbanisierung
tax advantage	–	Steuervergünstigung
tax reduction	–	Steuerentlastung

Helpful words and phrases for task 1:

Between ... /From ... onwards ...
Since ...
The number of ... grew/fell/dropped steadily/constantly/slowly/rapidly/slightly
... remained constant

M1 The city also offers attractive surrounding areas

General framework conditions:
- Growing affluence and the desire for more and larger living space.
- The wish to live "in a house in the countryside".
- Growth of individual transport facilities, e.g. car.

... and political/urban planning factors:
- Tax advantages (subsidies for building owner-occupied homes and tax reductions for commuters).
- Expansion of utilities, waste disposal, roads and public transport outside the core cities.

Source: Claaßen, Klaus & Thilo Girndt: Stadt-Umland-Wanderungen. In: Ders.: Stadt- und Stadtentwicklung. (=Diercke Spezial), 2019, p. 49, amended

M2 Framework conditions for suburbanisation since the 1950s

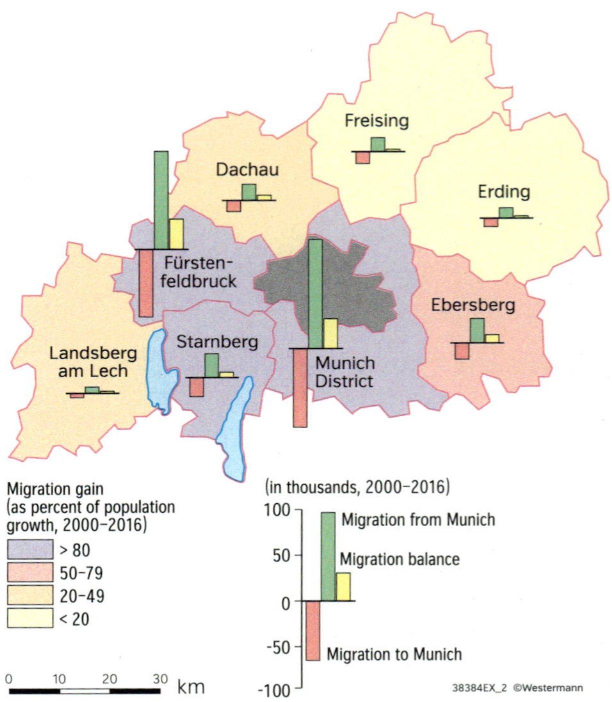

M3 Migration balance between Munich and its surrounding districts (2000 – 2016)

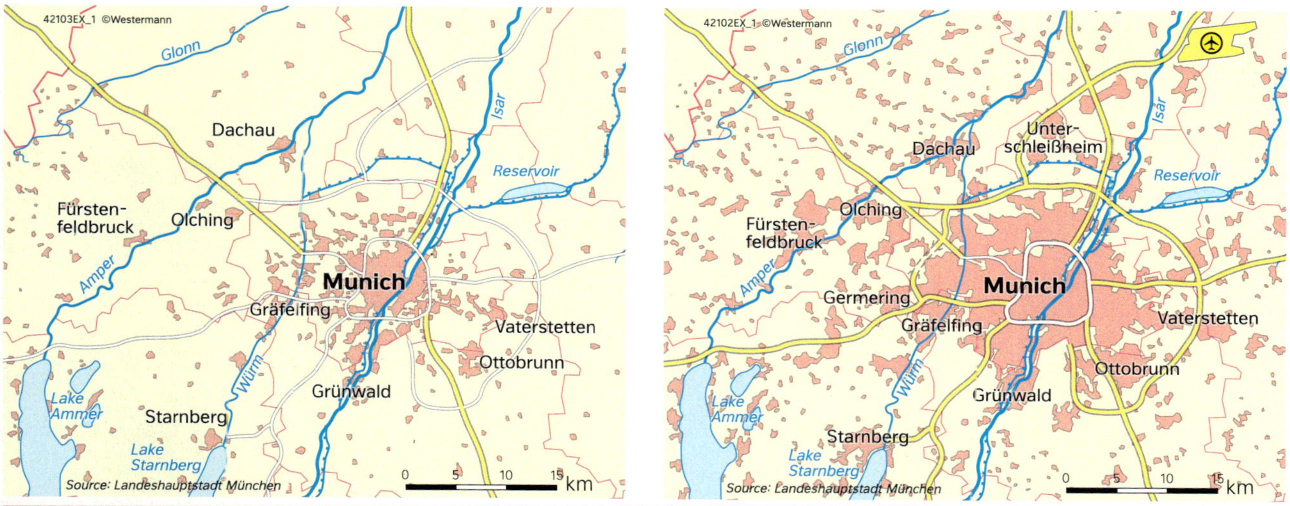

M4 Settlement area in the Munich Metropolitan Region, 1950 and 2000

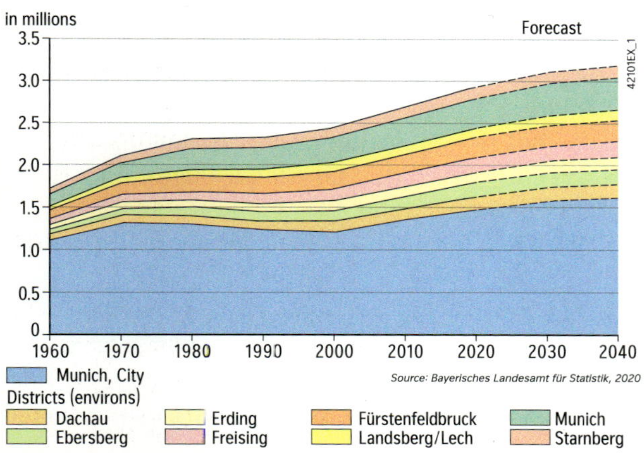

M5 Development of the population in Munich and its eight surrounding districts, 1950 – 2038

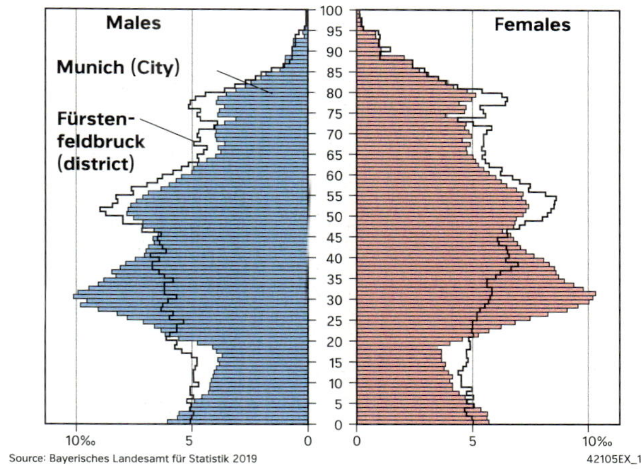

M8 Age structure of Munich and the rural district of Fürstenfeldbruck in comparison (2018)

Munich is one of the most popular cities in Germany and worldwide. Many people would like to live here. This has an impact on the housing market. The supply of living space is limited and as a result rental and purchase prices are rising. This has led to more and more people moving to the outskirts of Munich. Due to the strong demand, prices there have also continued to rise.

M6 Report on house building in Munich

INFO

M7 The relocation of population and urban functions to the edges of cities or their surrounding areas is referred to as **suburbanisation**. Relocation back into the city is called **reurbanisation**..

The Munich region is growing inexorably, its economic development is stable. The other side of the coin is the rising demand for housing, the high cost of living and the increasing traffic. [...] Most of them [emigrants] merely change their place of residence within southern Bavaria and continue to work in Munich. Frequently cited reasons for moving away are the intention to increase the size of one's household or the desire to live more in the countryside. As the survey showed, one third of the households only moved away when they could not find a suitable place to live in Munich. People appreciate the advantages of the big city in everyday life.

Source: Landeshauptstadt München: Wanderungsmotivuntersuchung. www.muenchen.de, accessed: 01.12.2020

M9 From the City of Munich's survey of motives for migration

Wenn du diese Aufgaben erfolgreich bearbeitet hast, kannst du ...

- *... am Beispiel der Stadt München Wachstum der Stadt in das Umland beschreiben und erklären.*
- *... die Fachbegriffe **suburbanisation** (**Suburbanisierung**) und **reurbanisation** (**Reurbanisierung**) erklären.*

How are population groups distributed in the city?

In cities, there is a great diversity in the inhabitants' lifestyles and living situations. There are often social differences in cities: different groups do not live together spatially, but rather separately from one another. However, there are always some neighbourhoods that are mixed in social terms.
How can social differences be described? How are different social groups distributed in cities?

1. a) Explain what conclusions you can draw from the indicators in M6 and M9.
b) Describe more precisely the situations in three different parts of the city, and then look for further parts of the city that have similar characteristics.

2. Compare the different living situations in Essen with one another, and discuss the future prospects of the people (M2 – M6, M8, M9). 217

C **3.** **A** Explain how the A40 in Essen can be described as a "social equator".
B Make a sketch of the social-spatial division of Essen. 204

4. Comment on the suggestions in M7.

decline	–	Niedergang
in-commuter	–	Einpendler
integration trap	–	Integrationsfalle
refugee	–	Flüchtling
segregation	–	Segregation
social disparities	–	soziale Disparitäten
social-spatial	–	sozialräumlich
stakeholder	–	Interessenvertreter/-in
top-selling	–	umsatzstark
turnover	–	Umsatz
unemployed	–	arbeitslos/ Arbeitslose/-r

Helpful words and phrases for task 1:

The ... indicate/s a/that ...
It becomes clear that ...
This may suggest a/that ...
From the fact that ... it can be concluded that ...
It is obvious that ...
Compared to ...

- The biggest city in the Ruhr Area – 590,000 inhabitants.
- 42 % of its total surface area is not built up.
- Since the 19th century, the city has been marked by coal mining and the steel industry. The last mine closed in 1986.
- Of 328,000 people who work there, some 45 % commute into the city. 86.1 % work in the tertiary sector and 13.8 % in the secondary sector.
- Essen is the headquarters of 7 of the 100 companies with the highest turnover in Germany.

M1 Profile of the city of Essen

She [...] graduated from high school and is planning to study law. "I grew up here. This cliché house, happy family, dog, horse, holiday, a nice car. The school is obviously a grammar school. [...] You can be sure that your life will be successful, because there are people supporting you."

Source: Pippa Kolmer: Diercke Praxis Band 3, Westermann, Braunschweig, 2018, translation: John Atfield

M2 Pippa K., 17 years old, grew up in the south of Essen

Our children come from 25 different nations. They have very different ethnic origins. We also have 70 refugee children who cannot even speak German yet. [...] We must compensate for many things that their parents can no longer do. There is a lack of language role models. It is not an ethnic or geographical problem, but [...] a social problem.

Source: Betül Durmaz: Diercke Praxis Band 3, Westermann, Braunschweig, 2018, translation: John Atfield

M3 Betül D., head of a primary school in the north of Essen

The decline of coal and steel has mainly affected the "working-class districts": In the districts where most workers lived in the 1987 census, the proportion of unemployed is particularly high, and there are also the largest number of children living on social benefits. In places where 20 years ago most of the inhabitants had no more than a secondary school leaving certificate, there are still particularly high proportions [...] who leave school without a certificate today.

M4 Background – from the specialist literature

M5 Life in Essen-Bredeney

M8 Life in Essen-Altenessen

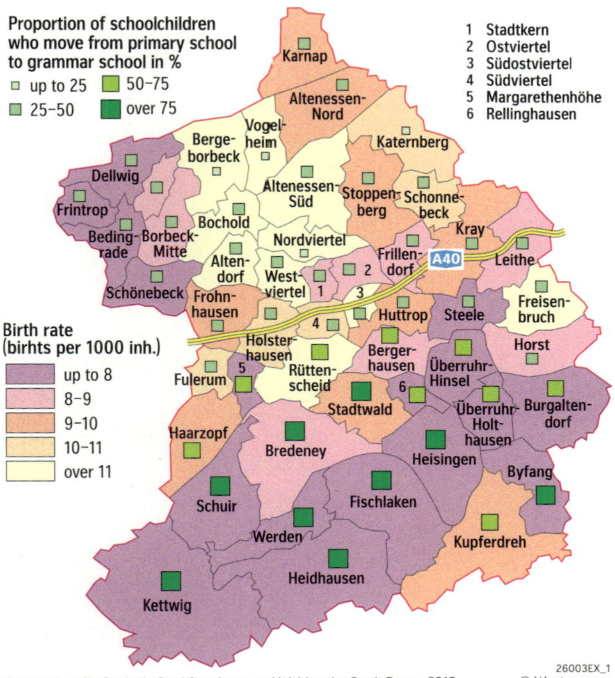

Source: Amt für Statistik, Stadtforschung und Wahlen der Stadt Essen, 2018 26003EX_1 © Westermann

M6 Social characteristics of the districts of Essen

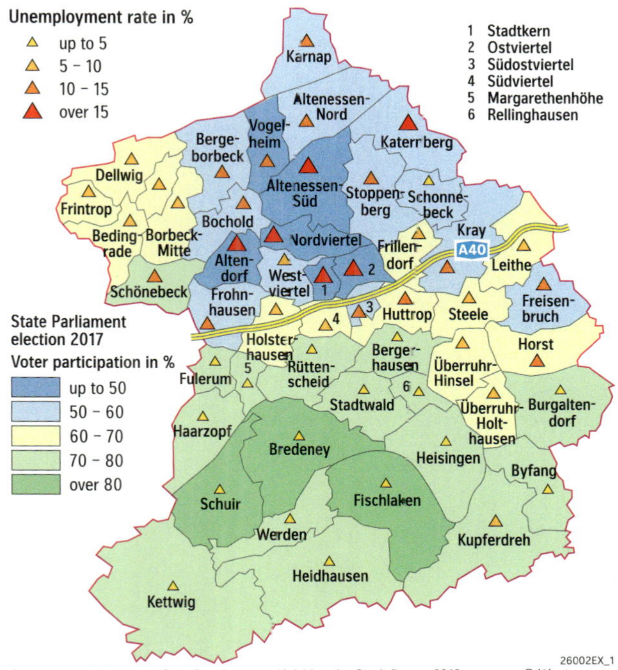

Source: Amt für Statistik, Stadtforschung und Wahlen der Stadt Essen, 2018 26002EX_1 © Westermann

M9 Social characteristics of the districts of Essen

If the education and health of the next generation are determined by their address, then bad addresses in the cities must be turned into better ones, and as good as possible. [...]
Segregation is the demographic and social challenge of urban society. The municipalities in the Ruhr Area are not able to cope with it on their own. It requires the cooperation of many stakeholders. It is not city mayors, but companies that create jobs. Housing companies are important allies when it comes to stabilising social conditions in disadvantaged neighbourhoods.
Without promoting education for the children of immigrants and the unemployed, ethnically segregated neighbourhoods become integration traps. Above all, education enables mobility and innovation. Education and work are the keys to the social integration of the next generation; without them there is no perspective for the Ruhr Area.

Source: M4, M7 from Kersting, Volker, Meyer, Christian, Strohmeier, Peter und Terpoorten, Tobias: Die A40 – Der Sozialäquator des Ruhrgebiets. Vielfalt und Wandel des Ruhrgebiets im Kartenbild. In: Prossek, Achim et al. (ed.): Atlas der Metropole Ruhr. Emons, Köln 2009, p. 144/145, amended

***M7** Perspectives – from the specialist literature

Wenn du diese Aufgaben erfolgreich bearbeitet hast, kannst du ...
... mit verschiedenen Indikatoren Bevölkerungsgruppen beschreiben.
... die soziale Differenzierung von Städten erläutern.
*... den Fachbegriff **segregation** (**Segregation**) erklären.*

Why do towns and villages lose population?

Many towns and cities are growing, but there are also shrinking towns. Shrinking means a population loss. Both of these phenomena occur not only in towns and cities, but also in rural areas. What are the causes and effects of a population loss in towns and villages?

Choose an area for yourself:

Urban area (Group 1)

1. Locate Altena (atlas). Describe the urban change and explain its causes (M1, M2, M7, M8).

2. Present clearly the consequences for the town of the population loss in a diagram.

3. The measures taken by the town of Altena are a classic example of a shrinking town. Research what ideas and measures have been taken there in recent years (internet). 213

Rural area (Group 2)

4. Locate Voigtsdorf and classify it (M3, M5, atlas).

5. **A** Calculate the percentage shares of the different age-groups in Voigtsdorf (M3).
B Present the age structure of Voigtsdorf in an age diagram (M3).

6. Outline push and pull factors of Voigtsdorf from the viewpoint of a schoolgirl or schoolboy and of a pensioner (M6).

deterioration	– Verschlechterung
municipality	– Kommune
purchasing power	– Kaufkraft
remote region	– abgelegene Region
vacancy rate	– Leerstandsquote
well-funded	– kaufkräftig

Helpful words and phrases for task 1:

... is located/situated ...
... east/west/north/south of ...
... in the northern/southern/eastern/western part of ...
near/close to ...
in a region/area that is ...
densely/sparsely populated

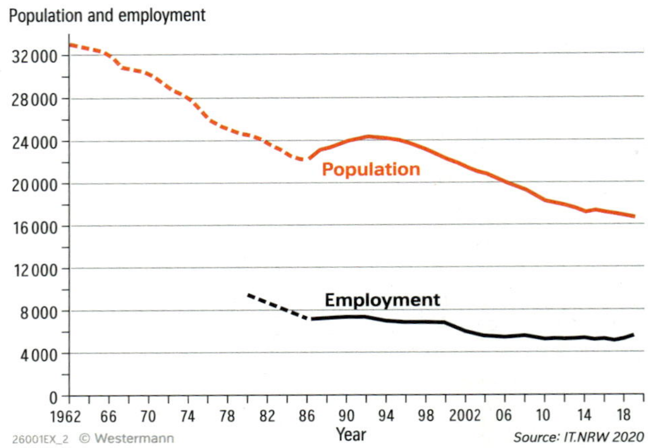

M1 Development of population and employment in Altena

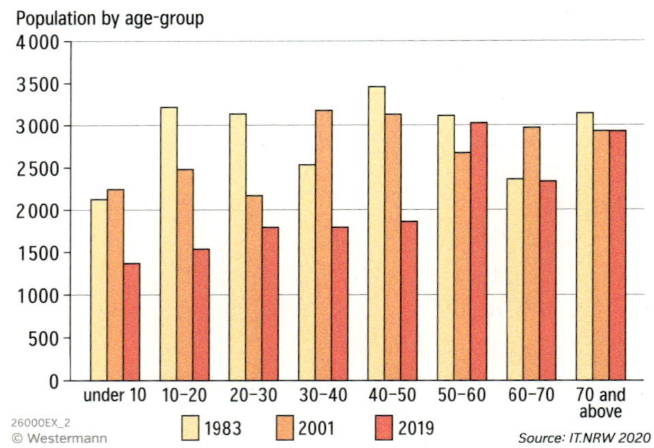

M2 Population of Altena by age-group

Voigtsdorf, founded 750 years ago, is the smallest independent municipality of the Woldegk district (Mecklenburgische Seenplatte district). The nearest town is Strasburg (Uckermark).

	1990	2018
Population	180	104
Age-groups:		
under 15	45	21
15 to under 30	35	7
30 to under 45	36	15
45 to under 60	29	31
60 to under 75	27	20
75 and above	8	10

M3 Data and facts about Voigtsdorf

M4 Altena

Especially the end of several large industrial operations in the 1980s is still causing problems for the town. At that time, many young residents moved away because of the poor job opportunities. Today the town has the same problem as many municipalities in the region: it is too far away and at the same time too close to the conurbations around Cologne and the Ruhr Area. Too far away to benefit from the attraction of the big cities as a wealthy suburb. Too close because many young people move from the small towns to the big cities.

Source: Jacobsen, Lenz: Altena versucht's eine Nummer kleiner. www.zeit.de, Hamburg, 12.05.2012, accessed: 16.12.2020

M7 Background to the development

Growing villages..
- are in the surrounding area of towns or cities with a short commute to places of work,
- have attractions for tourists,
- have larger companies based there.

Shrinking villages...
- lie in remote, rather thinly populated regions,
- offer few places of work and poor infrastructure,
- suffer from emigration, resulting in empty buildings and further deterioration in infrastructure (e.g. schools, pubs, shops)..

M5 Characteristics of growing and shrinking villages

We have a very high vacancy rate. Now the first primary school is being closed. We have already lost sports facilities and one of two outdoor swimming pools. Important infrastructure providers are relocating their service facilities to the medium-sized centres in the immediate vicinity. Churches, associations and clubs are suffering just as much from the town's population development. All this does not make the locality more worth living in. In the retail sector, there is a noticeable decline in purchasing power. The well-funded households with high incomes are missing! The fixed costs for the urban infrastructure are increasing per capita.

Source: Hollstein, Andreas: Gesellschaftliche Aspekte des demographischen Wandels am Beispiel der Stadt Altena. Vortrag auf Schloss Ahaus, 2005, p. 7–9, amended.

M8 Speech to entrepreneurs by Andreas Hollstein, Mayor of the town of Altena

M6 Statements from Voigtsdorf

Wenn du diese Aufgaben erfolgreich bearbeitet hast, kannst du ...
... Ursachen und Auswirkungen von starkem Bevölkerungsrückgang in Städten und Dörfern erläutern.

M1 Cartoon

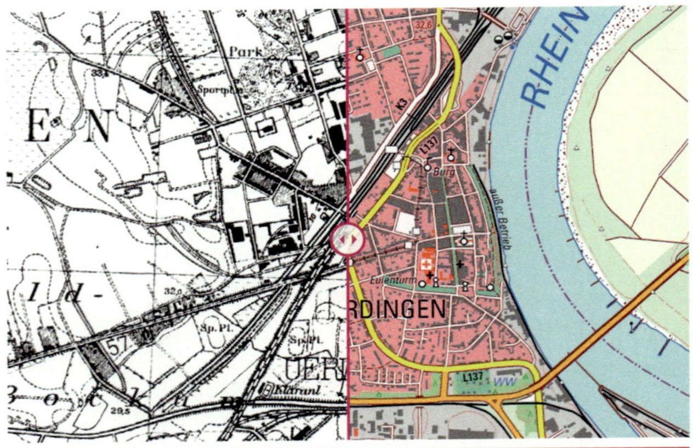

M2 Uerdingen, 1936 and 2020 (scale 1:35,000; based on: geoportal.nrw.de, Geoviewer)

WHAT WE HAVE LEARNED

How do towns and cities develop?

The city centre – core of the city

110 The central business district (CBD) is located in the **city centre**. There is a lot of office and commercial space here, as well as many jobs. The city centre has excessive importance. Many functions are concentrated here, for example the business area with pedestrian zones, administrative and banking districts or museum districts. While the daytime population is very large, few people are there at night. The importance of a city depends on its functions and facilities. Inner-city retail shops are under great competitive pressure from shopping centres and online retailers. That is why the inner cities are changing.

What are the characteristics of the different centres?

112 In cities, amenities and companies in the service sector often form centres where they are spatially concentrated. Service providers often depend on direct customer contact. That is why retailers, for example, choose locations close to the customer. Their business needs to be easily accessible for many customers. In many locations, businesses have advantages due to the proximity of other businesses because, for example, they attract more customers together than alone. Business-oriented service providers also benefit from such **agglomeration advantages**.

How have towns and cities developed?

114 Cities have always been economic centres with many jobs in production and services. At the same time, cities have constantly changed. Urban life, the appearance of cities and their structure have changed with the situation in each respective era. Cities reveal their development in their townscape. In their centres, there are the oldest buildings and structures. Connected to these are the extensions of different era. In today's cities, features of their era are often preserved.

How do towns and cities spread into their surrounding area?

After the Second World War, many cities grew into the 116 surrounding countryside and the conurbations increased greatly in population. With **suburbanisation**, extensive new housing estates were built in older places in the surrounding areas. More recently, there have also been movements in the opposite direction, a return to the core city (**reurbanisation**).

How are population groups distributed?

Different population groups are often more or less clear- 118 ly separated from each other in cities (**segregation**). The reason for this is their respective financial possibilities. For example, the more affluent have moved "into the countryside" on the outskirts of the city or into the surrounding countryside. Neighbourhoods close to the inner city have also become attractive for a wealthy population. However, the composition of the population of neighbourhoods can vary. Neighbourhoods can be described by the social characteristics of their inhabitants. Differences can be indicated in terms of income, education or household size, for example.

Why do towns and villages shrink?

Towns or cities in which the number of population as a 120 whole is declining are usually those that have lost part of their economic base and are undergoing structural change. Shrinking villages generally tend to be located in remote regions with few jobs and poor infrastructure compared to the larger population centres. In towns and regions for example,, the consequences of population decline are empty houses and closed shops, declining tax revenues, and difficulties in maintaining the infrastructure.

PROJEKT VOR ORT

Eine Stadtexkursion vorbereiten

Recherchiert zum Beispiel ...
... beim statistischen Amt der Stadt zu zwei unterschiedlichen Stadtvierteln Angaben wie Haushaltsgröße, Einwohnerdichte, Migrationshintergrund, Altersstruktur usw. und vergleicht. Findet aussagekräftige Orte, an denen man die Unterschiede zeigen kann.

M3 Recherchieren 213, 221

Kartiert zum Beispiel ...
... die Nutzung des Erdgeschosses in einer Straße der City und anderen Zentren in eine Karte eintragen (siehe M6). Stellt dann die Straße, ihre Funktion und Nutzung vor.
Erforscht zum Beispiel ...
... die Veränderung eines Stadtteils in alten Karten (M2) und zeichnet die Veränderung in einer topographischen Karte ein. Stellt dann den Wandel des Stadtteils an aussagekräftigen Standorten vor.

M4 Kartieren 201, 222

Fotografiert zum Beispiel Motive zu Themen wie
- was eine lebenswerte Stadt ausmacht,
- unterschiedliche Merkmale von Stadtteilen,
- Wachsen und Schrumpfen von Städten,
- Gebäude, die für bestimmte Epochen der Stadtentwicklung stehen,
- städtischer und ländlicher Lebensraum.

M5 Fotografieren 205

Hier findet ihr Anregungen, die verschiedenen Themen des Kapitels in eurer Stadt mit den Methoden Fotografieren, Befragen, Recherchieren und Kartieren zu erforschen. Die Arbeitsergebnisse können zu einem kurzen Referat oder einer Präsentation mit geeignetem Anschauungsmaterial als Beitrag zu einer Exkursion vorbereitet werden. 211, 218, 222

Tipps zum Vorgehen:

Planung: Bildet Gruppen und entscheidet euch für ein Thema und eine Untersuchungsmethode. Legt fest, wo ihr eure Untersuchung vornehmen wollt.

Durchführung: Besorgt euch die nötigen Materialien und führt eure Arbeit durch. Beachtet die Vorschläge.

Auswertung: Bereitet ein kurzes Referat vor. Schlagt einen oder mehrere Standorte vor, an denen euer Beitrag zur Exkursion vorgetragen werden soll.

Befragt zum Beispiel
... Besucher einer Fußgängerzone, wie attraktiv sie diesen Einzelhandelsstandort finden, mit welchem Verkehrsmittel sie hingekommen sind oder nach ihrem Wohnort. Erstellt einen Fragebogen. Formt Ergebnisse in Tabellen und Diagramme um. Vergleicht mit eurer eigenen Einschätzung der Einkaufsstraße.
... Einwohner eines Stadtteils. Lasst euch berichten, wie es dort „früher einmal war" und was sich verändert hat, was sie hier hin gezogen hat, was ihnen gut gefällt und was sie verändern würden.

M7 Befragen 222

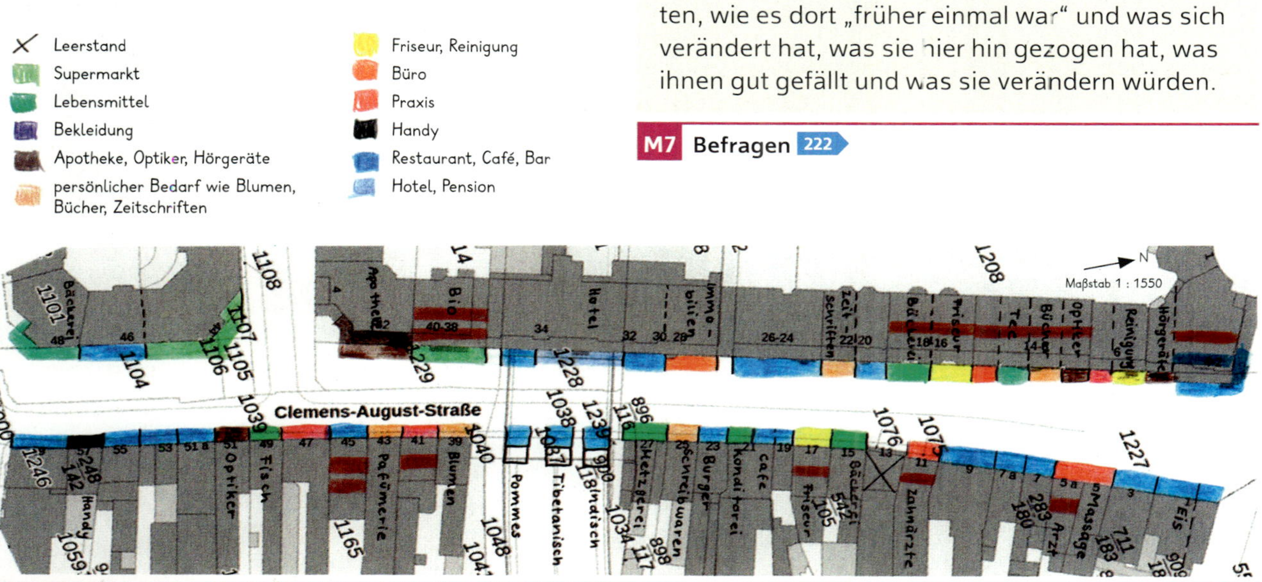

M6 **Nutzung der Gebäude in der Clemens-August-Straße, Bonn-Poppelsdorf (Grundlage: geoportal.nrw.de, Geoviewer: TopPlus-Web-Open) – Kartierung von Mara aus Bonn**

Towns and cities of the future – made for people?!

Suitable and affordable housing – a challenge

Housing is one of the basic functions of towns and cities, because it is a basic need for everyone. Housing means protection and security. People identify with it, they have their privacy here, and sometimes the house is also a status symbol. What are the causes of this housing shortage? How can it be countered in the future?

1. a) Describe the situation shown in the picture in M4 and use this example to explain the function of a house or apartment.
b) Everyone wishes to have suitable living space. Describe what "your" housing should be like.

2. a) Describe the household structure in Germany and how it has changed (M1, M2).
b) Give reasons for why this has an influence on the housing market.

3. Summarise the background to the housing shortage in a flow chart (M1 – M3, M5, M6).

4. Is it better to live in the city or in the suburbs? Discuss (M6).

Ⓐ **5.** Prepare a contribution for an excursion on the subject of this double page, referring to your town or city. Select suitable locations.

accelerated designation	beschleunigte Ausweisung/Kennzeichnung
building site	Baugebiet
bureaucracy	Bürokratie
conurbation	Ballungsraum
council housing unit	Sozialwohnung
market for housing	Wohnungsmarkt
municipal house building	kommunaler Wohnungsbau
suburb	Vorort

Helpful words and phrases for task 1:

On the one hand, ...
On the other hand, ...
In fact, ...
advantage/disadvantage
It depends on ...
Considering ...

Source: Destatis	Area	
	per dwelling	per person
	in m²	
Total households	93	46
Household size with ...		
1 person	68	68
2 persons	99	49
3 and more persons	122	33
Households ...		
with child(ren) under 18 years	116	32
without children under 18 years	87	54

42106EX_1

M1 Sizes of housing in Germany

42107EX_1

Households	total	in %				
Number of persons		1	2	3	4	5 and more
Essen (2019)	310010	49.8	28.5	11.0	7.2	3.5
Munich (2019)	832628	53.9	24.9	10.4	7.7	3.1
Germany (2017)	in 1000					
Conurbations	26735	43.9	32.5	11.6	8.8	3.2
Rural areas	14569	38.0	35.5	12.8	10.0	3.7
Germany (2019)	41506	42.3	33.2	11.9	9.1	3.5
1999	37795	35.7	33.2	14.9	11.8	4.4
1989	27793	35.3	30.1	16.8	12.6	5.3
1969	22287	25.8	27.3	19.6	15.0	12.3
1961 Source: Destatis	19460	20.6	26.5	22.6	16.0	14.3

M2 Household size and number of households

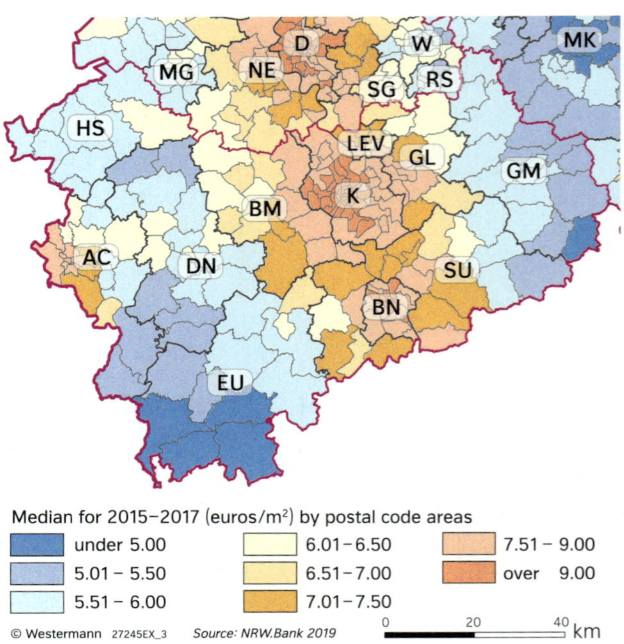

M3 Rents charged in parts of North Rhine-Westphalia (euro/m², average 2015 – 2017, post code areas)

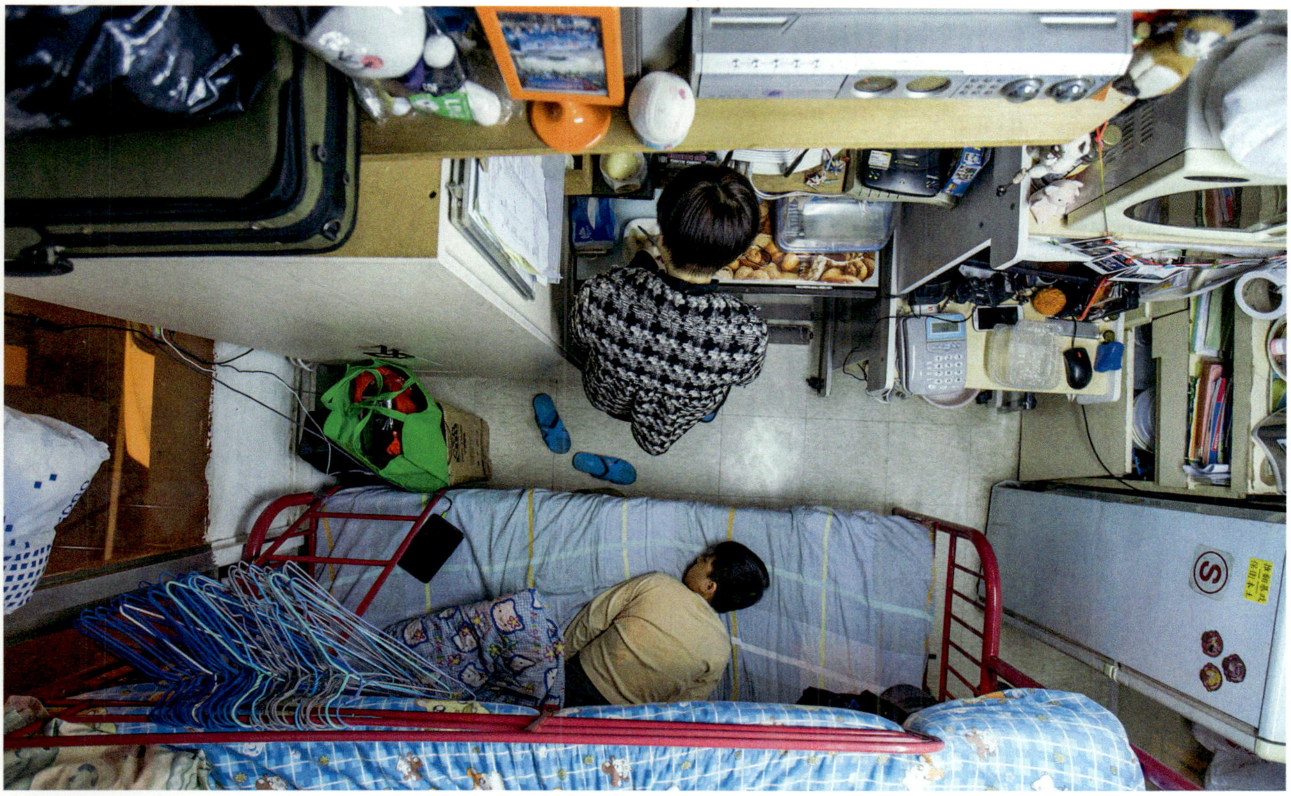

M4 **5.6 m² for 427 euros per month. In Hong Kong, the dramatic shortage of living space forces many citizens to get by in the tiniest of spaces. Here, an average of 71 percent of household income is spent on housing. In Frankfurt it is 35 percent.**

Does Germany have too little housing?
The high demand for real estate has led to an increase in prices. To make their investment pay off, the new owners of the property often demand higher rents. So there are fewer and fewer affordable homes. Another reason for the housing shortage is that municipal housing construction has been reduced in many cities. For example, the number of council housing units has fallen sharply. These are homes for which low rents are guaranteed. Thirty years ago, there were almost three million council housing units, today there are only about one million.

Source: Anon: Fragen zur Wohnungsfrage. In: Fluter, Bonn, 74/2020, p. 10, amended

M5 **Background: the market**

How can more affordable housing be created in cities?
To promote house building, some people call for less bureaucracy: among other things a simplification of building laws or the accelerated designation of building sites. In some cities, housing companies that are permitted to build in good locations are obliged to offer a proportion of the homes for low-income tenants. Another proposal for a city is to sell land more cheaply to investors who are not out to make maximum profits [...]. The call for more municipal house building is also becoming stronger.

Source: Anon: Fragen zur Wohnungsfrage. In: Fluter, Bonn, 74/2020, p. 10

M7 **Measures for affordable housing**

Especially the university cities, and cities in successful economic regions, continue to show population growth, particularly among the young population. At the same time – due to the price gap that still exists a lot of young families move from the cities to the surrounding areas. As a rule, most people looking for housing must adapt to the conditions and in the more expensive cities they must either lower their housing expectations (size, location, quality) or accept to pay higher rents.

Source: Kaltenbrunner, Robert et Matthias Waltersbacher: Besonderheiten und Perspektiven der Wohnsituation in Deutschland. In: bpb (Hg.): Gesucht! Gefunden? Alte und neue Wohnungsfragen. Bonn 2019, p.52/53, amended

M6 **Back to the cities**

Wenn du diese Aufgaben erfolgreich bearbeitet hast, kannst du ...
... die Funktion und Bedeutung von Wohnungen erklären.
... Faktoren, die auf den Wohnungsmarkt wirken und Ursachen der Wohnungsknappheit erläutern.

Mobility and a city made for people – a contradiction?

When one talks about traffic in cities, it is usually about cars: climate change, bad air and pollutant emissions in cities, noise, and too many cars that clog up cities and take away space from others. Yet cities have been planned and built for decades with motorised traffic in mind. In the whole of Europe, cities are arguing, experimenting and searching for solutions. What future-looking ideas are there for urban transport?

1. Explain briefly the idea of the "car-friendly city". Give examples of how this idea has been realised in your town or city, or in one that you know.

2. Compare the use of means of transport in urban and rural areas. Give reasons for the differences (M2, atlas). 217

3. Select three means of transport and list their respective strengths and weaknesses (M4).

C **4.** **A** Point out how the different measures in the examples should work. Describe how they would change the traffic and the cities (M5, M6).
B Characterise these measures in keywords and show their intended effect in a diagram.

A **5.** Prepare a contribution for an excursion on the subject of this double page, referring to your town or city. Select suitable locations.

basic parts of living	–	Grunddaseinsfunktionen
car-friendly city	–	autogerechte Stadt
clog up	–	verstopfen
pollutant emission	–	Schadstoffemission
postwar period	–	Nachkriegszeit
season ticket	–	Dauerkarte
superfluous	–	überflüssig
synchronised traffic lights	–	grüne Welle
traffic jam	–	Verkehrsstau

Helpful words and phrases for task 2:

The pie charts show ...
The percentage of ... is as large as/smaller than/larger than ...
The second/third largest ...
The biggest/smallest share of ...
In Bonn, the share of ...

M1 **Traffic in Münster, along with Karlsruhe the "bike capital" of Germany. The bicycle is an important means of transport, especially for young people.**

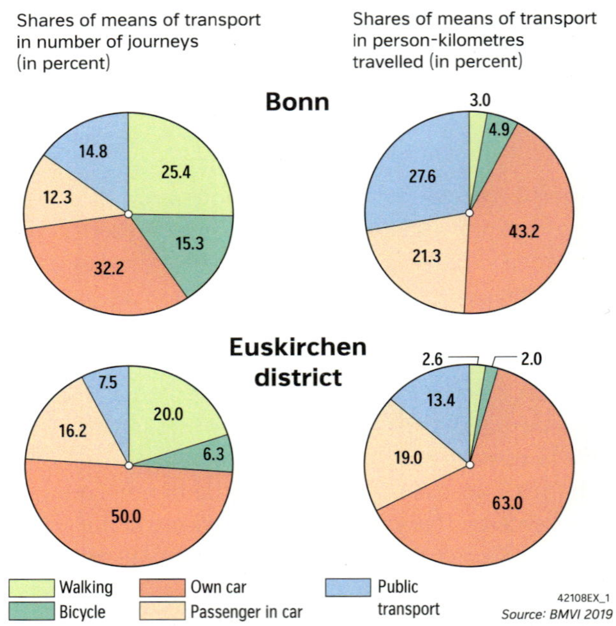

M2 **Distribution of means of transport in Bonn and in the Euskirchen district**

INFO

M3 Designing cities in Europe to be "car friendly" is an idea from the post-war period. In the **car-friendly city**, **basic parts of living** such as work, home and leisure were separated from one another. They were to be connected mainly through the car. Cities were designed to meet the needs of car drivers.

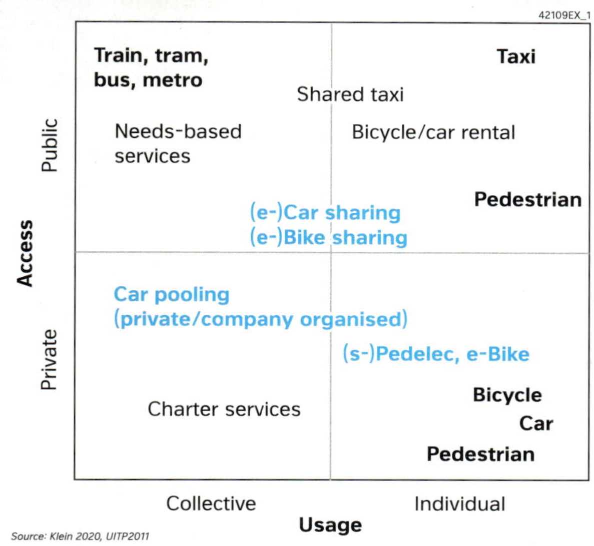

M4 **Means of transport – can be used in a variety of interlinked ways**

If there are not enough parking spaces, more should be created. Where there are traffic jams, cars need even more space! That sounds logical, but it is wrong. For a long time, the guiding principle was that the more traffic increased in cities, the more roads were built. And the more roads were built, the more car traffic appeared, and more traffic jams were created. The "always more and more" principle was misleading. This mechanism has now even been scientifically researched. It is called the Downs-Thomson Paradox. The two researchers [...] discovered that improvements in the road network often result in more traffic jams. As simple as it sounds, the solution to traffic jams is not to drive. Instead, take the bus or train, cycle or walk.

Source: Pinzler, Petra: Laufen und laufen lassen. In. DIE ZEIT, Hamburg, 39/2020, p. 2

M5 **What helps against traffic jams.**

Copenhagen

"A city [must] offer enough space for people to meet, and the traffic must have a speed that suits people. [...] The development of the Danish capital, Copenhagen, is strongly influenced by this philosophy. Copenhagen has become world-famous for its exemplary bicycle infrastructure. The cycle lanes are wide, the synchronised traffic lights are set for cycling speed, there are fast cycle lanes and cycle bridges. For people living there, the fastest way to make most daily journeys is by bike. The bicycle is now considered to be the main means of transport in Copenhagen. Its share in 2017 was 41 percent of work-related trips and 29 percent of all trips."

Vienna

"Parking became more expensive for residents, so space was created. The revenue went into the expansion of public transport. This and a cheap annual ticket for 365 euros, which is valid on buses and trains, have more than doubled the number of these season tickets since 2012. There are now more of them than cars. Some car parks have become small urban oases with seating and greenery. In 2019, Vienna made it to the top of two international city rankings for urban quality of life, also thanks to its transport policy."

Source: Heinrich-Böll-Stiftung & VCD Verkehrsclub Deutschland e. V. (editors): Mobilitätsatlas 2019. 2nd edition, 2020, p. 13

Helsinki

"One goal [...]: more space for more people, but less space for cars. Their numbers should decrease despite growth [...]. However, the local public transport network is being greatly expanded. The city hopes that this will make cars completely superfluous for commuters. [...] In any case, Helsinki sees itself as a test area for a new concept: Mobility-as-a-Service (MaaS). This describes how public and private providers can work together to get travellers to their destination in the best possible way. In Helsinki, for example, one can book trips via an app and choose whether to travel by bus, train, taxi, rental bike or rental car, or with a mix of transport modes. You pay for everything in one ticket within the same app."

Source: harl (SZ authors): Vier europäische Metropolen, vier Verkehrsmodelle. In: Süddeutsche Zeitung, München, 45/2018, 23.02.2018, p. 14

M6 **Models for new approaches**

Wenn du diese Aufgaben erfolgreich bearbeitet hast, kannst du ...

... Vor- und Nachteile unterschiedlicher Verkehrsmittel bewerten.
... zukunftsträchtige Ideen für den Stadtverkehr erklären.
*... die Fachbegriffe **car-friendly city** (**autogerechte Stadt**) und **basic parts of living** (**Grunddaseinsfunktion**) erklären.*

Smart Cities – a perspective?

When people talk about cities of the future, they often refer to the development of cities into so-called Smart Cities. Smart Cities are supposed to be the solution to some of the problems of modern cities. What are the characteristics of a "Smart City"? How will a development in that direction change urban life?

1. Assign the events in M4 to the illustration in M3. For each event, describe how it will change people's lives. Give a rating to the events, using a scale from -5 = very bad to +5 = very good.

2. Develop your own three events that could take place in a Smart City and evaluate them.

3. Name three problems or challenges of cities that are solved by the development into a Smart City.

Ⓒ **4.** **A** Explain how far the examples from Mönchengladbach can support high-street retailing (M1).
B Explain the intention and objective of the model projects in M2. How should they change life in the cities and the cities themselves?

Ⓐ **5.** Prepare a contribution for an excursion on the subject of this double page, referring to your town or city. Select suitable locations. 203

cisterne	–	Zisterne
charging station	–	Ladesäule
consumer	–	Verbraucher
dustbin	–	Mülltonne
electricity meter	–	Stromzähler
energy supplier	–	Energieversorger
high-street retailing	–	stationärer Einzelhandel
land-use plan	–	Flächennutzungsplan
power network	–	power network
public participation	–	Bürgerbeteiligung
roof garden	–	Dachgarten
rubbish truck	–	Müllwagen
ticket machine	–	Fahrkartenautomat
traffic management	–	Verkehrssteuerung
video monitoring	–	Videoüberwachung
waste management	–	Abfallwirtschaft
water leak	–	Wasserleck
weather conditions	–	Witterung

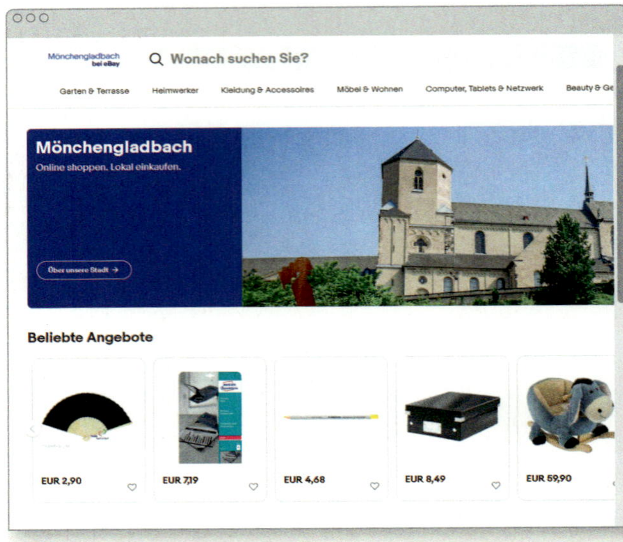

Portal ***Mönchengladbach bei eBay***: Here, local shopkeepers can also offer their goods on the online marketplace eBay. Ordered items can be collected from the seller's location or delivered to the buyer's home. The traders thus also become centrally visible on the internet as Mönchengladbach traders and can be advertised together.

mg.fashion box: A "Click & Collect" service aims to link together online trade and high-street retailing. Thus, clothes bought online can be tried on at a central location in the city centre, taken home or returned directly.

M1 Smart inner city, examples from Mönchengladbach

Freiburg: Development of a central data and exchange platform with data on climate, housing, society, commerce, infrastructure, nature and landscape, transport or tourism for urban planning, mobility; public participation.

Iserlohn: Establishment of a "town laboratory", digital forms of public participation and urban development; to facilitate cooperation between residents, business, science and administration.

M2 Smart City – Model project examples, 2020

M5 ***Smart City*** *–*
a city heavily networked with modern technology, which can therefore use resources efficiently and in which the standard of living should rise ... beep ... beep ... low battery ...

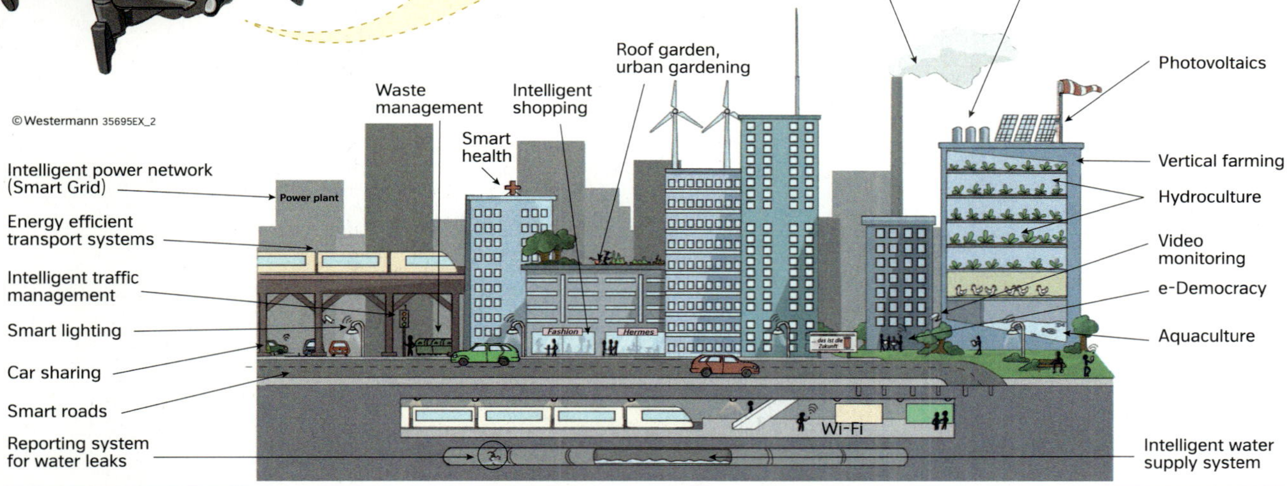

M3 **Intelligent, networked, sustainable – this is generally associated with "Smart City"**

1 On the way to his office, Klaus D. is directed away from his normal route. This is to avoid a traffic jam.

2 Sensors notify the public utilities that the plants need to be watered so that gardeners can be sent out.

3 Some people already find it difficult to operate an automatic ticket machine. Erik B. now has to cope with a multiplicity of computers.

4 Through the on-board computer in her car, Lisa K. receives the news that the carton of milk in the refrigerator is missing. She is on her way to the fitness studio.

5 The self-propelled electric rubbish truck is notified by the dustbin when it is 95% full.

6 The street lighting can be controlled depending on the weather. The municipality switches off the lighting in Julian's street when less than five pedestrians per hour are counted.

7 Electricity meters inform the energy suppliers about consumers' energy use at all times. The company can even recognise which TV programme Carla P. is currently watching.

8 Data from smartphone apps can be used to adapt public transport to demand. Thomas B. is asked to change trains because his connection is overcrowded today.

9 In the inner city, parking spaces are centrally allocated. When Inga W. comes home, she is directed to a parking space in her neighbourhood.

10 More and more supermarkets close, because fewer and fewer people visit the supermarkets.

11 A sensor identifies that nobody has been in the apartment for a long time. The heating is turned down.

12 The refrigerator always re-orders 1 kg of tomatoes when the last tomato has been used.

13 By analysing the images from the surveillance cameras, it is possible to count people, register conspicuous behaviour and monitor objects. Anika's route to school is thus precisely recorded.

14 The municipal authority invites people to view the proposals for changing the land-use plan and, if necessary, to comment and join in the discussion.

15 Andreas B. loads the washing machine in the morning. It will switch on when the electricity price is at its lowest.

16 Server breakdown: the food drones cannot come today because the network is out of action.

17 On cycle paths, green LEDs and time indications on illuminated signs guide the drivers. Emre C. does not have to stop at red lights so often.

M4 **Events in the Smart City**

Wenn du diese Aufgaben erfolgreich bearbeitet hast, kannst du ...

- *... Möglichkeiten, die die Entwicklung zu Smart Cities bietet, erläutern und bewerten.*
- *... erläutern, wie diese Entwicklung das Leben in Städten verändert.*
- *... den Fachbegriff* ***smart city (Smart City)*** *erklären.*

Sustainable city – the city of the future?

What should the city of the future be like? Today, there is one answer above all: sustainable! Of course, cities contribute decisively to climate change because the majority of people lives in cities. Therefore, combating the causes of climate change is an important contribution to sustainability. In addition, we must also respond to the consequences of climate change. What makes cities sustainable?

Ⓒ **1. A** Explain how modern transport concepts (see pp. 126/127) contribute to the sustainability of a city.
B Outline the sustainability of Smart Cities (see pp. 128/129).

2. Explain the core goals of sustainable urban development (M1).

3. a) Explain why cities often become "heat islands" in summer (M2 – M6).
b) Derive from the causes, what measures could be taken to avoid extreme temperatures (M6) in summer months.

Ⓒ **4. A** Explain the services that a tree performs for the climate. Give reasons for the role that parks, courtyards with greenery, gardens or avenues play in the city (M7).
B Explain the function of cold air corridors for cities in view of climate change (M6).

Ⓐ **5.** Prepare a contribution for an excursion on the subject of this double page, referring to your town or city. Select suitable locations. 203

city dweller	–	Stadtbewohner/-in
cold air corridor	–	Kaltluftschneise
mitigate	–	entschärfen
sealing	–	Versiegelung
sealed	–	versiegelt
urban heat island	–	Wärmeinsel in der Stadt
water expanse	–	Wasserfläche

Helpful words and phrases for task 3:

Due to ...
As a result of ...
A result is ...
As a consequence ...
Because of ...
Therefore ...
This leads to ...

- Urban building density:
  Dense building, keeping spaces free.
- Mix of uses:
  Housing and work as well as amenities and leisure in the urban neighbourhood.
- Social mix:
  Varied housing options.
- Polycentricity:
  Concentration on selected settlements in the surrounding areas of the cities.

M1 **Core goals of sustainable urban development**

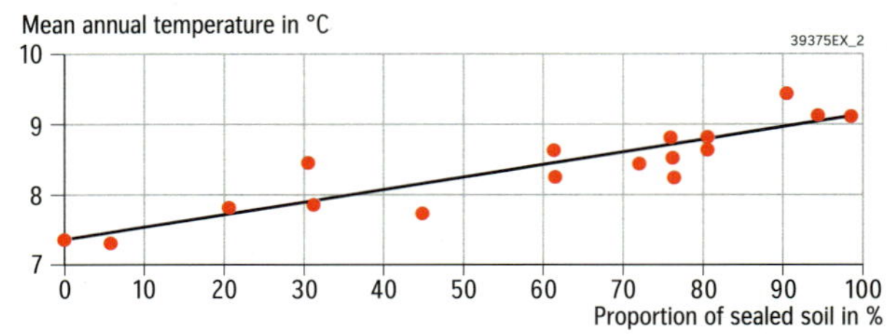

M2 **The relationship between sealing and air temperature**

INFO

M3 **Urban heat island**
Phenomenon of often higher air temperatures in cities compared to the rural hinterland

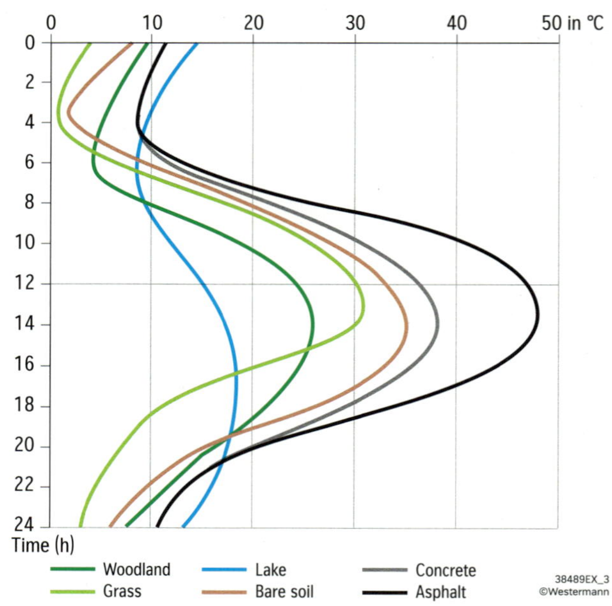

M4 **What contributes to warming of the city**

M5 Vertical forest (Bosco Verticale) in Milan, completed in 2014. The exterior facade of the two buildings, 80 and 112 m high, improves biodiversity in the district and the microclimate, as well as providing shade and reducing dust and heat for the residents. The idea has since been implemented frequently in other places.

The summer heat is causing more and more problems, especially for city dwellers. And the forecasts predict more and more extreme temperatures in the coming decades. Is it still possible to change this trend?

Simone Sandholz: We cannot reverse the climate change that has taken place so far, we can only mitigate its consequences and prevent it from getting worse in the future. What we can do is take action to ensure that things remain reasonably tolerable.

Can the consequences of the global climate crisis be limited at all by local measures?

Sandholz: In recent years, we have been involved in a project in which small-scale climate scenarios were calculated for the city of Bonn. This clearly showed that there are cold air corridors through which cooler ambient air reaches the city centre at night. Keeping these free is absolutely necessary if the city is not to become a heat island. At the same time, it was also seen that greened streets are significantly cooler than sealed open areas. Even small expanses of water spread across the city can have an amazing effect.

Source: Wein, Martin: Stadtplanerin plädiert für klimagerechten Umbau Bonns. In: General Anzeiger Bonn, 17.08.2020, p. 17, amended

M6 From an interview. Simone Sandholz is a senior researcher at the Institute for Environment and Human Security at the United Nations University in Bonn.

Ein Baum!

Diese etwa 100 Jahre alte Buche sollten Sie sich etwa 20 m hoch und mit etwa 12 m Kronendurchmesser vorstellen. Mit mehr als 600.000 Blättern verzehnfacht sie ihre 120 qm Grundfläche auf etwa 1.200 qm Blattfläche. Durch die Lufträume des Blattgewebes entsteht eine Gesamtoberfläche für den Gasaustausch von ca. 15.000 qm, das entspricht etwa zwei Fußballfeldern! 9.400 l = 18 kg Kohlendioxid verarbeitet dieser Baum an einem Sonnentag. Bei einem Gehalt von 0,03 % Kohlendioxid in der Luft müssen etwa 36.000 cbm Luft durch diese Blätter strömen. In der Luft schwebende Bakterien, Pilzsporen, Staub und andere schädliche Stoffe werden dabei größtenteils ausgefiltert. Gleichzeitig wird die Luft angefeuchtet, denn etwa 400 l Wasser verbraucht und verdunstet der Baum an demselben Tag. Die 13 kg Sauerstoff, die dabei vom Baum durch die Fotosynthese als Abfallprodukt gebildet werden, decken den Bedarf von etwa 10 Menschen. Außerdem produziert der Baum an diesem Tag 12 kg Zucker, aus dem er alle seine organischen Stoffe aufbaut. Einen Teil speichert er als Stärke, aus einem anderen baut er sein neues Holz. Wenn nun der Baum gefällt wird, weil eine neue Straße gebaut wird, oder weil jemand sich beschwert hat, dass der Baum zu viel Schatten wirft oder gerade dort ein Geräteschuppen aufgestellt werden soll, so müsste man etwa 2.000 junge Bäume mit einem Kronenvolumen von jeweils 1 cbm pflanzen, wollte man ihn vollwertig ersetzen. Die Kosten dafür dürften etwa 150.000,- € betragen.

www.die-gruene-stadt.de

M7 The services that a tree performs

Wenn du diese Aufgaben erfolgreich bearbeitet hast, kannst du ...

- ... *Prinzipien einer nachhaltigen Stadtentwicklung erläutern.*
- ... *Möglichkeiten, die zukunftsträchtige Stadtentwicklungsideen bieten, hinsichtlich der Nachhaltigkeit bewerten.*
- ... *Maßnahmen erläutern, mit denen in Städten Folgen des Klimawandels entgegengetreten werden kann.*
- ... *den Fachbegriff* ***urban heat island*** *(**Wärmeinsel**) erklären.*

WHAT WE HAVE LEARNED

Towns and cities of the future – made for people?!

M1 **Cable cars are important means of transport in La Paz, Bolivia. They can be used to connect poorer neighbourhoods on the mountain slopes with public transport.**

M2 **Sustainable urban planning: The Vauban district in Freiburg has long been regarded as exemplary, even internationally.**

Housing – a challenge

124 "Housing", as a basic need for all people, is one of the important functions of cities. Therefore, it is problematic when adequate and affordable housing is not available for many. Currently, the housing market is under great pressure. This is because investors are putting money into real estate and municipal housing construction has declined. Also, the demand for housing from increasingly smaller households has changed. In addition, population growth is concentrated in certain cities and regions.

Mobility and a city made for people – a contradiction?

126 After the Second World War, cities were rebuilt and designed primarily according to the principle of the **car-friendly city**. However, the future of mobility in cities has long been seen in the intelligent linking of different means of transport – from private, individually-used vehicles such as the bicycle, to shared public transport, such as trains or buses. Especially the bicycle is often seen as playing an important role. After all, its use makes no noise, it is cheap and takes up little space.

Smart Cities – a perspective?

128 The development of cities into so-called **Smart Cities** is associated with the solution of various problems, because the strong networking of a wide range of functions allows resources to be used more efficiently. Attempts are also being made to promote inner-city retail shops and to enable a more ecologically tolerable urban life overall, for example through new forms of mobility.

Sustainable city – the city of the future?

The city of the future should be sustainable. Therefore, it should be dense and compact, to consume little space. In addition, it should be mixed in terms of use, to enable short distances, and mixed in terms of population, to maintain neighbourhoods that are good to live in. Building in the surrounding areas should be limited to spatial focal points. Furthermore, measures should be taken to prevent the **"urban heat island"** from becoming unbearably hot in summer. 130

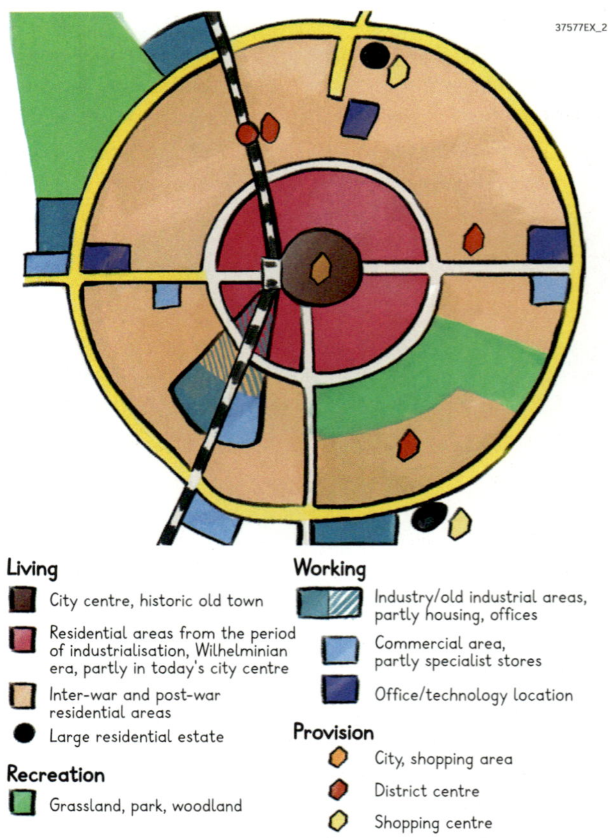

M3 **Model of the Central European city** 210

ORIENTATION

European agglomerations

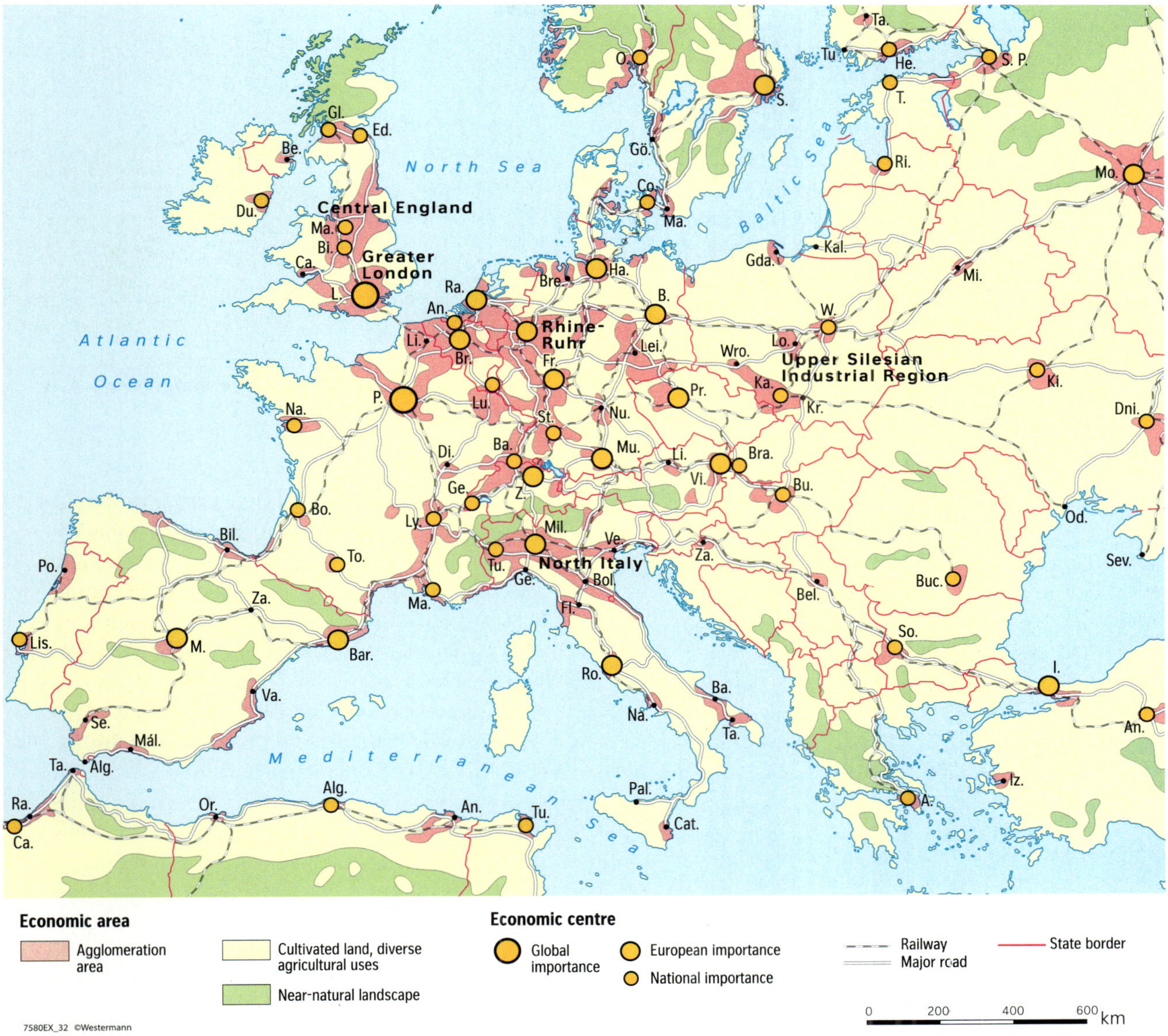

M4 European agglomerations

1 Work with the atlas (map: Europa, physische Übersicht). List the European agglomerations that are larger than 5 million and larger than 3 million inhabitants.

2 Compare it with the illustration showing agglomerations at night (map: Ballungsräume bei Nacht).

3 Work with the model (M3) 210
a) Explain how historical structures from the Middle Ages and the period of industrialisation are considered in the model.
b) Give reasons for three locations of the service sector in the model.
c) Explain the arrangement of the urban functions in the model.
d) Explain whether the model also reflects your town or city.

4 Evaluate photos M1, M2 and page 20 M3 and explain whether they are examples of a sustainable city.

5 Work with the atlas (map: Freiburg – nachhaltige Stadtentwicklung).
a) Describe the construction, infrastructure, traffic and greening concept.
b) Evaluate whether "core goals of sustainable urban development" (p. 130 M1) have been achieved here.

Apply and practise

PROJEKT VOR ORT

Debatte – welche Stadt wollen wir?

Wie die Stadt in Zukunft sein wird, kann man heute kaum sicher sagen. Mit Blickrichtung Zukunft ist es aber wichtig, zu überlegen, wie die Stadt sein soll, die wir wollen. Schließlich handelt es sich um unser direktes Lebensumfeld, das wir mitgestalten können. Die Frage, ob Städte zu Smart Cities entwickelt werden sollen, ist beispielsweise sehr umstritten. Welche Ansprüche hast du an die Stadt? Welche Ansprüche haben Angehörige anderer Gruppen? Wie soll sie sein, die Stadt der Zukunft?

1. Wie schätzt du deinen Schulort ein? Begründe, ob du ihn für lebenswert hälst. Knüpfe an deine Bearbeitung von S. 114 Aufgabe 2 an.

2. a) Liste, ausgehend von den Grunddaseinsfunktionen (M2) auf, welche Ansprüche du an die Stadt hast.
b) Welche Ansprüche an die Stadt haben ältere Menschen, berufstätige Singles, Abiturientinnen und Abiturienten und Familien mit kleinen Kindern? Begründe.
c) Plane ein Stadtviertel, das deinen Vorstellungen von der lebenswerten Stadt der Zukunft entspricht und das zugleich Ansprüchen unterschiedlicher Gruppen gerecht wird.

3. Vergleiche deine Vorstellungen von einer lebenswerten zukünftigen Stadt mit der Wortwolke (M1).

4. Führt eine Debatte. 220
a) Analysiere M3 und M5: Notiere in Stichworten Pro- und Kontra Argumente zur Entwicklung von Städten zu Smart Cities.
b) Bereite dich auf eine Debatte zum Thema „Smart Cities – wünschenswert oder Schreckensszenario?“ vor.

M1 Vorstellung von der Stadt der Zukunft

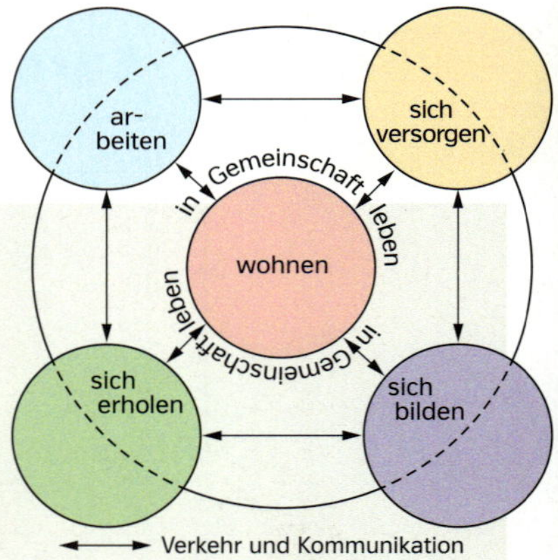

Grunddaseinsfunktionen sind Tätigkeiten und Leistungen, die notwendig sind, damit Menschen leben können. Jede dieser Funktionen hat auch Auswirkungen auf den Raum und zeigt sich in der Stadtstruktur.

M2 Aus einem Lexikon: Was sind Grunddaseinsfunktionen? 210

Aktuelle Prognosen zu den Lebensbedingungen in Städten gehen noch nicht davon aus, doch es ist möglich: ein gesünderes, angenehmeres, entspannteres Leben in Metropolen weltweit. Die Voraussetzung dafür ist, dass diese smart, also regelrecht schlau werden. [...] Die Vorstellung, dass sich diese Verstädterung fortsetzt, ist erschreckend und alarmierend. Aber es gibt die berechtigte Hoffnung, dass sich diese Trends umkehren lassen. Saubere Luft statt Smog; entspannte Mobilität statt verstopfter Straßen; genug sauberes Trinkwasser statt Wasserquellen, in denen sich Krankheitserreger finden; ein nach Bedarf verfügbarer und bezahlbarer Strom aus regenerativen Quellen [...]. Der Weg dorthin führt [...] nach vorn, in ein Zeitalter der Digitalisierung, in dem Städte wie große Computer funktionieren. [Die Städte] sind also im Begriff, im modernsten Sinne schlau zu sein: auf Basis von Daten – und mithilfe einer Vielzahl ausgeklügelter Technologien. Diese übermitteln beispielsweise die Information, wo Parkplätze frei sind und reduzieren damit Verkehrsaufkommen und Emissionen bei der Suche danach [...]. Präzise Vorhersagen, basierend auf sorgfältig erfassten Daten, sind der Kern fast aller Bereiche einer schlauen Stadt. So werden beispielsweise intelligente Stromnetze Stromschwankungen als Folge wechselnder Wetterlagen ausbalancieren können. Schon jetzt sind Aussagen dazu möglich, wie sich eine Flotte von E-Autos in das Gebäudemanagement integrieren lässt und dort als Energiespeicher fungieren kann. [...] Mit solchen Ansätzen öffnen sich völlig neue Märkte für Technologien und Dienstleistungen.

Quelle: Zistl, Sandra: Smart Cities. Schöne neue Stadtwelt. www.siemens.com, 11.07.2017, Zugriff: 26.04.2018.

M3 Smart Cities, Schöne neue Stadtwelt – der Technologiekonzern Siemens auf seiner Homepage

M4 Schriftzug des Künstlers Merlin Bauer in der Kölner Innenstadt. Mit seinem Projekt wollte er auf die Geschichte der Nachkriegsarchitektur in der Stadt aufmerksam machen und Menschen dazu motivieren, sich in der eigenen Stadt für eine lebenswerte Stadt zu engagieren.

Mit Blick auf die Zukunft der Städte machen mir drei Dinge Sorgen: Datenschutz, Ausgrenzung und Sicherheit. So wie „smarte“ Städte aktuell gestaltet werden, führt der Einsatz von intelligenten Datensammlern dazu, dass wir nicht nur im öffentlichen Raum, sondern auch in unseren eigenen vier Wänden permanent überwacht werden. Das ist vor allem besorgniserregend, da in den Städten demokratische Teilhabe stattfindet, politische Gruppen sich treffen, Proteste organisiert werden. Wir müssen uns fragen, für welche Bevölkerungsgruppen die „smarten Städte“ intelligent sein sollen. Funktionieren sie auch für die Schwächsten? Was passiert mit den Menschen, die keinen Zugang zur Technologie haben? [...] Wenn die Regierungen dabei versagen, sensible Daten zu schützen, taumeln wir in eine Smart-City-Ära, ohne auf die möglichen Konsequenzen vorbereitet zu sein: Was passiert, wenn die digitale städtische Infrastruktur gehackt wird? Dann könnten wir nachts zum Beispiel im Dunkeln sitzen [...]. Smart Cities ermöglichen dem Staat und Großkonzernen gleichermaßen, Daten zu sammeln und zu verkaufen. [...] Anstatt alle Städte „smart“ zu machen, sollten die Bürgerrechte das wichtigste Kriterium sein, wenn wir über Urbanisierung nachdenken. Wir müssen auch die benachteiligten Stadtbewohner miteinbeziehen und diejenigen, die keinen Zugang zur Technik haben. Systematische Datensammlung sollte nicht die Antwort auf alle Herausforderungen der Zukunft sein.

Quelle: „Gläserne Menschen in Filterblasen“ (Auszug, gekürzt, FAZ.NET vom 03.03.2018 von Judith Lembke) © Alle Rechte vorbehalten. Frankfurter Allgemeine Zeitung GmbH, Frankfurt. Zur Verfügung gestellt vom Frankfurter Allgemeine Archiv

M5 **Statement von Eva Blum-Dumontet (NRO „Privacy International“)**

Wer den Traum vom Leben in einer leisen, sicheren und sauberen Stadt verwirklichen will, muss also zuerst und vor allem den Verkehr stoppen. Konkret: Verkleinert die Straßen! Pflanzt Bäume auf Parkbuchten! Und hört auf, alle Dieselfahrzeuge durch E-Autos ersetzen zu wollen. Die sind zwar leiser, aber wenn jeder eines fährt, brauchen sie genauso viel Platz. Lasst saubere Busse fahren, am besten elektrisch betrieben, gespeist mit Solarstrom. Investiert in die Digitalisierung, damit Taxen mehrere Passagiere sammeln und von Ort zu Ort bringen. Bevorzugt Radfahrer und Fußgänger. Wagt den Krach mit den Autofahrern – und wenn die etwas von Freiheit erzählen, schenkt ihnen Freifahrkarten für Busse und Bahnen. Zeigt ihnen die Kinder, die dort spielen, wo früher Parkplätze waren. Die Menschen, die auf breiten Boulevards vor Cafés sitzen. Lasst sie die Blumen riechen und die Vögel hören, die dann auch Platz in der Stadt haben.

© Petra Pinzler für DIE ZEIT, Hamburg, 1/2018 (27.12.2017): „Wir brauchen Utopien für die Stadt“.

M6 **Stadt bürgerfreundlicher machen**

- Ein aktuelles Stadtentwicklungsprojekt am Schulort – Beitrag zur nachhaltigen Stadt?
- Schrumpfende Dörfer – unterstützen oder aufgeben?

M7 **Themenvorschläge für weitere Debatten**

Wenn du diese Aufgaben erfolgreich bearbeitet hast, kannst du ...

... mithilfe der Grunddaseinsfunktionen verschiedene Ansprüche an den Stadtraum erläutern und eigene Vorstellungen begründen.
... unter Beachtung des richtigen Vorgehens Rollen in einer Debatte einnehmen.

IN FOCUS

Amsterdam – a European metropolis

Gabled houses, the canals (Grachten) that shape the city centre like a spider's web, tolerance that characterises the city – these are the ideas coming into people's minds when they think of the city. Yet Amsterdam has become a pioneer of progressive urban transport, and it is in the headlines because people in Amsterdam are resisting mass tourism.

1. Work out
C **A** A short excursion guide.
B An excursion programme for a class trip.
C A virtual excursion to Amsterdam.
Use the two double pages (p. 136 – 139).

Tips for preparation:

- Compare Amsterdam in 1450 (M1) with the model of a medieval town. Point out special features of the town.
- Describe the Canal District (Grachtengordel) and its function (M2, internet).
- Illustrate the growth of the city of Amsterdam (M3, M5 – M7).
- Almere is characterised as an "overflow city". Explain (M7).
- Describe the IJburg project. Consider the type of use (M8).
- Explain how far the measures taken for mobility in Amsterdam can be considered as exemplary (M9 – M12).
- Based on M18, explain why so many people visit Amsterdam and the effects of tourism in the city (M13 – M19).

M2 Old Amsterdam, Amsterdam's Canal District (Grachtengo

M3 Today's Amsterdam Metropolitan Region in 1950 ...

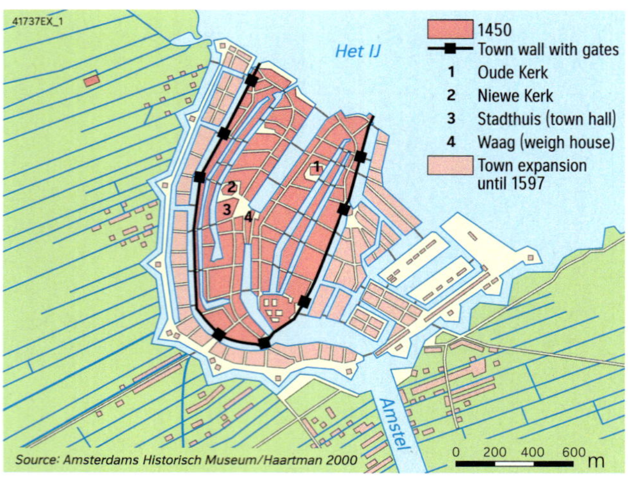

M1 Origin of the town, Amsterdam in 1450 and 1597

Around 1000: Fishing village built on dams in the marsh that protect against flooding

Middle Ages: Growth on both sides of the Amstel, favourable location for shipping, trade to Hamburg and the Baltic region, shipbuilding

17th century: Founding of the East India Company: trade with Asia, especially in spices, silk, slaves; in 1680: 205,000 inhabitants, one in two is an immigrant or refugee, mainly from Germany; expansion around the Canal District (Grachtengordel), "Golden Century"

18th century: Slow decline

19th century: Boom begins with the construction of the North Sea Canal, railway connections, industrialisation: diamonds, textile, engineering and shipbuilding industries; World Fair 1883; city expansions

M4 From fishing village to world city

gabled houses

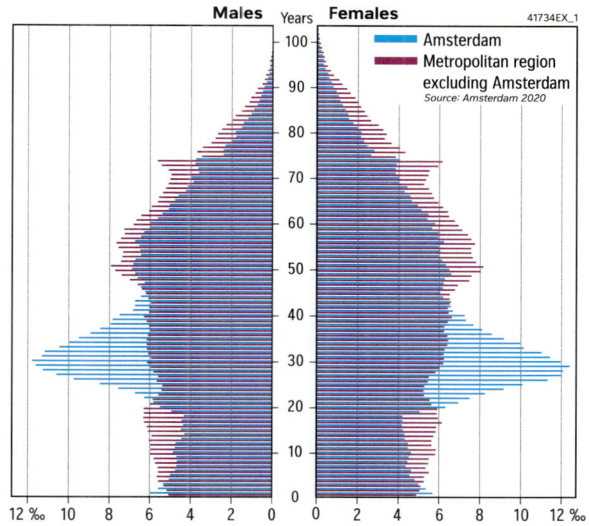

M6 Age structure of Amsterdams and the Metropolitain Region

... and 2015

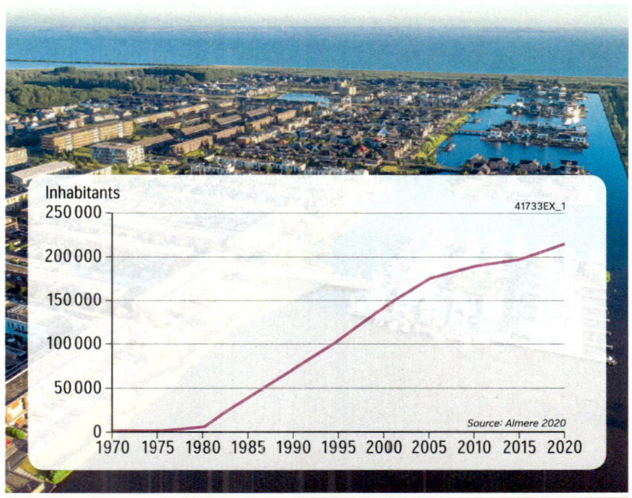

M7 Almere – residents moved into the first houses on the newly created polder in 1976; today it is the seventh largest city in the Netherlands

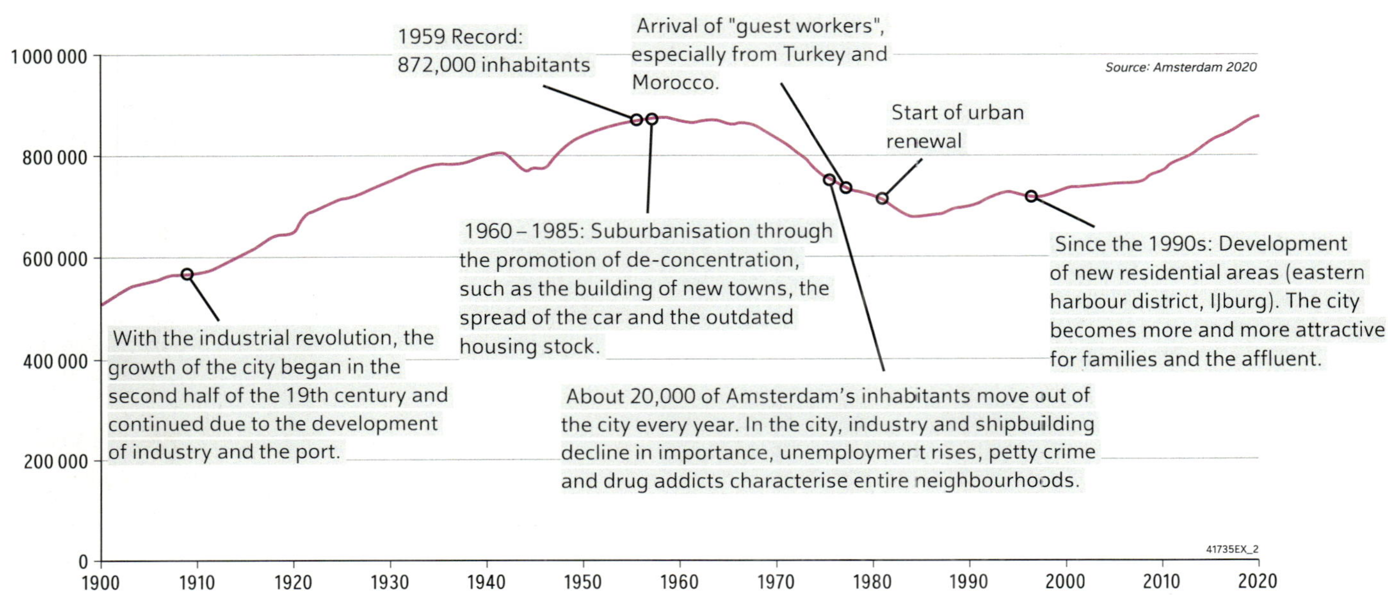

M5 Population development of Amsterdam (1900 – 2020)

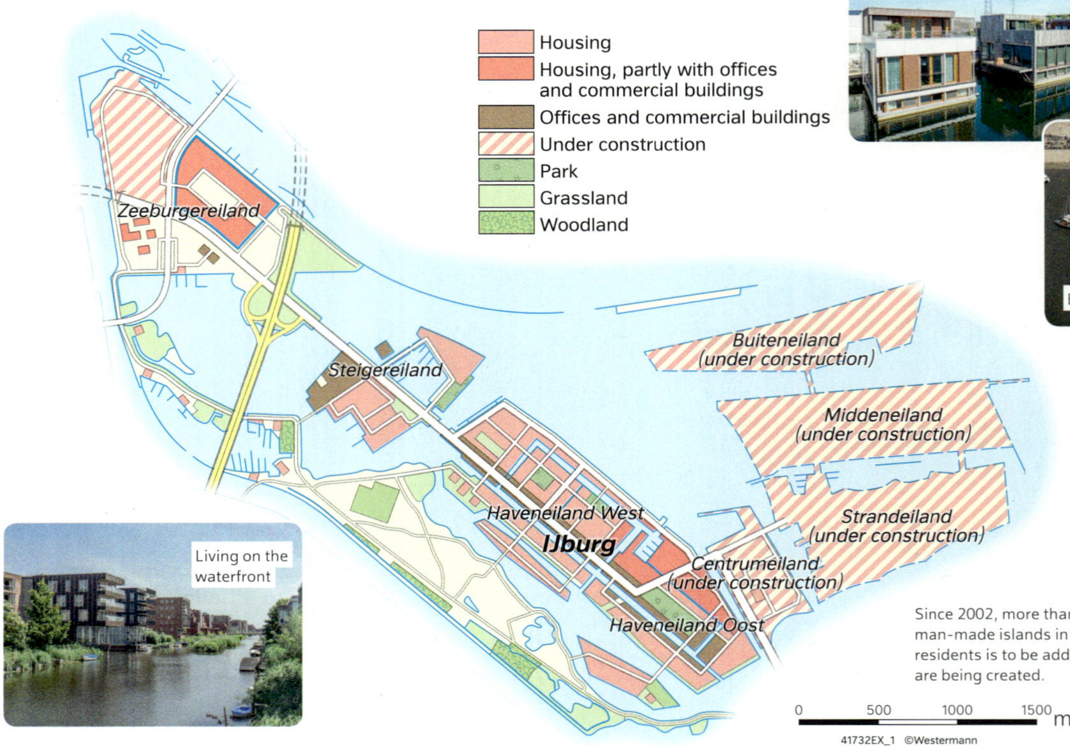

Since 2002, more than 25,000 residents have moved to the man-made islands in the IJ. Housing for about 20,000 more residents is to be added. In the south-east, more islands are being created.

M8 The city is to become a compact city again. It is growing along the waterfront, but also in IJburg in old port areas to the east.

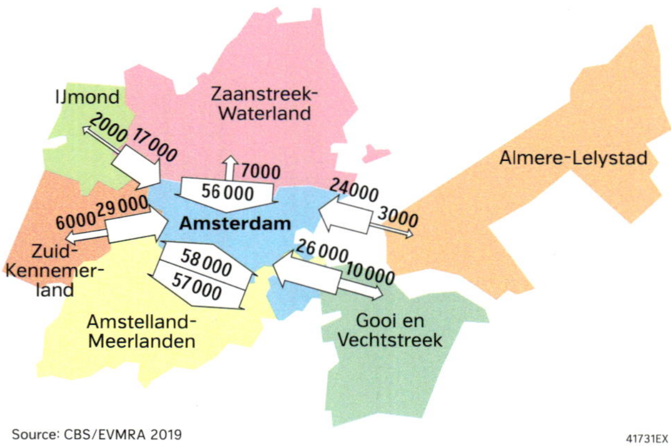

M9 Commuters into and out of the Amsterdam Metropolitan Region (2018)

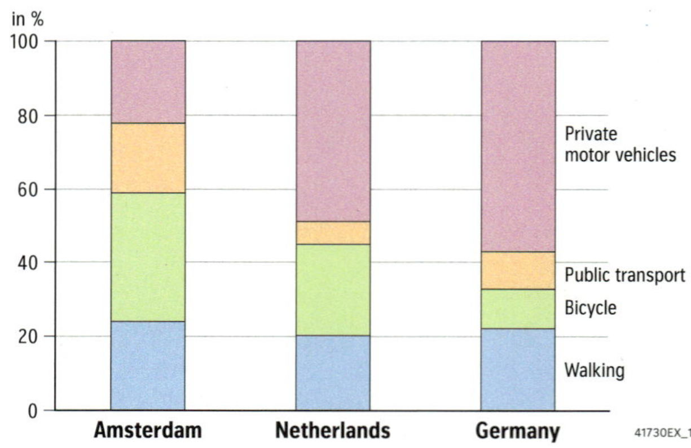

M11 Shares of means of transport in Amsterdam in comparison

M10 Bicycle parking at Centraal Station. Around 60 percent of people in Amsterdam cycle every day. Wide, modern cycle paths and lanes take one wherever one wishes to go.

- A cable car will connect northern districts with the city centre via the harbour.
- Safe and clearly marked pedestrian and cycle paths
- More and wider cycle lanes, new cycle bridges
- Expansion of bicycle parking spaces and garages
- Intelligent "smart" traffic guidance systems (Smart Mobility)
- Regular reductions of car parking spaces: widening of pavements, cycle paths and bicycle parking spaces

M12 Innovative solutions for urban transport

M13 New tourist attraction: The Film Museum (left) in a new urban district with A'DAM-Tower (centre; lit with a heart in 2020 for the Corona pandemic)

M17 Selfie hotspot in front of the Rijksmuseum. The "must" for every Amsterdam tourist has since been dismantled, a symbolic act against overtourism.

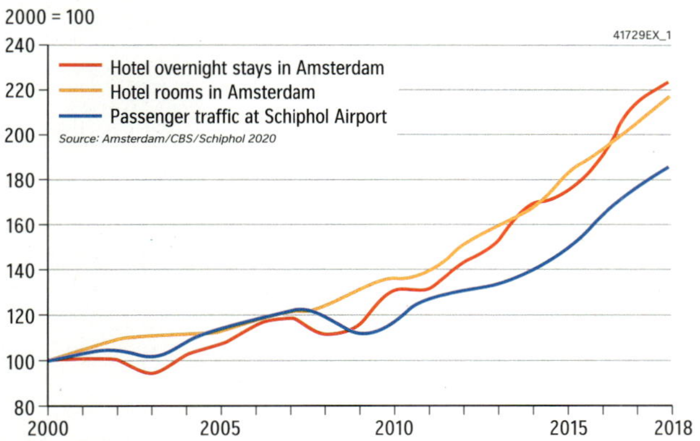

M14 Development of tourism in Amsterdam. In 2018, almost 20 million visitors were counted in Amsterdam.

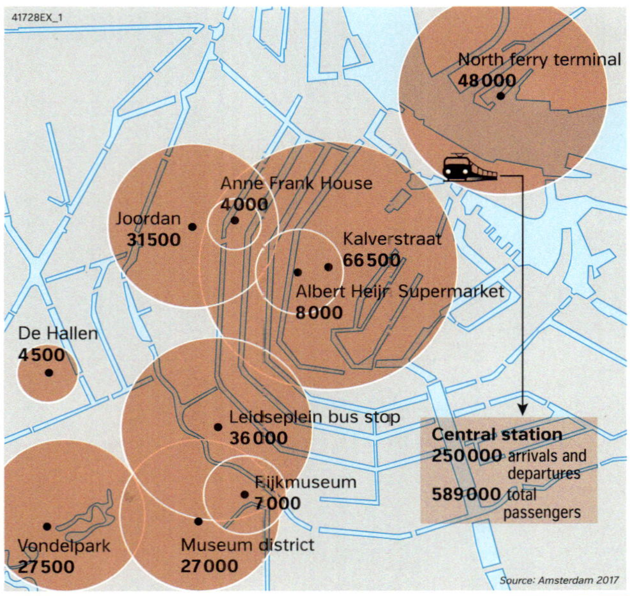

M18 Visitors on an average day in August (2016)

INFO

M15 Overtourism – meaning that a place becomes overburdened by mass tourism. Local people feel massively disturbed and their normal daily lives endangered.

To resist the increasing number of visitors, Amsterdam has reacted with countermeasures. Since July 2020, the rental of private holiday flats in the old city centre has been prohibited. In the other parts of the city, private individuals are only allowed to rent out accommodation for a period of up to 30 days. There are heavy fines for non-compliance. Improper behaviour and alcohol consumption on the street are also subject to fines. [...] So-called "B-marketing", the marketing of supposedly second-ranking tourist destinations, seems to work successfully. The seaside resort of Zandvoort, about an hour's drive from Amsterdam, is advertised as "Amsterdam Beach".

Source: Astor, Ellen: Touristenhochburg Amsterdam. In: Geographische Rundschau 10/2020, p. 50

M16 Countermeasures

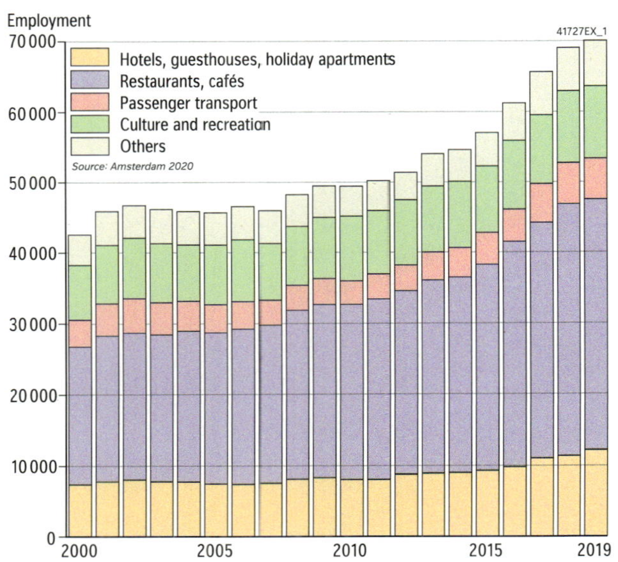

M19 Development of jobs in the tourism industry

4. Regions in transition due to globalisation and digitalisation

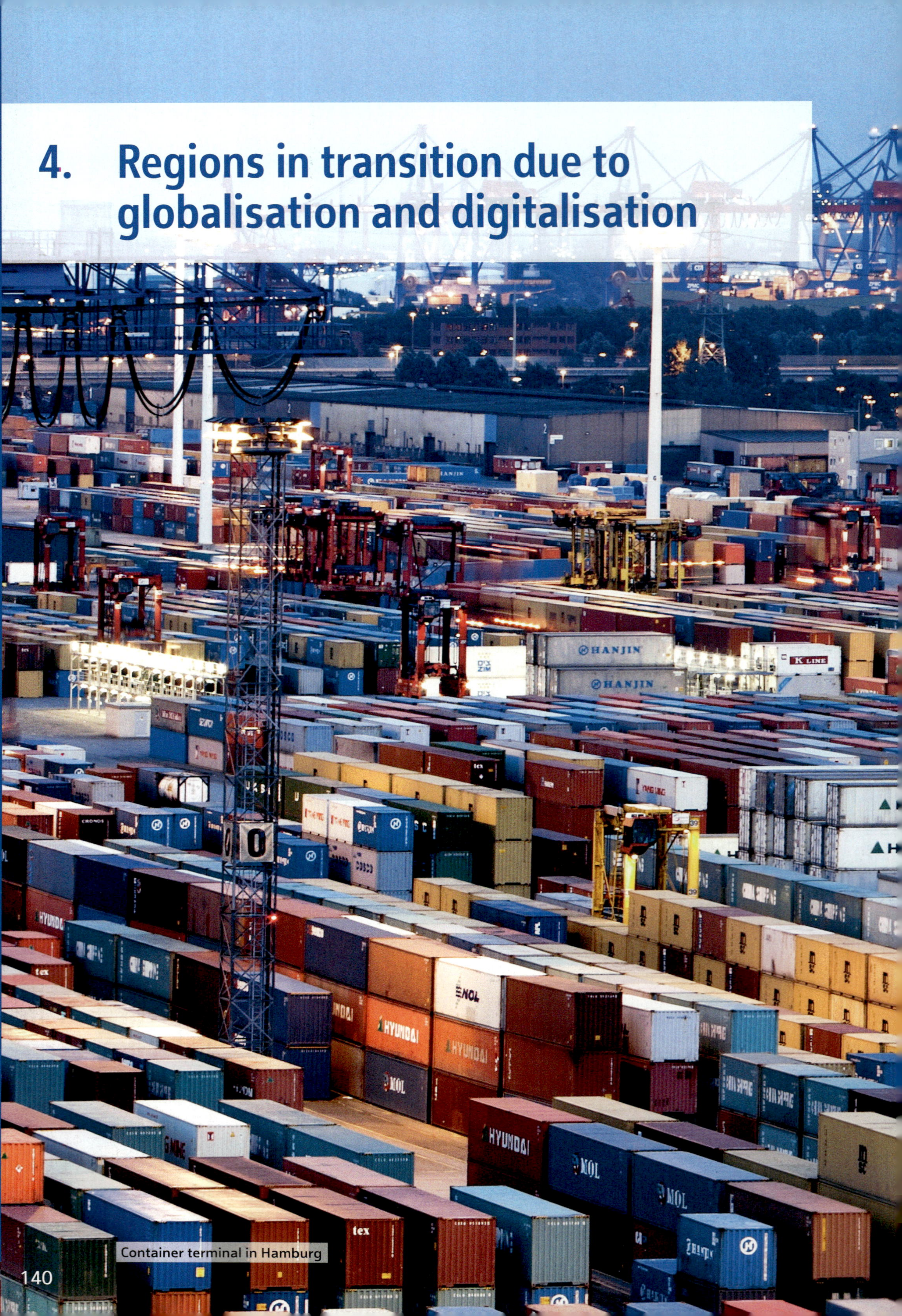

Container terminal in Hamburg

HYUNDAI
HYUNDAI
UASC
COSCO
K LINE
YANG MING
YANG MING

In the age of globalisation – is it all only virtual?

What does globalisation mean?

It is said that "we live in a globalised world". It makes our world ever smaller.

But what exactly is meant by "globalised" or "globalisation"? How does it make our world smaller?

1. Explain which dimensions of globalisation the photos exemplify (M2, M5). 205
2. a) Characterise the YouTube video as a globalised product (M5).
   b) Refer to selected terms in the mind map (M4).
   c) Describe the effects that the YouTube video has caused.
3. (C) **A** The world is getting smaller. Explain wether this sentence is true (M1 – M3).
   B Summarise M3 in a few sentences.
4. (A) Add more examples for the branches in the mind map of globalisation (M4).

accelerate	–	beschleunigen
interconnection	–	gegenseitige Verbindung
production chain	–	Produktionskette
production site	–	Produktionsstandort
rapid	–	schnell
vulnerable	–	anfällig

Helpful words and phrases for task 1:

... the internationalisation of production and services ...
... the globalisation of production ...
Advances in information and communication technologiy ...
... is an example of ...
... is marked by ...
... can be characterised by ...
trade barriers
international networks of production and services
global player

M2 Examples of globalisation

International relations have existed for a long time. At the latest since Marco Polo travelled the Silk Road to China around 1300 or Columbus reached America in 1492. But we only started talking about **globalisation** around 1990, when major technological innovations (e.g. the internet) and political changes (the end of the "Iron Curtain" between East and West) led to enormous changes in the economy and society. The world suddenly became more interconnected. Travelling became easier, cheap textiles and electronics came from all parts of the world. Trade grew rapidly, and global prosperity also increased. Today, globalisation leaves its traces all over the world. However, its economic, social and ecological consequences are being criticised. The Corona (Covid-19) pandemic also highlighted how vulnerable the global interdependencies are.

M3 Development of globalisation

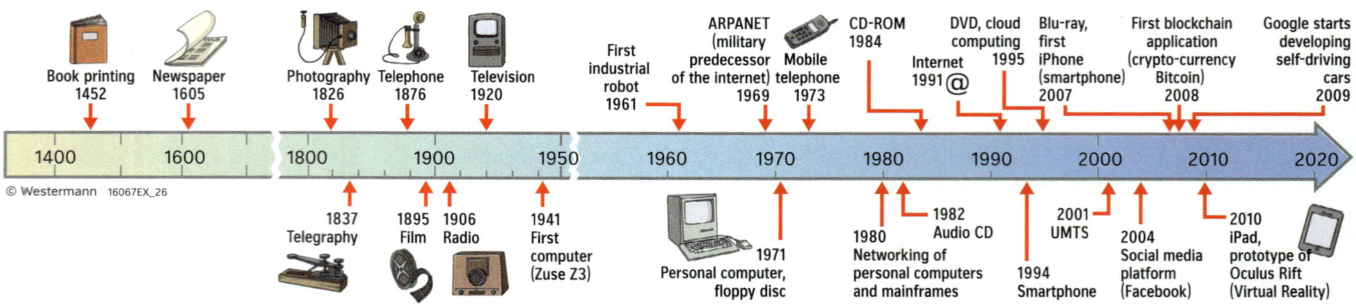

M1 Inventions of important media – the path to digital globalisation

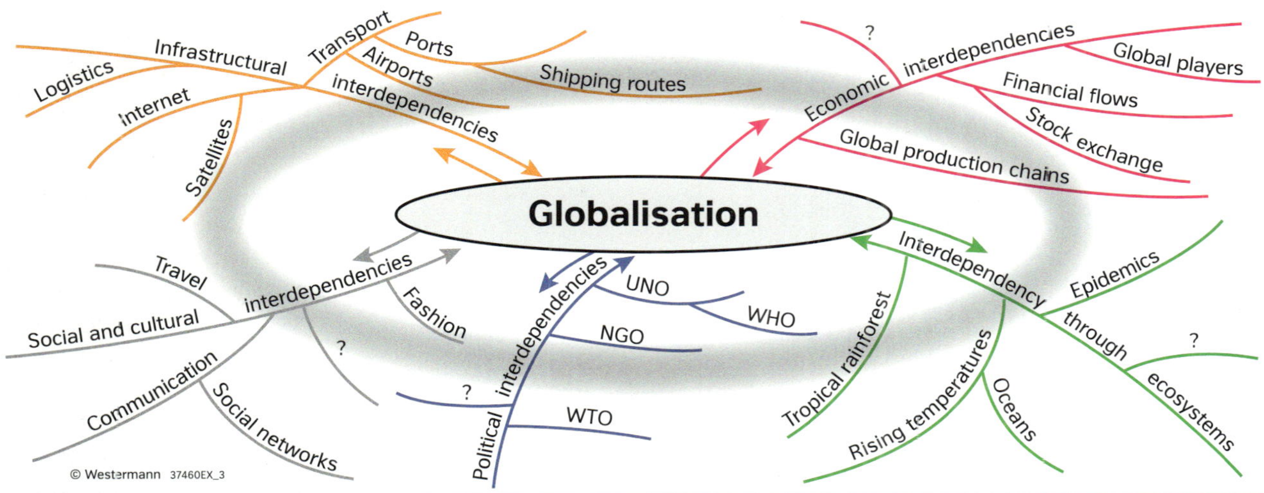

M4 Worldwide areas of interdependence that shape and reinforce globalisation (diagram)

The video "Despacito" has been viewed seven billion times worldwide since 2017 (10/2020). The song has been in the charts for weeks in over 20 countries. The music label is based in California and is a subsidiary of the French media group Vivendi. The singer lives in Florida and sings in Spanish. The video was recorded in Puerto Rico. Since then, interest in the country on travel websites has increased significantly. Most people watch the video with a smartphone that was "Made in China". Huge amounts of data are sent around the world in the process.

M5 Digital globalisation – interdependencies and effects

Wenn du diese Aufgaben erfolgreich bearbeitet hast, kannst du …
- *… Erscheinungsformen der Globalisierung mit Beispielen beschreiben.*
- *… erläutern, welche Bereiche die weltweite Verflechtung bewirken.*
- *… den Fachbegriff **globalisation** (**Globalisierung**) erklären.*

COMPONENT 1 What is the role of trade?

Trade is seen as the engine of the global economy. Raw materials and finished products are imported or exported from one region of the world to another. Services and financial products are also traded globally. However, trade flows are distributed across the globe in different ways, and the global economy is changing.

1. a) Assign some countries to each of the world's three large economic areas (M4, atlas).
   b) Compare trade within Europe with trade in Asia and Africa (M4).
2. Analyse the phases of globalisation from 1990 until 2020. Consider the relationship between world trade and the world economy as a whole (M1).
3. Ⓒ Name possible ways to influence trade flows:
   A Generally as a citizen and consumer.
   B You personally in your everyday life
4. List which industries are heavily dependent on exports or imports (M3).
5. Ⓒ Do some research (internet): 213
   A The current "largest" export countries (M5 A).
   B The development of worldwide exports since 2020 (M1).

be dependent on	–	abhängig sein von
economic performance	–	Wirtschaftsleistung
flow of goods	–	Warenstrom
global economy	–	Weltwirtschaft
merchandise	–	Handelsware
raw material supplier	–	Rohstofflieferant
shares	–	Anteile
trade agreement	–	Handelsabkommen
trade flows	–	Handelsströme

Helpful words and phrases for task 1:

World trade ist dominated by ...
The leading region is .. / leading regions are ...
Trade between ... and ...
... plays / does not play an important role.
... has only limited significance

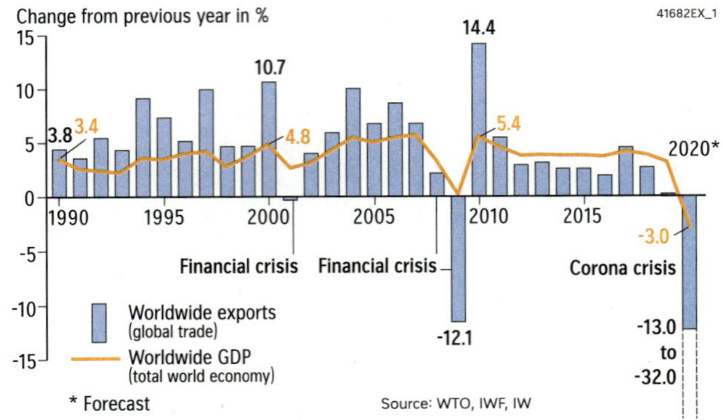

M1 Globalisation since 1990: the development of world trade and the global economy as a whole

Vor 5000 Jahren fanden drei Viertel des Handels in einem Umkreis von 20 Kilometern statt. Heute hat ein Smartphone bis zu uns 20000 Kilometer Seeweg zurückgelegt.

*Der **Welthandel** hat seit 1990 dramatisch zugenommen.*

Der Nutzen daraus ist aber ungleich verteilt?!

Die reichen Länder profitierten überproportional, während vielen armen Länder weiterhin nur die Rolle als Rohstofflieferanten bleibt. In der Weltwirtschaft haben nicht alle die gleichen Chancen.

Wie sehen Sie die Zukunft des Welthandels?

Diese ist angesichts zunehmend nationalistischer Wirtschaftspolitik deutlich unsicherer geworden. Die internationale Politik muss für verlässliche Regeln durch Handelsabkommen sorgen, die so gestaltet sind, dass sie auch ärmeren Ländern eine faire Chance geben und die Ökosysteme schützen.

M2 Interview mit dem Wirtschaftsgeographen Professor Rauch aus Berlin (2021)

Merchandise	Export in bill. euros	Import in bill. euros
Food and animal feed	56.7	49.7
Oil and natural gas	11.9	63.5
Chemical products	118.6	86.7
Pharmaceutical products	83.3	58.9
Metals	57.6	56.9
Data processing devices, electronic and optical products	118.8	118.6
Electrical equipment	89.8	66.8
Machinery	196.4	88.2
Vehicles and components	224.6	128.5
Other goods	79.6	99.5

Source: Destatis 2020

M3 Germany's foreign trade – the most important export and import goods, 2019

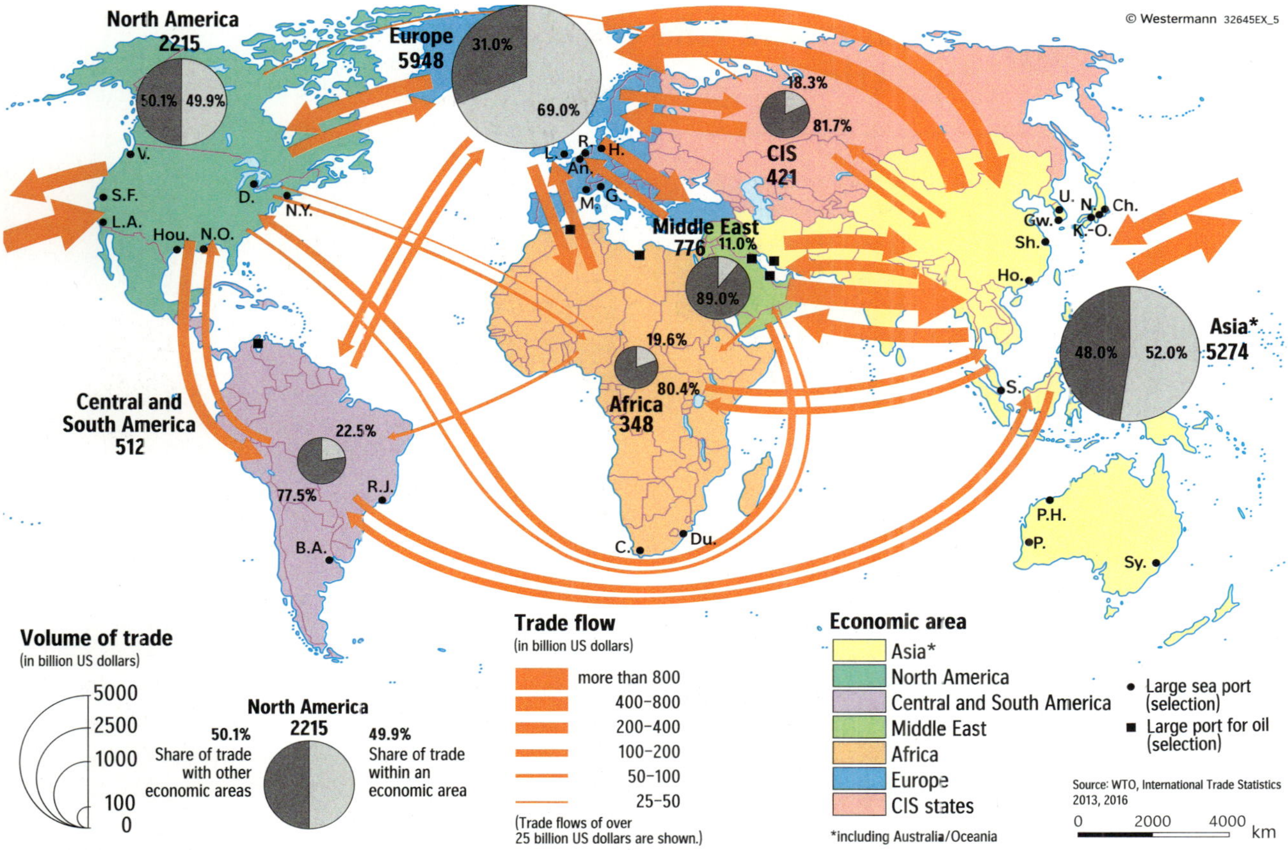

M4 Regional and international flows of goods in the major world economic regions. 90 percent of the goods (approx. 7 billion tonnes in 2019) are transported by sea and 10 percent by rail or air.

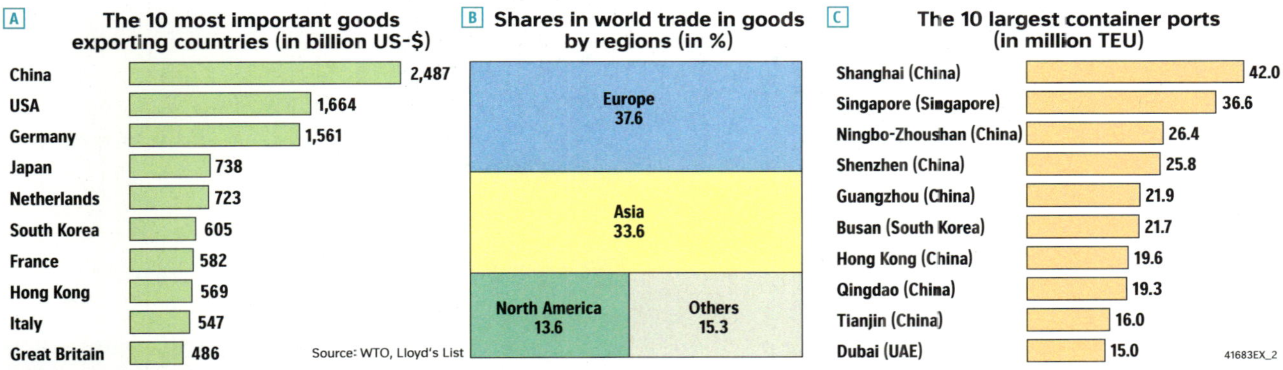

M5 Data and facts about world trade (2018)

M6 Merchandise in Africa: Old clothes from Europe at a market in Gambia (example)

INTERNET

M7 At Marinetraffic you can track worldwide shipping traffic in real time and also differentiate, for example, by sea region or ship type.

Wenn du diese Aufgaben erfolgreich bearbeitet hast, kannst du …

- … die Entwicklung des Welthandels darstellen.
- … die Verteilung der globalen und regionalen Handelsströme sowie die regionalen Unterschiede charakterisieren.
- … die Bedeutung des Handels und der Weltwirtschaft für die Globalisierung und für einzelne Länder oder Regionen erläutern.
- … den Fachbegriff **world trade** (**Welthandel**) erklären.

COMPONENT 2

What is the role of logistics?

What stages have my new smartphone or bicycle passed through when they arrive at my home? Logistics plays a major role in the processes behind this.

1. a) Note down all the stations in the supply chain for Lily's rucksack purchase. For each station, add the countries and means of transport that are mentioned (M2).
b) Colour the stations according to the phases in M7.

2. Trace the railway line from Duisburg to China. Note down the countries and their capitals and the number of time zones (M4, atlas).

3. Give reasons why Duisburg is a good transshipment and logistics location for Germany and Europe (M1, M4, M6, atlas).

Ⓒ **4.** Explain with the help of the satellite image and map, how it can be shown that it is a logistics location (internet): 203, 206
A Duisburg Trimodal Terminal (D3T).
B Köln Eifeltor.

Ⓐ **6.** Assign the production steps of the smartphone to the diagram in M7 (M5).

central warehouse	–	Zentrallager
component	–	Bauteil
distribution centre	–	Verteilzentrum
final assembly	–	Endmontage
freight transport	–	Gütertransport
logistics centre	–	Logistikzentrum
procurement	–	Beschaffung
stackable	–	stapelbar
supply chain	–	Lieferkette
transshipment	–	Warenumschlag

Helpful words and phrases for task 3:
transshipment point – Umschlagplatz
transport connections
traffic infrastructure
logistical infrastructure
waterway
motorway
railway
supply chain management
means of transport
A change from ... to . .

The first train from Weihai (China) to Duisburg, 2017

M1 ***duisport*** **is a logistics specialist for the 10,000 km long rail transport to China, with up to 50 trains running per week. The company began in 2011 with a connection to Chongqing. Electronics, textiles and toys are delivered for Central Europe.**

Lily wants to buy a new rucksack. She orders one on the internet from an Irish shop that meets her expectations. Two days later it is delivered. A very complex process has made the order possible. For example, the rucksack order was sent digitally to a central warehouse in Neuss, where the rucksack was packed and sent to a parcel centre in Lily's home town. There, the delivery driver takes over. Before that, however, the rucksack has already gone through several stages of a **supply chain**: designed in London, made of Indian leather in Ambur (India) with dye from China, then by truck to the container port, in the container firstly to Rotterdam, then to Duisburg and from there by truck to Neuss, where it was finally stored in the logistics centre until it was ordered. In this way, a mouse click triggers orders all over the world.

M2 Lily buys a rucksack – with a lot of **logistics**

M3 **Warehouses are very important for logistics. Goods handled there are first stored and then distributed to the merchants. In large distribution centres, about 3,000 shipments can be handled in just under three hours.**

How does a delivery of mountain bikes from a factory near Shanghai get to Duisburg? Which transport route and which means of transport are the cheapest?
The logistics company responds to the customer's needs. How urgently and at what time does the customer need the delivery? What costs are involved on each route to the logistics centre in Duisburg (e.g. freight costs, insurance, customs duties, warehousing)? And increasingly important: what is the environmental impact?

Note: Sea freight costs in container per bicycle: 11 euros; Sea freight costs in container per smartphone: 0.1 euros; by air freight it is five times that amount in each case

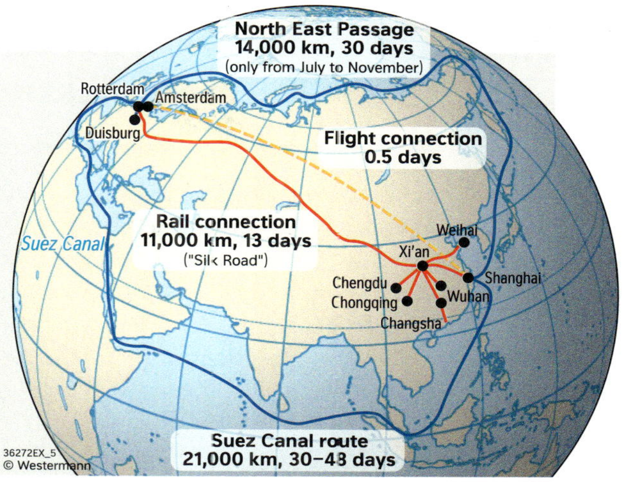

M4 Possible transport routes from Shanghai to North Rhine-Westphalia

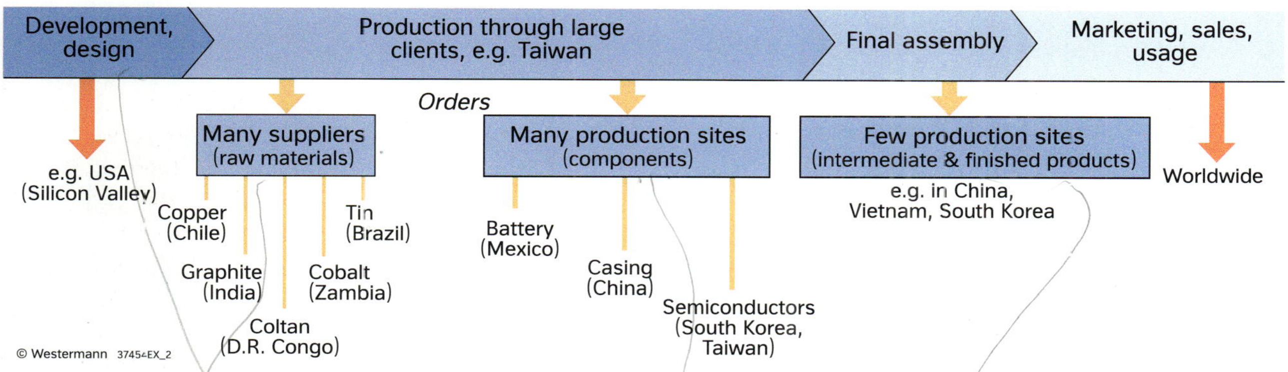

★M5 Production steps and global delivery chain (Example: Smartphone)

INFO

M6 **Logistics** is part of the service sector. Its task is the planning, control, coordination, and implementation of all operational processes in the transport of goods and data traffic, both locally and globally. In North Rhine-Westphalia, more than 600,000 people are employed in logistics.

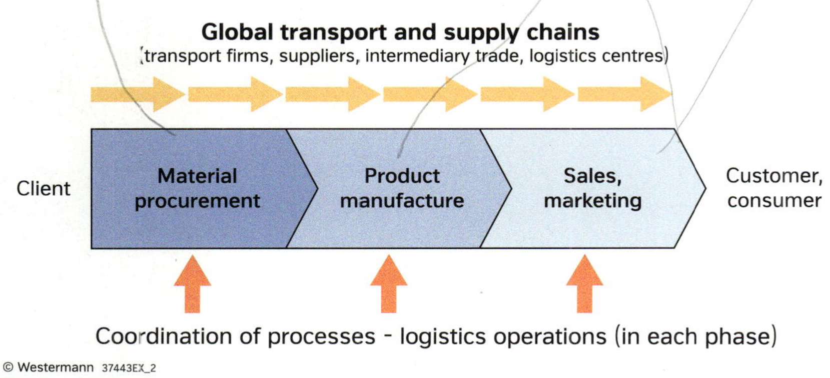

M7 Logistics and its importance in global production of goods (diagram)

Advantages: Worldwide standard dimensions; standardised for ship, rail, truck; easily stackable; saves warehouse costs; simplifies logistics in trade.

M8 Containers – "Boxes of globalisation"

Wenn du diese Aufgaben erfolgreich bearbeitet hast, kannst du ...

- *... Beispiele für globale Lieferketten beschreiben (Rucksack, Smartphone).*
- *... die Logistik konkret auf mehreren Ebenen sowie im Modell erläutern.*
- *... die Fachbegriffe **supply chain** (**Lieferkette**) und **logistics** (**Logistik**) erklären.*

COMPONENT 3

What is the role of digitalisation?

"Alexa, please tell my fridge to tell my TV to tell the lawnmower to ask my smartwatch what time it is." Admittedly, this scenario is an exaggeration. It could take place in Bielefeld, but also in Venice, Tokyo or Cape Town. Digitalisation permeates our lives, the economy and society all over the world.

C **1.** Report about your digitalised everyday life:
 A What data infrastructure is necessary for your use of data (M2, M5, M8, M9)?
 B Where can be seen links to globalisation (M1, M5, M9, M10)?

2. Explain what location factors make Frankfurt a globally important digital hub (M2, M3, atlas).

3. Show how digitalisation influences the global ecology and society by giving examples (M1, M5 – M10, internet).

4. Start a virtual tour through a data centre, e.g. Google (M4, internet).
What has surprised you? 213

data centre	–	Rechenzentrum
data throughput	–	Datendurchsatz
digital map service	–	digitaler Kartendienst
fibre optic cable	–	Glasfaserkabel
garden hose	–	Gartenschlauch
internet exchange	–	Internetknoten
permeate	–	durchdringen
power grid	–	Stromnetz
storage space	–	Speicherplatz
underwater cable	–	Unterseekabel
transmission	–	Übertragung
transshipment point	–	Umschlagplatz

Helpful words and phrases for task 4:
Without modern information and communication technology ...
... would not exit/be available.
... would be more difficult.
Information and communication technology ...
... has improved ...
... has enabled to ...
But, considering ...

Hanging out on the sofa with my girlfriends in the evening watching a stream, taking a selfie with my smartphone, uploading it to social media. Milliseconds later, it appears on the smartphones of my girlfriends sitting next to me while the episode continues.

M1 Instagram and Co – widely travelled: It is hard to believe that the photo has travelled via Frankfurt, across the Atlantic to North Carolina, to Facebook in California, and finally to Facebook servers in Ireland.

Frankfurt is a gigantic digital hub. The German Commercial Internet Exchange, or DE-CIX for short, is the largest internet exchange in the world in terms of data throughput. Frankfurt's position as an international financial and service hub is highly dependent on the smooth functioning of this digital infrastructure. A decisive role is played by the reliable supply of electrical energy to the data centres.

Source: Alsheimer, Constantin H. (Mainova): Stromnetz – Zuverlässige Versorgung. (IHK WirtschaftsForum July 2012), www.frankfurt-main.ihk.de, accessed: 21.11.2020, amended

M2 Data centres in Frankfurt – location factors

AMAZING

M3 Frankfurt is one of the largest internet exchanges worldwide. The data throughput is up to 10 terabits per second (11/2020). This corresponds to the transmission of 2.2 billion printed A4 pages or 2.2 million HD videos per second. DE-CIX in Frankfurt is located in close to 40 data centers with more than 1,000 networks connected. Companies like to locate near large data centres.

INTERNET

M4 Use the search term "data centre" to find virtual tours of data centres on video portals.

Behind every use of data, there are real places with a real **digital infrastructure** all over the world. Without them, there would be no digitisation and no internet. This includes power grids and power stations, data centres or submarine cables in the oceans and satellites in space.
All data traffic passes through data centres, the core of the digital infrastructure. These are highly secured buildings with sealed server rooms where customers (e.g. internet service providers, companies) use computing power, storage space and direct access to many other network operators or large cloud providers such as Amazon or Microsoft. In the process, ever larger amounts of data must be transported, e.g. through video streaming. Larger volumes of data also need ever more gigantic amounts of storage space.
Data security and a high-performance, stable power grid are immensely important for the location. In Germany, there are several thousand small and medium sized data centres. There are about 100 larger data centres (over 5,000 m²) most of them located in and around Frankfurt.

M5 **Digital infrastructure – worldwide, without borders**

M8 **There were around 2000 satellites in 2020, and countless more are launched into space every month to enable high-speed internet via satellite anywhere in the world in the future. Hazards arise from space debris.**

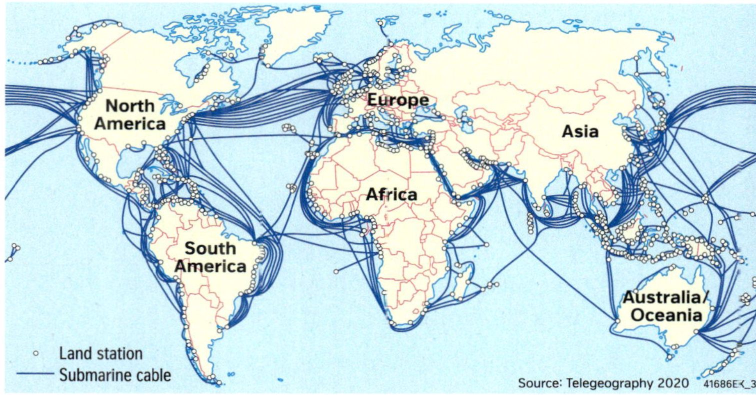

M9 **Underwater cables transport most data worldwide. These connect the continents via land stations. The cable pipes, more than 400 glass fibre cables (2020), are over 1.2 million kilometres long and hardly thicker than a garden hose. The cables can be damaged by ship anchors, earthquakes or sabotage, among other things.**

M6 **The downside of digitalisation: robots could put 500,000 workers in the textile industry out of work in Cambodia, and 1.7 million office workers in Indonesia. Low-wage countries run the risk of losing their locational advantages.**

AMAZING

M10
- The CO_2 emissions of the internet correspond to those of global air traffic in 2019.
- Video streaming accounts for 80 per cent of data traffic (2020).
- The waste heat from an average data centre could heat 10,000 homes.
- Cybercrime is the biggest business risk for companies worldwide for the first time in 2020.
- The world's first website (1990) can still be accessed today: info.cern.ch

– Flugscham? Streamingscham! –
Wer den Computer anschaltet, wer per Smartphone oder Tablet Filme konsumiert, verhält sich nicht „grüner" als Autofahrerinnen oder Flugreisende. Es sieht nur sauberer aus.

Quelle: Peitz, Christiane: Wie sehr die Digitalbranche das Klima belastet. www.tagesspiegel.de, Berlin, 19.10.2019, Zugriff: 21.11.2020

Digitalisation

41685EX_2

Social and leisure
- Social networks
- Video games
- ...

Education, learning
- e-Learning
- Hybrid teaching
- Distance learning
- ...

Home
- Smart devices
- ...

Transport, spatial
- e-Mobility
- Digital card services
- Autonomous driving
- ...

Work
- Home office
- IT
- Robotics
- e-Commerce
- ...

Supply
- e-Commerce
- ...

M7 **Impact of digitalisation on the economy and society worldwide (diagram)**

Wenn du diese Aufgaben erfolgreich bearbeitet hast, kannst du ...
... die globale digitale Infrastruktur beschreiben und ihre Bedeutung am Beispiel Frankfurts erläutern, z. B. Internetknoten.
... Beispiele für die digitale globale Vernetzung anführen und Auswirkungen aufzeigen (Umwelt, Wirtschaft, Gesellschaft).
*... den Fachbegriff **digital infrastructure** (**digitale Infrastruktur**) erklären.*

COMPONENT 4 What is the role of the global players?

Coca-Cola, Apple, Toyota and Co – who does not know them? One can hardly avoid them, anywhere in the world.

1. Characterise the worldwide Nestlé Group as a global player (M1, M4, M5, M7).
2. Illustrate the differences between a global player (e.g. Nestlé) and the Krah company in a chart (M1 – M4, M9). 207
3. Research a global player. Introduce it to the class (M6, internet; Tip: make a profile). 213
4. a) Analyse M6.
   b) Compare the composition of the top 10 of 2020 with that of 2008 (e.g. countries, industries) and derive from it the emerging development (M6, M8, internet).
5. Record a short commentary (3 minutes) on:
   A the business methods of global players (M1).
   B the title "Help! Global players rule my life!"
6. Research how the company *Nießing Anlagenbau GmbH* from Borken is a hidden champion (internet).

be on the cutting edge	–	auf der Höhe der Zeit sein
confectionery	–	Süßwaren
denounce	–	anprangern
marketing manager	–	Marketingleiter/in
niche product	–	Nischenprodukt
nutritional information	–	Angaben zum Nährwert
shareholder	–	Gesellschafter/in, Anteilseigner/in
unscathed	–	schadlos
trade market	–	Absatzmarkt
wire resistor	–	Drahtwiderstand
world market leader	–	Weltmarktführer

Helpful words and phrases for task 5:

Firstly, I'd like to explain/look at ...
Next, I'll come to ...
Finally, ...
Therefore, ...
An example is ...
I'd like to give an example.
To sum up, ...

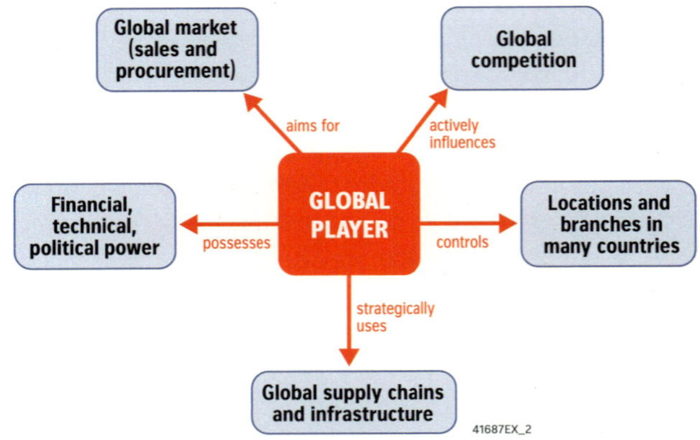

M1 Importance and economic role of the global players

Mrs. Mason-Hermann, you belong to the management of the world market leader Krah. What makes Krah a world market leader?

We are an automotive supplier and specialise in wire resistors for manual air conditioning systems. These are specially manufactured for each customer.
We operate in nine countries worldwide, focused on Europe, East Asia and South America. Our export quota is over 70 percent, our annual turnover is 175 million euros. All well-known car brands worldwide drive with our resistors.

Do you think globalisation is an advantage?

Definitely! It is necessary for our survival! The national market is too small for us.

M2 Interview with Mrs. Mason-Hermann, partner and marketing manager of a world market leader from Westphalia

INFO

M3 **Hidden champions** operate like the big **global players** on the world market. They are mostly medium-sized businesses. Their turnover is between 50 million and three billion euros. They hold a top position in their industry worldwide and work on niche products - highly specialised, innovative, and efficient. They are often family businesses. In Germany, there are many world market leaders (about 1,600). There are a particularly large number of hidden champions in Baden-Württemberg, South Westphalia and Münsterland.

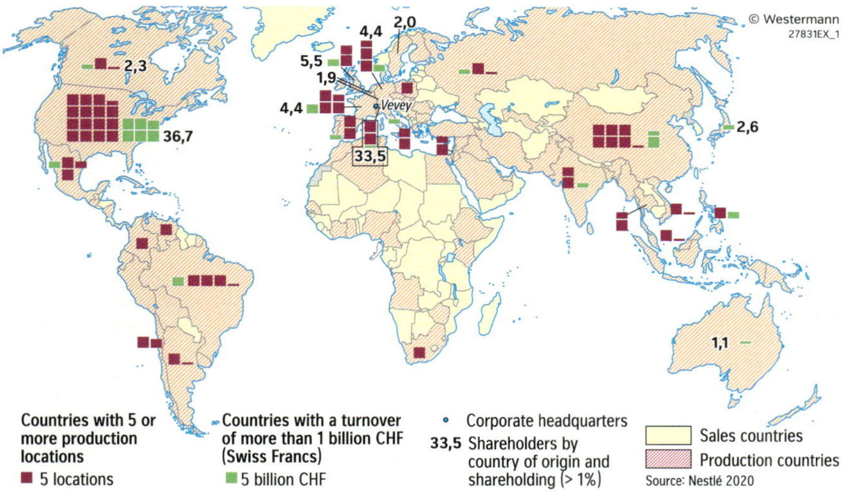

M4 **Nestlé: The largest food company in the world with 291,000 employees**

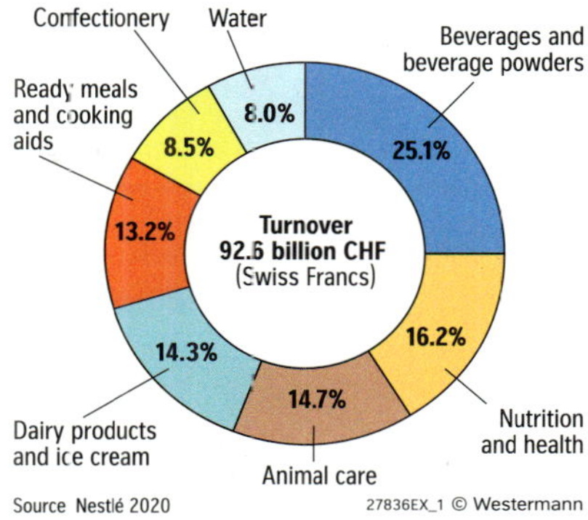

M9 **Nestlé produces more than 2,000 brands (e.g., Maggi, Nesquik, Vittel). Some 100 factories are fully automated and paperless.**

The world's largest food company has often been denounced by environmentalists in the past, because it pumps water even in places where it is scarce, such as in Africa, because it uses palm oil for confectionery and thus promotes deforestation of the rainforest, because it tolerates animal testing with mice for its cosmetic products. But in the meantime, the Swiss corporation is increasingly showing itself to be on the ecological cutting edge. [...] The change can be seen in many innovations: clear nutritional information on food packaging, refillable containers for cat food and coffee to go, more and more burgers, sausages and schnitzels on a vegetable basis [...]. Nestlé's sheer size means that it can weather crises and criticism relatively unscathed. In 187 countries, the world's largest food company sells a total of more than one billion products a day. [...] The Corona crisis showed how such a solid company [...] knows how to hold its own.

Source: Klemm, Thomas: Nestlé ist unverwüstlich. In: FAZ, 24.05.2020, p. 26

M5 **"Nestlé is indestructible".**

M7 **Nestlé presents new products as well as digital marketing innovations and new technologies at a trade fair in Beijing.**

Rank 2020	Rank 2008	Company	in billion US-$
1.	–	Saudi Aramco (Saudi Arabia)	1741
2.	41.	Apple (USA)	1568
3.	7.	Microsoft (USA)	1505
4.	–	Amazon.com (USA)	1337
5.	36.	Alphabet (USA)	953
6.	–	Facebook (USA)	629
7.	–	Tencent (China)	599
8.	–	Alibaba (China)	577
9.	13.	Berkshire Hathaway (USA)	430
10.	111	Visa (USA)	372
...	...	...	...
48.	112	SAP (Germany)	169

Source: PWC 2013, 2020

M6 **Largest companies in the world by stock market value (June 2020; best-placed German company)**

AMAZING

M8

- Amazon founder Jeff Bezos is the richest person in the world with a fortune of more than 200 billion dollars (09/2020). This is equivalent to the economic output (GDP) of Peru (211 billion with a population of 35 million), four times that of Bulgaria (50 billion, population 7.5 million) and a hundred times that of the Central African Republic (1.9 billion, population 5 million).
- Google (Alphabet), Amazon.com, Facebook, Apple and Microsoft are among the winners in the 2020 pandemic, because their digital services were in particularly high demand during the pandemic.

Wenn du diese Aufgaben erfolgreich bearbeitet hast, kannst du ...

- *... Bedeutung und Wirtschaftsweise der Global Player beschreiben und kritisch betrachten.*
- *... den Erfolg von Hidden Champions erläutern und sie mit Global Playern vergleichen.*
- *... die Fachbegriffe **Hidden Champion** (**Hidden Champion**) und **Global Player** (**Global Player**) erklären.*

COMPONENT 5 What is the role of online trading?

Buying and selling anywhere in the world and around the clock - online trading makes it possible. How does it work? What impact does it have?

Ⓒ **1.** Online trading or stationary retailing? How do you shop? What do you personally think makes sense? Produce
A newspaper article.
B a simulated interview.

2. Give possible reasons for the location of Amazon's logistics and distribution centres in Germany (M5).

3. "Online trading is typical of digitalisation and globalisation." Explain this statement (M2, M3, M6, M7).

4. Complete M10.

5. Evaluate the importance of shopping events (M4, M7 – M9).

6. Using Eurostat, report on the use of e-commerce in the countries of the European Union. 212

Ⓐ **7.** Compare and evaluate online trading and stationary retailing from the point of view of
a) retailers,
b) transport service providers,
c) environmental protection,
d) the mayor of a small town.

arts and crafts	–	Kunsthandwerk
commission	–	Provision
on the internet	–	im Internet
stationary retail trade	–	stationärer Einzelhandel
parcel shipment	–	Paketsendung
payment services	–	Zahlungsdienste
state-of-the-art	–	neuester Stand der Technik
turnover	–	Umsatz

Helpful words and phrases for task 2:

As you can see, ...
It is obvious that ...
Most of the ...
That is why ...
... are closely connected to ...
... is/are located/situated ...
... near/close to ...
Whereas the ...
To sum up, ...

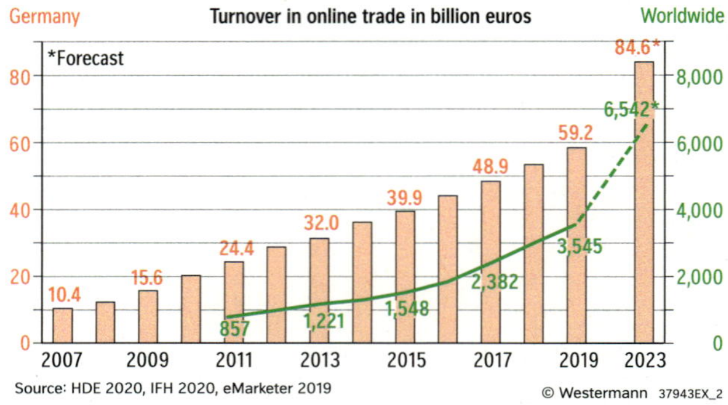

M1 **Turnover in online trade: Germany and the world**

The Cologne retailer Johanna B. has been running a shop with arts and crafts from Eastern Europe for years. She noticed that customers were shopping less in her inner-city shop. To counteract this, she got training and set up a professional online shop at a large online marketplace. She was able to draw on her existing networks and suppliers. When, during the Corona crisis, the high-street retail sector declined to the point where its existence was threatened, the online retail platform became a lifeline. She even increased her turnover significantly. Wooden handicrafts as garden ornaments were particularly in demand. Her customers now come from all over Germany and the world, and the wood workshop in Poland had to hire new workers. Johanna B. is happy that she can continue to run her shop, and she has even found a way into the export business. The only thing that concerns her is the high commissions that she must pay to the online trading platform for every sale.

M2 **Retailers and the online marketplaces of global players**

INFO

M3 **Online trading**

- Digital commerce on a global scale
- Digital payment options
- Emergence of digital marketplaces (eBay, Amazon Marketplace) for professional and private traders
- Offer of digital services (e.g., payment services, website design)
- Deliveries and returns with customised logistics - increase in parcel shipments
- Flexible shopping, anytime, anywhere
- Competition for the stationary retail trade

M4 Logistics centre in Wuxi on Singles Day. Traders from all over the world sell their goods with discounts.

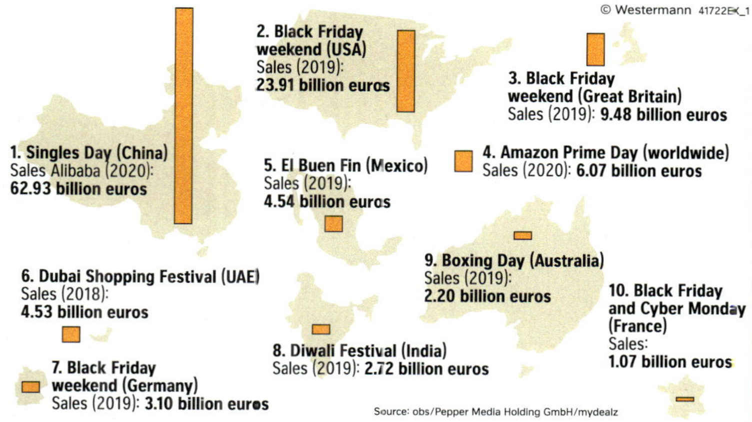

M8 The biggest global shopping events with sales of 1 billion euros or more

M5 Amazon logistics locations in Germany, 2020

Number	Mastercard (USA)	Paypal (USA)	Alipay (China)
Countries	210	c. 200	c. 110
Currencies	150	25	15
Users in millions	2400	c. 350	1000

Source: Alipay 2020, Mastercard 2020, Paypal 2020

M6 Data on online payments

AMAZING

M7
- Lipsticks sold are equivalent to the height of 47 Eiffel Towers (Singles Day 2020).
- DHL transported 1.8 billion parcels in Germany in 2020, 15 % more than in 2019.

- 3 million additional workers
- More than 3,000 aircraft and cargo ships for goods imports, 700 charter flights for export deliveries
- 10,000 mobile lockers and packing stations
- State-of-the-art digital infrastructure; global supply chains including logistics infrastructure

M9 Logistics company (Cainiao) from Alibaba (China) on Singles Day 2020

M10 Effects of online trading (diagram)

Wenn du diese Aufgaben erfolgreich bearbeitet hast, kannst du …

… die Entwicklung des Onlinehandels beschreiben sowie seine Infrastruktur am Beispiel von Bezahldiensten und Logistikzentren in Deutschland erläutern.
… die Vor- und Nachteile des Onlinehandels für unterschiedliche Interessengruppen analysieren und erörtern.
… die Auswirkungen eines globalen Shopping-Events kritisch bewerten.
*… den Fachbegriff **online trade** (**Onlinehandel**) erklären.*

M1 Apple's corporate headquarters "The Ring" in Cupertino (Silicon Valley, California) measures 461 metres in diameter, the inner park is 12 hectares in size. More than 12,000 employees work on four floors. The construction costs amounted to 5 billion US dollars (the sum corresponds to the value of the GDP of Malawi or five times that of Gambia). The design was produced by the English architect Norman Foster, the glass construction was supplied by a company from Augsburg.

WHAT WE HAVE LEARNED

In the age of globalisation – is it all only virtual?

What is globalisation?

142 **Globalisation** is a process of increasing interconnection of different regions and cultures of the world. This is mainly evident in the economic sphere (trade, production in goods supply chains, transport, logistics). In addition, there are other dimensions: global interconnections in culture (fashion, media, eating habits), politics (organisations such as the UNO, WTO) as well as in communication (internet) and ecology (world climate). Globalisation affects everyone of us and directly or indirectly influences every part of the Earth with its effects.

What is the role of trade?

144 International economic relations between countries have always existed, but since around 1990 they have grown with more and more countries and people. **World trade** in particular has increased. Trade flows are unevenly distributed; 80 percent of them are between Europe, Asia and North America. Important trade goods include machinery, electronics, energy raw materials and textiles. Crises (political upheavals, pandemics) can endanger global trade.

What is the role of logistics?

146 Globalisation is inconceivable without **logistics**. The industry uses state-of-the-art technology to plan and coordinate production and transport in global **supply chains**, from the manufacturer to the consumer. In addition to digitalisation, central instruments are globally standardised containers and transport hubs.

What is the role of digitalisation?

The increased digitalisation in all areas of life has boosted digital globalisation. Processes are changing; supply chains are becoming simpler, online trade is growing and robots are already part of everyday life, which in turn threatens jobs. Data flows need a real **digital infrastructure**, regionally through strong power grids and data centres, globally through internet nodes (e.g. Frankfurt) or submarine cables between continents and satellites in space. Digitisation causes high electricity demand and CO_2 emissions. 148

What is the role of the global players?

Global players are corporations on a world scale with worldwide production facilities and markets. They supply the world market and exploit their financial and technical power and the advantages of global supply chains. Their stock market value often exceeds the GDP of entire countries. Smaller **hidden champions** also serve the world market, mostly with highly specialised niche products. Both large and small global players are in fierce global competition. 150

What is the role of online trading?

Online trade is booming. It requires both things: globalisation (e.g. supply chains) and digitalisation (e.g. platforms, payment services). Private individuals as well as traders can buy and sell digitally at any time. This has displaced analogue (offline) trade and created new logistics facilities and jobs. 152

	Production of raw materials	Production phase		Marketing, selling and consumption phase
Countries	Kazakhstan, India	Ethiopia, Bangladesh	Indonesia, Bangladesh	EU, USA
Operations, jobs	- Plantations - Weaving - Low wages	- Dyeing operations - Sand blasting - Low wages	- Sewing - Packing - Low wages	- Wholesaling - Textile shops - Discounters - Minimum wage
	 Transport	 Transport	 Transport	
Health hazards	- Pesticides - Fertilisers	- Poisonous chemicals	- Long working hours - Polluted air - Lack of safety standards	- Fumes from the textiles - Working hours also at evenings/weekends
Environmental pollution	Monoculture	Disposal of chemicals in rivers, lakes, soil	Harmful exhaust air in factories	Clothing waste

M2 The textile manufacturing supply chain (diagram) – from a sustainable viewpoint

M3 Shop sustainably – why?

M6 Robots in an Amazon warehouse

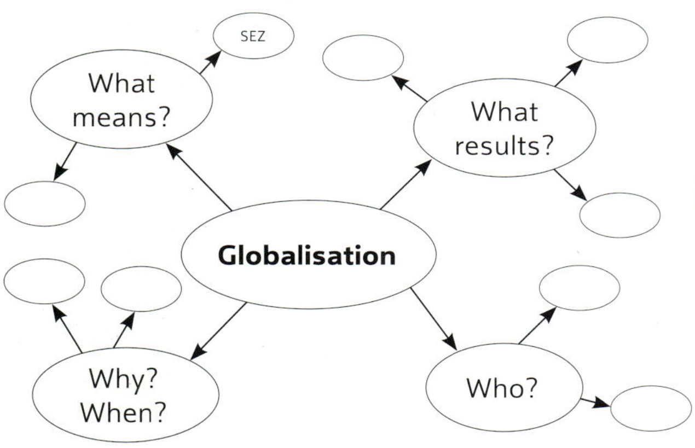

M4 Suggestion for a mind map

container – digital infrastructure – digitalisation – globalisation – global players – global competition – internet nodes – supply chain – logistics (location) – online trade – world trade

M5 Technical terms - do you know them?

1 The headquarters of the global player Apple is a symbol of both globalisation and digitalisation. Justify this thesis (M1).

C 2 **A** Match M3 and M6 to a phase of the supply chain (M2). Explain your decision. Where would M1 be placed?
B Research on the internet or in the book: Find photos of each phase of the supply chain (M2). Give reasons for your choice in keywords.
C What needs to be changed for goods supply chains to be sustainable? Suggest examples (M2, M3).

3 Complete the mind map on globalisation (M4). Also use terms from M5. 199

4 Let your classmate explain three of the terms from M5.

What are the effects of globalisation pressure? – in Germany

Industry 4.0 – the digitalisation of work

Global competitive pressure forces companies in Germany to continually develop new production concepts and strategies in the context of digitalisation. The goal is to reduce costs and increase productivity. This increases the competitiveness of companies. What concepts and strategies are companies pursuing today to remain competitive?

1. a) Describe what a modern industrial company looks like in your opinion.
b) Compare your results with the characteristics of Industry 4.0 (M1 – M3, M8). Note similarities and differences.

2. Work out the development towards Industry 4.0. Draw a timeline (M1 – M3, M8).

Ⓒ **3.** Explain the principle of just-in-time production by making
A a diagram (M4 – M6). 199
B a short explanatory video (M4 – M6). 219

4. Explain what problems can arise from just-in-time production (M4 – M7, internet).

5. Evaluate whether just-in-time production is a sustainable concept (M4 – M7, internet).

assembly – Montage
body shell – Rohkarosserie
competitiveness – Wettbewerbsfähigkeit
control system – Steuerungssystem
dashboard – Armaturenbrett
division of labour – Arbeitsteilung
mechanical loom – mechanischer Webstuhl
outsource – auslagern
precondition – Voraussetzung
press plant – Presswerk
production line – Produktionsband
programmable logic controller – speicherprogammierbare Steuerung
slaughterhouse – Schlachthof
timeline – Zeitstrahl

Helpful words and phrases for task 5:

Just-in-time production means ...
The advantages of ... are ...
So, on the one hand, ...
But ...
On the other hand, ...
This leads to ...

M2 An engineer controls the automatic maintenance of a robotic arm in a smart factory

INFO

M3 In **Industry 4.0**, industrial production processes are networked with state-of-the-art information and communication technology. This makes it possible to produce more economically and flexibly. At the heart of Industry 4.0 is the **smart factory**.
In a smart factory, intelligent machines coordinate production processes independently, robots cooperate with humans in assembly, and driverless transport vehicles handle logistics orders independently.
The essential precondition are powerful computers. This also changes the world of work. In the future, humans will work hand in hand with robots and computer systems. They do not even have to be directly present.

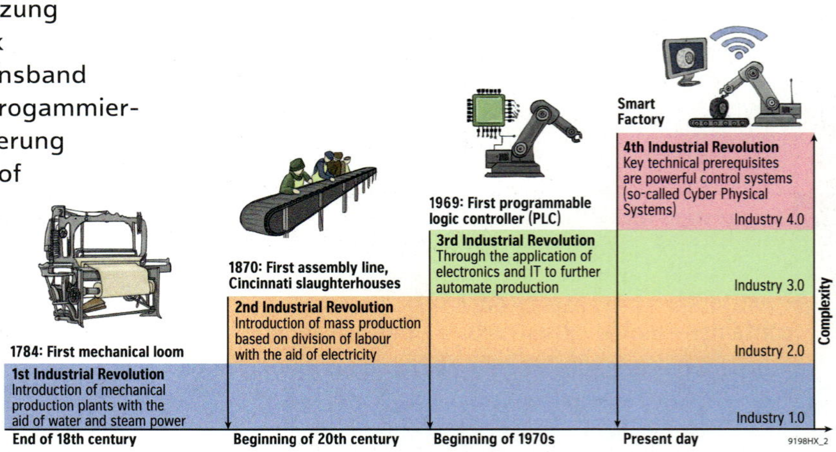

M1 The development towards Industrial Production 4.0

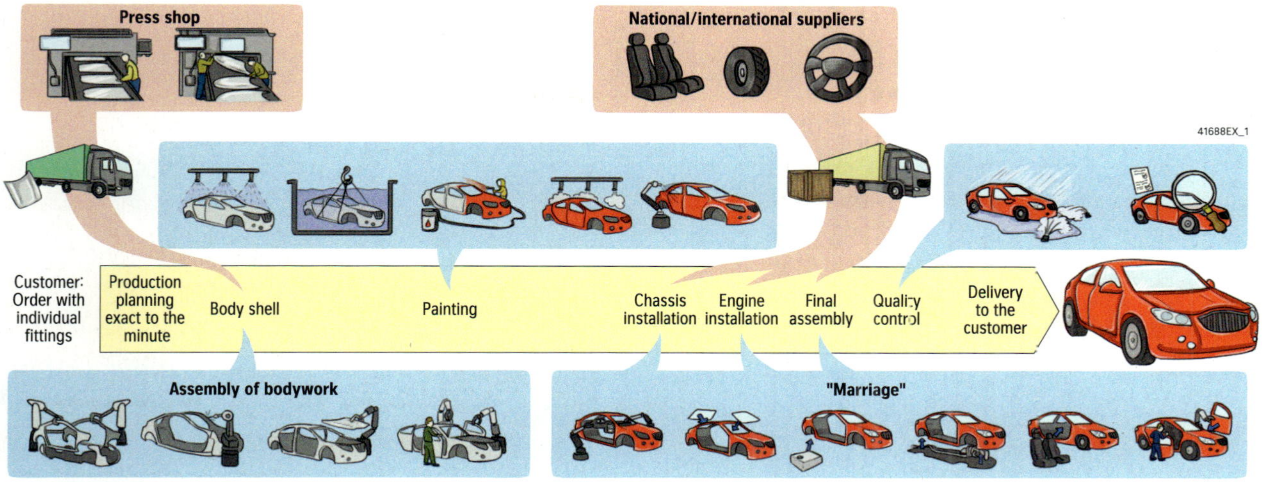

∗M4 Each automobile is produced individually – entirely according to the customer's wishes (e.g. with different equipment). It is produced according to the principle of just-in-time production.

INFO

M5 Nowadays, many production steps are no longer done by the companies themselves, they are **outsourced**. This means that they are subcontracted to other companies. The individual parts, but also complete components (e.g. the dashboard), are then delivered by the subcontractors to the production line at the time they are needed, **just-in-time**. This saves on warehousing costs. As a result, there is a high degree of mutual dependence between supplier and customer. The car manufacturers are dependent on the very punctual delivery of car parts. Outsourcing, even to far-away companies, increases the volume of traffic and thus also the environmental impact.

M6 "Smart Transport Robot" from BMW and the Fraunhofer Institute. Autonomous transport robots deliver car parts just-in-time to car plants.

Until now, many companies wanted to be supplied with parts for production according to the just-in-time principle, i.e. only when they were needed. The Corona crisis showed that this principle is very risky because it severely restricts the security of supply. If, for example, China cannot deliver because it is in the middle of a pandemic, no parts arrive in Germany and production comes to a halt. This has recently cost the German automotive industry a lot of money. "We are currently experiencing a move away from just-in-time in many sectors," says logistics consultant Neumeier.

Source: „Platz für Pasta", Bärbel Brockmann, SZ, München, 20.06.2020, amended

M7 Just-in-time – a production concept with a future?

Artificial intelligence (AI) is a branch of computer science. Its goal is to imitate and even improve upon human intelligence. AI systems are software, machines or robots that perform tasks on their own continuously, without each step being programmed by humans. They are characterised by the fact that they can learn, programme and develop themselves and others.

M8 Artificial intelligence in Industry 4.0

Wenn du diese Aufgaben erfolgreich bearbeitet hast, kannst du ...

... die Entwicklung zur Industrie 4.0 herausarbeiten.

... das Verfahren Just-in-time erklären.

... Probleme erläutern, die durch eine Just-in-time-Fertigung auftreten könnten.

*... die Fachbegriffe **artificial intelligence – AI** (**künstliche Intelligenz – KI**), **Industry 4.0** (**Industrie 4.0**), **smart factory** (**Smart Factory**), **outsourcing** (**Outsourcing**) und **just-in-time** (**Just-in-time**) erklären.*

What will the industry of tomorrow look like?

In the course of digitalisation and globalisation, industrial production in Germany faces constant changes and challenges. The digital possibilities in industry in particular ensure that the world of work is constantly changing. How will we work in the future? Will robots make humans superfluous? What technical innovations will shape industry in the future?

1. Name at least three examples where 3D printing is used (M1, M6, M7, internet).

2. Make a chart showing the advantages and disadvantages of 3D printing (M1 – M3, M6, M7, internet). 207

C **3.** Comment on one of these questions:
A Will digitalisation make people unemployed (M1 – M7, p. 156/157 M1 – M8, internet)?
B What will the world of work of the future look like in the context of digitalisation (M1 – M7, p. 156/157 M1 – M8, internet)?

4. Report on two modern professions that have emerged in the course of digitalisation (M4, M5, M7, internet).

A **5.** Make a digital collage on the industry of the future (M1 – M3, M7; p. 156/157 M1 – M8, internet).

harden	–	aushärten
one-off product	–	Einzelstück/Unikat
on the verge	–	an der Schwelle
qualitative leap	–	Qualitätssprung
resin	–	Harz
trigger	–	auslösen
unemployment	–	Arbeitslosigkeit

Helpful words and phrases for task 3:

The impact of digitalisation ...
The working environment ...
This effects ...
Due to ...
Therefore ...
This leads to ...
As a result of ...
As a consequence ...
I agree/disagree with ...
From my point of view ...

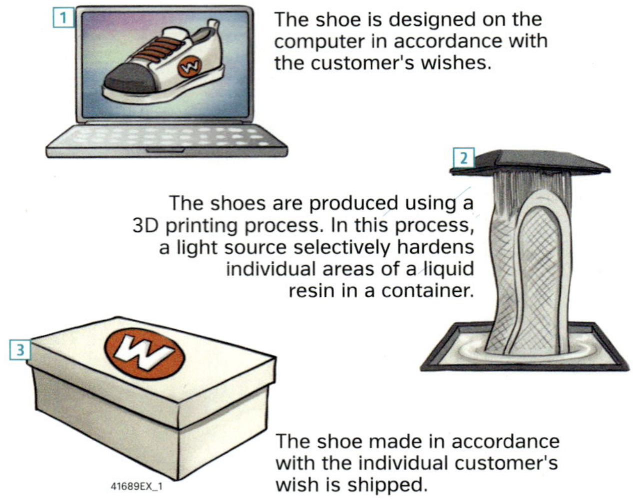

M1 3D printing in shoe manufacturing. 3D printing makes it possible to develop and manufacture one-off products on a computer according to the wishes of individual customers.

Many experts believe that 3D printing will trigger a new technological revolution. Goods will no longer have to be transported on land, water or by air. They can be sent to any place in the world as a digital 3D model via the internet. This does not only save transport costs, but it could make the automated production of goods possible anywhere in the world at the same price. This could bring the production of many goods back to Germany. 3D printing does not only save valuable time, but also saves warehouse space and raw materials. Goods are produced as and when they are required, and only with the raw materials needed.

M2 A revolutionary process – 3D printing

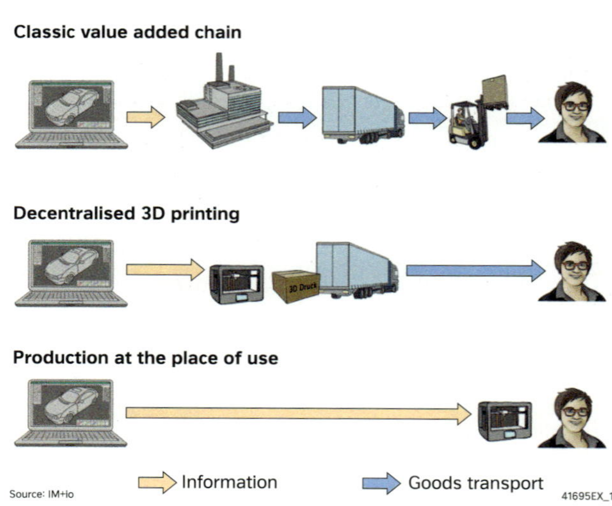

M3 Companies need to ask themselves where 3D printing can be employed

Headlines

Digitalisierung schafft Arbeitsplätze: Neue Technologien ermöglichen neue Produkte und Dienstleistungen [...]

Source: Schaible, Stefan et al. (Roland Berger GmbH): 12 Thesen zu Auswirkungen der Digitalisierung auf die Arbeitswelt der Zukunft. Stuttgart 11/2017, S. 4

Arbeit der Zukunft: Roboter vs. Mensch?

Source: Mahanty, Wolfgang: Arbeit der Zukunft : Roboter vs. Mensch? T4M Transofmations-Magazin.com, www.transformations-magazin.com 09.12.2019, Zugriff: 12.11.2020

Fabrik der Zukunft – Mehr Möglichkeiten für gut ausgebildete Ingenieure

Source: Wallner, Philipp: Frünf Trends für die Fabrik der Zukunft. www.industr.com, 10.12.2019, accessed: 12.11.2020

Die voranschreitende Digitalisierung wird die Wirtschafts- und Arbeitswelt der Zukunft auch weiterhin stark verändern.

Source: Altmaier, Peter (BMWi): KI und Robotik im Dienste der Menschen. Grußwort von Bundesminister Peter Altmaier. Berlin 09/2019, S. 2

Industrie 4.0: Roboter sind die zentralen Elemente in der digitalen Produktion

Source: Industrieanzeiger: Roboter sind die zentralen Elemente in der digitalen Produktion. www.industrieanzeiger.industrie.de, Leinfelden-Echterdingen, 28.02.2020, accessed: 12.11.2020

Corona als Stresstest – wie die Krise die Industrie 4.0 pusht

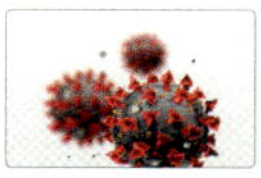

Source: Lina, Stephan (BR): Corona als Stresstest – wie die Kriste die Industrie 4.0 pusht. www.br.de, München, 12.06.2020, accessed: 12.11.2020

M4 **Selection of headlines – the future of employment**

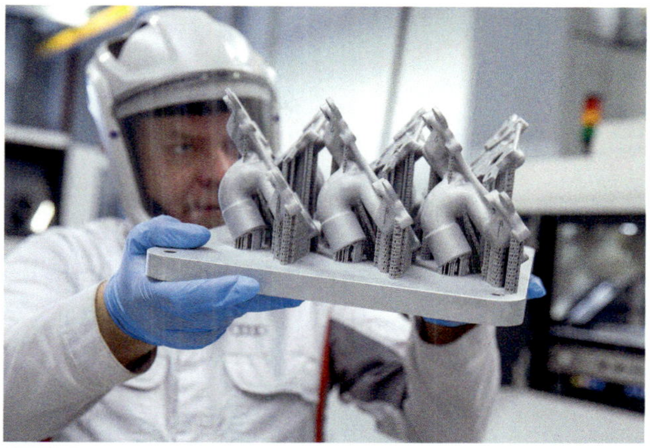

M6 **Car parts from a 3D printer. In this way, complex components can be produced quickly and in small quantities. 3D printing is increasingly finding its way into industry. This will also change the demands on the workforce in the future.**

INTERNET

M7 You can find further information on Industry 4.0, the working world of the future and application examples for 3D printing here:

- acatech
- BMWi
- Plattform Incustrie 4.0
- 3D-grenzenlos Magazin

tagesschau.de: Which jobs are most at risk from digitalisation?
Schwentker: *The jobs that are most likely to be affected are those for which no training or only a short training is needed, so-called helper jobs. The better the training, the less risk there is that the job will be replaced. That is the general trend. If you have studied and are a specialised expert, then the danger is less.*
tagesschau.de: What happens to the people who are replaced? Will new jobs be created or are we threatened with a huge increase in unemployment?
Schwentker: *[...] There are researchers who are very pessimistic and believe that many jobs will be lost. And there are others who say the exact opposite will happen. Personally, I tend to be more optimistic. This is from a historical perspective. At the latest, since Henry Ford has invented the production line in car maufacturing, we have had this effect that machines can take over more and more of what people do. Since then, however, employment has continued to rise at the same time because the economy has also continued to grow. This is a trend that - apart from a few crises – has actually continued throughout time. And one would have to provide arguments as to why we are now on the verge of a qualitative leap that we have never experienced before in history.*
tagesschau.de: Do fully automated factories, which are often mentioned, not represent a new dimension?
Schwentker: *[...] There are very few factories in Germany where this is being implemented as you might imagine. And even if such a development were to happen, industrial manufacturing is by no means the only sector of the economy that determines the labour market. With increasing automation, jobs are shifting further towards the service sector. This is a trend we have been observing for some time. In this sector, the number of employees is also growing. I do not tend to be very worried. There is always change. You must go along with the change. And best of all, you must shape it yourself.*

Source: Abdulaziz-Said, Aimen (tagesschau.de): „Man muss den Wandel mitgehen". www.tagesschau.de, Hamburg, 26.10.2016, accessed: 12.11.2020

M5 **Extract from an interview with the data journalist Björn Schwentker**

Wenn du diese Aufgaben erfolgreich bearbeitet hast, kannst du ...
... Vor- und Nachteile des 3D-Drucks erläutern.
... Stellung nehmen zum Thema „Die Arbeitswelt der Zukunft".

A region reinvents itself again and again – the Ruhr Area

Where research is carried out today, in the past everything revolved around coal. The Ruhr Area was once the largest coal-mining region in Germany. Why has the Ruhr Area changed? How did the Ruhr Area develop?

1. What do you associate with the Ruhr Area? Write down three keywords.

2. a) Describe the development of coal production in the Ruhr Area (M1, M6).
b) Calculate the index values for the data in chart M6. 217
c) Present the index values in a diagram.

3. Make a timeline that illustrates the development from Bochum 1.0 to Bochum 4.0 (M1 – M10).

4. Report to your class on the current status of the Mark 51°7 project in Bochum. Also add a satellite image (internet). 203, 206

C **5.** Explain the structural change in the Ruhr Area by making (M1 – M10)
A a mind map. 199
B a concept map. 199
C an explanatory video. 219

A **6.** Describe the structural change in Saarland (atlas). 201

M2 *Bochum 1.0:* **Hard coal was mined in the first deep shaft at the Dannenbaum Colliery in Bochum from 1860. Coal mining ceased in 1958.**

The economic importance of the Ruhr Area can be traced back to the rich coal deposits and the iron and steel industry. The period before and during the Second World War and the post-war period proved to be boom times for the region.

The import of low-cost coal and the switch to alternative energy sources such as oil or gas led to an economic crisis in the Ruhr Area from 1957 onwards. Numerous mines and iron and steel works had to be closed. In the mid-1970s, an oversupply of cheap steel on the world market led to falling steel prices. In addition, plastics and aluminium were introduced as alternatives to steel products. In addition to the **coal crisis**, the Ruhr Area also suffered from a **steel crisis.**

M3 **The rise and the crises of the Ruhr Area**

colliery site	–	Zechengelände
diversification	–	Diversifizierung
hard coal mining	–	Steinkohleförderung
iron and steel works	–	Hüttenwerk
iron ore	–	Eisenerz
mine	–	Bergwerk
one-sided orientation	–	einseitige Ausrichtung
shaft	–	Schacht
structural change	–	Strukturwandel

Helpful words and phrases for task 2a:

From ... to ...
Since ...
The number of ...
The development of ... indicates the ...
... the decline of ...

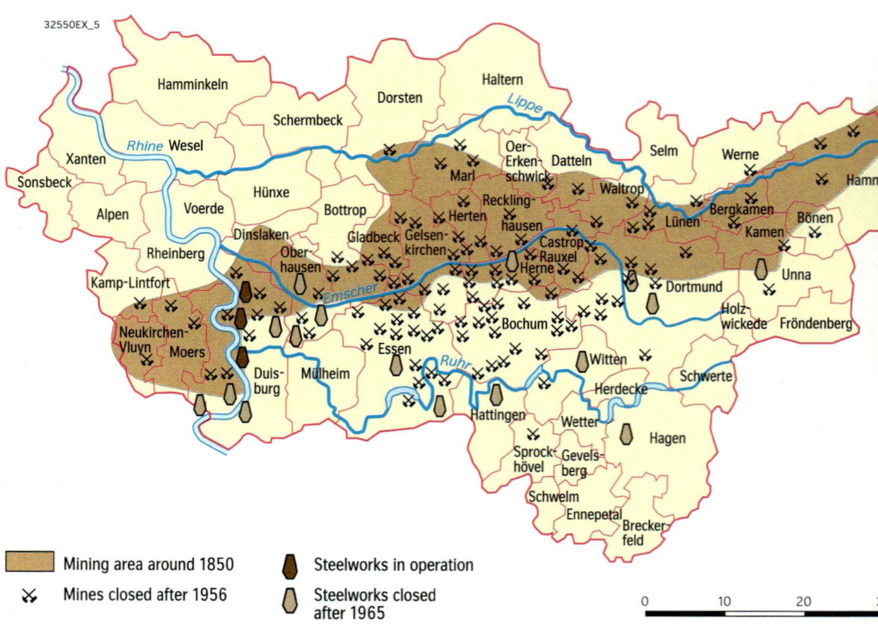

M1 **The Ruhr Area with locations of steelworks and coal mines, in the past and today**

M4 *Bochum 2.0:* **The former Dannenbaum colliery site in changing times ... The Opel company opened a car plant on the former colliery site in 1962.**

M7 *Bochum 4.0:* **The former Dannenbaum colliery site in changing times ... In 2019, the logistics company DHL opened a parcel centre on the former Opel factory site.**

The coal crisis and the steel crisis led to a profound **structural change** in the Ruhr Area. The problem of the region at that time was primarily the **monostructure** of the economy, i.e. the one-sided orientation towards mining and the iron and steel industry. Therefore, from 1960 onwards, politicians promoted **diversification**. New industries were established: mechanical engineering, chemicals, vehicle construction and electrical engineering. The Finnish mobile phone manufacturer Nokia started producing mobile phones in Bochum in 1989 and created more than 2,300 jobs there (*Bochum 3.0*). Jobs in the service sector were also created in newly built universities and in technology parks as well as in shopping centres and leisure facilities.

M5 The Ruhr Area in transition

With the help of "Bochum 4.0", the city of Bochum is trying to successfully master the next structural change. "MARK 51°7" is the new name of the 70-hectare industrial and commercial area of the Opel plant that closed in 2014. The *Industrie-, Technologie- und Wissens-Campus* is being developed here that will address the requirements of future worlds of work. Today, 60 percent of the available industrial and commercial space has already been planned. It will offer more than 6,000 new jobs in the future.

M8 The structural change continues

INFO

M9 **Structural change** refers to the change in the structure of a region (e.g. **economic structure**) or a business (company structure).

	Employment in hard coal mining (in 1000)	Hard coal extraction (in million t)	Number of coal mines	Employment in the iron and steel industry (in 1000)
1960	398	115	122	264
1970	202	91	56	227
1980	143	69	29	184
1990	99	55	19	123
2000	45	26	9	51*
2010	19	10	3	47*
2018	3	2	1	47*

Source: Destatis, IT.NRW, Regionalverband Ruhr, Statistik der Kohlenwirtschaft e. V. *Change in the statistics

M6 The development of the mining and steel industries in the Ruhr Area

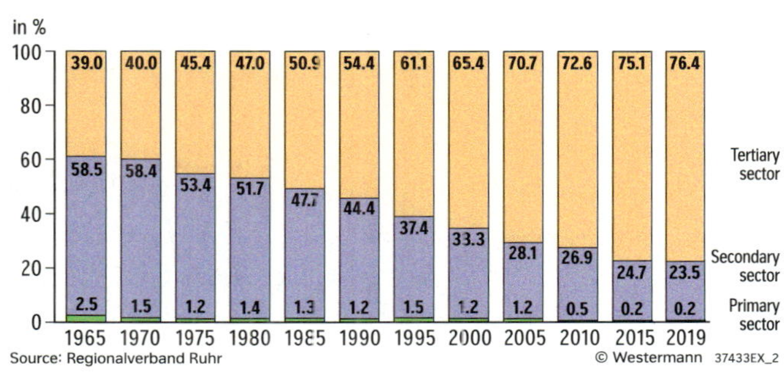

M10 Employment by economic sector in the Ruhr Area

Wenn du diese Aufgaben erfolgreich bearbeitet hast, kannst du ...
- *... die Entwicklung der Steinkohleförderung beschreiben.*
- *... die Ursachen und Folgen des Strukturwandels erläutern.*
- *... die Fachbegriffe **coal crisis** (**Kohlekrise**), **steel crisis** (**Stahlkrise**), **structural change** (**Strukturwandel**), **monostructure** (**Monostruktur**), **diversification** (**Diversifizierung**) und **economic structure** (**Wirtschaftsstruktur**) erklären.*

The Ruhr Area – modern and innovative

"City of cities" - the Ruhr Area gets a new slogan. This campaign is intended to draw attention to the Ruhr Area around the world. The aim is to attract managers and young innovative companies and to bind them to the region in the long term. What does the economy in the Ruhr Area look like today and in the future?

1. a) Describe the characteristics of a start-up company (M2).
b) Note down which start-ups you know (M2).

Ⓒ **2.** You are the mayor of Bochum. Explain why the Ruhr Area is an attractive region for start-ups (M1, M2, M7).
A Write a press release.
B Record a voice message.

3. a) List the characteristics of Dortmund as a technology location (M3 – M6).
b) Give reasons why Dortmund is a good location for a technology centre (M3 – M6).

4. "The Ruhr Area - an economic area with a future". Comment on this statement (M1 – M7, internet).

Ⓐ **5.** Develop an advertising slogan for the modern and innovative Ruhr Area.

B2B	–	business-to-business (Unternehmen zu Unternehmen-Beziehung)
business model	–	Geschäftsmodell
Computer Science	–	Informatik
founder	–	Gründer/in
large cable railway	–	Großkabinenbahn
mayor	–	Bürgermeister/in
revenue groth	–	Umsatzwachstum
technology park	–	Technologiepark
University of Applied Sciences	–	Fachhochschule

Helpful words and phrases for task 3:

... can be characterised as ...
A characteristic is ...
Furthermore, it should be emphasised that ...
One reason is ...

With Masterplan.com, Stefan Peukert and Daniel Schütt are already building their second start-up in Bochum. Since 2017, the duo have been working on their mega-exciting e-learning concept on the topic of digitalisation, which is aimed at companies and their employees. [...]

Let's talk about the Ruhr Area. When it comes to start-ups in Germany, people immediately think of Berlin. What speaks in favour of the Ruhr as a start-up location?

Thanks to a large network of universities and colleges, there are a lot of well-educated young people with exciting ideas in the Ruhr Area, and that is the basic prerequisite for a good start-up location. [...] A good infrastructure also helps.

What exactly is the appeal of the start-up scene in Bochum?

The start-up scene in Bochum is small, but very well networked. Masterplan is in very good company in the city centre, with many technologically oriented start-ups are in the neighbourhood. We also keep very close contact with the universities in Bochum. In addition, many large companies are located here, which is ideal for B2B start-ups to gain their first customers.

What is easier in Bochum than in the rest of the country?

[...] The competition among start-ups is less than in Berlin, for example, so it is easier to attract talented applicants from the region for your own company.

What is missing in Bochum and the Ruhr Area?

More courageous founders who take advantage of the potential that undoubtedly exists here in the Ruhr Area.

Source: Hüsing, Alexander: #Interview. „Es fehlen mehr mutige Gründer, die das Potential, das im Ruhrgebiet vorhanden ist, ausnutzen". www.deutsche-startups.de, 12.07.2019, accessed: 09.10.2020

M1 Interview with Stefan Peukert, founder of the start-up company Masterplan

INFO

M2 Start-ups ...

- ... are not more than ten years old,
- ... are (highly) innovative with their technology and/or their business model,
- ... have – or aim for – significant revenue growth.

The world's most valuable start-ups (2019):

1. Chinese tech company ByteDance (including TikTok; US-$ 75 billion)
2. US-based ride-hailing company Uber (US-$ 72 billion)
3. Chinese ride-hailing company Didi Chuxing (US-$ 56 billion)

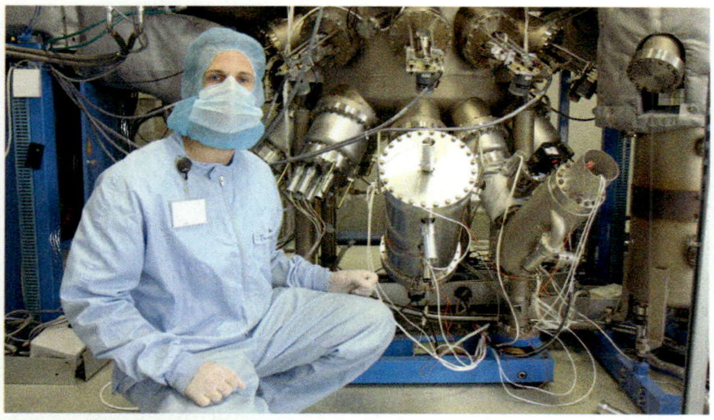

M3 TechnologieZentrumDortmund (TZDO) is one of the five leading technology centres in Europe. Around 300 technology-oriented companies with more than 10,000 employees have settled there.

M6 #DigitalGipfel19 – the German Federal Government's Digital Summit took place in Dortmund on 28 and 29 October 2019. The event focused on digital platforms.

Specialists
- A total of over 9,300 students in the Computer Science Department at the TU Dortmund and the Dortmund University of Applied Sciences.
- in Dortmund, 11,300 people work in the information and communication sector.

Local support
- *TechnologieZentrum Dortmund* and Dortmund Technology Park.
- Other sector-specific locations such as *Stadtkrone Ost* and *Phoenix West*.
- Locate IT: Support for foreign companies settling in the region.
- Start2grow start-up competition and growth initiative.

Know-how
- University of Dortmund Department of Computer Science
- Dortmund UAS Department of Computer Science
- IT Centre Dortmund.
- Fraunhofer Institute for Material Flow and Logistics
- Fraunhofer Institute for Software and Systems Engineering
- Research Institute for Telecommunications.

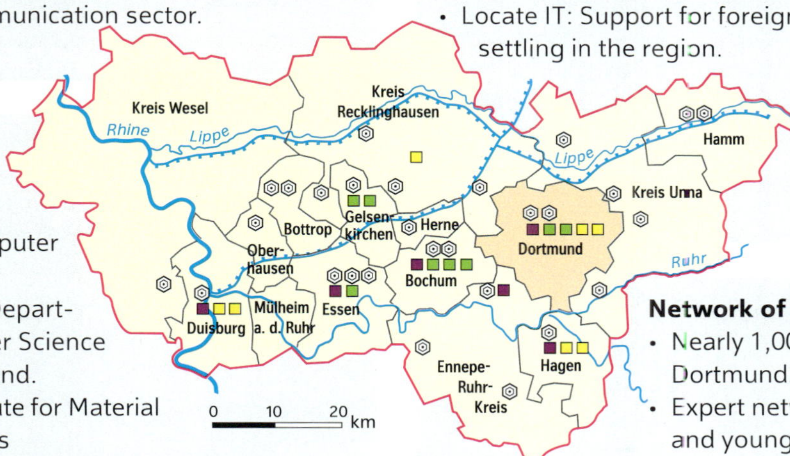

Network of partners
- Nearly 1,000 IT companies in Dortmund.
- Expert network for founders and young companies.

M4 Dortmund – an attractive location for the IT industry (IT = information technology) (as of 2019)

M5 The H-Bahn Dortmund is a fully automatically controlled large-cabin railway that connects the two campuses of the university and the adjoining technology park.

INTERNET

M7 *Gründerszene* and *Berlin Valley* are magazines from the German start-up scene and regularly publish reports, interviews and statistics on start-ups. *ruhr:hub* is the central contact point for the digitalisation of the economy in the Ruhr metropolitan region.

Wenn du diese Aufgaben erfolgreich bearbeitet hast, kannst du ...
- *... Gründe für die Ansiedlung von innovativen Unternehmen im Ruhrgebiet erklären.*
- *... Merkmale des Technologiestandortes Dortmund beschreiben.*
- *... den Fachbegriff **start-up (Start-up)** erklären.*

It is the location that counts!

Many of the cars produced in Germany are made in Saxony. There are numerous companies from the automotive industry there, for example BMW or Porsche. Why is that the case? Why do companies choose certain locations?

1. a) Describe the situation of Saxony as a location for the automotive industry (M4, atlas). 200
b) Explain why Leipzig is an attractive location for the automotive industry (M1 – M5, atlas).

2. Explain the term "cluster". Make a diagram (M4, M6). 199

3. a) Analyse how far Saxony as an automotive location fulfils the characteristics of a cluster (M3, M4, M6, atlas).
b) List possible advantages for a location in a cluster from the point of view of a company (M1 – M6, internet).

C **4.** Analyse the location factors that could be decisive in the choice of a location (M2, M5, internet, atlas): 204, 223
A for a high-tech company.
B for a logistics centre.
C for any company.

A **5.** Draw a sketch map showing the distribution of the locations of the automotive industry in Germany (atlas, internet). 204

business promotion – Wirtschaftsförderung
economic sector – Wirtschaftszweig
investment grant – Investitionszuschuss
hard location factor – harter Standortfaktor
research institution – Forschungseinrichtung
spacial proximity – räumliche Nähe
soft location factor – weicher Standortfaktor
supply industry – Zulieferindustrie

Helpful words and phrases for task 1:
The automotive industry in Saxony is situated ...
... extends from ...
There are several important cities ...
There are ... good reasons why ...

- **BMW and Porsche** established at the location
- **Trimodal transport hub:** excellent accessibility via road, rail and air
- **Worldwide rail connections:** especially to China
- Over 110 large-scale and developed **industrial sites** with over 10,000 square metres available
- **Excellent telecommunications infrastructure:** 5G standard in preparation
- Production centre for **electric and hybrid vehicles**
- **Daily freight traffic** to major European ports
- **24-hour airport** for freight flights
- **Outstanding potential for skilled workers** in a growing region
- **Investment grants** of up to 30 percent by state development bank

Source: City of Leipzig 41678EX_1

M1 Ten good reasons for Leipzig as a location

INFO

M2 Does every company fit in every location? Many factors play a role in searching for the right location for a business. These factors are called **location factors**. A distinction is made between hard and soft location factors. Hard location factors have a direct impact on a company's production costs. The soft location factors make up, in their sum, the quality of life in a region and are more difficult to measure. As a rule, it is not just one single factor that is decisive in the choice of a location.

Today, "Autoland Saxony" is one of Germany's top car-manufacturing locations, with five Volkswagen, BMW and Porsche vehicle and engine plants as well as around 780 component suppliers, equipment manufacturers and service providers in the industry. About one in eight cars built in Germany comes from Saxony. At the same time, the automotive industry is the driving force behind the manufacturing sector in Saxony. More than 95,000 employees, over 80 % of them in the component supply industry, produce more than a quarter of Saxony's industrial output. From powertrains to accessories, suppliers in Saxony can develop and manufacture almost all the components and parts needed for a vehicle, as well as the necessary production equipment. In total, more than 50 university and non-university research institutions in Saxony [...] deal with the topic of automobiles.

Source: Wirtschaftsförderung Sachsen GmbH (WFS): Das „Autoland Sachsen" im Überblick. www.standort-sachsen.de, Dresden, accessed: 12.11.2020

M3 The automotive industry is a key industry for Saxony

Porsche plant Leipzig: floor space 427 hectares, around 4,300 employees, daily production 550 vehicles.

Infineon Dresden: location for semiconductor production and research for the automotive industry, among others, more than 2,700 employees.

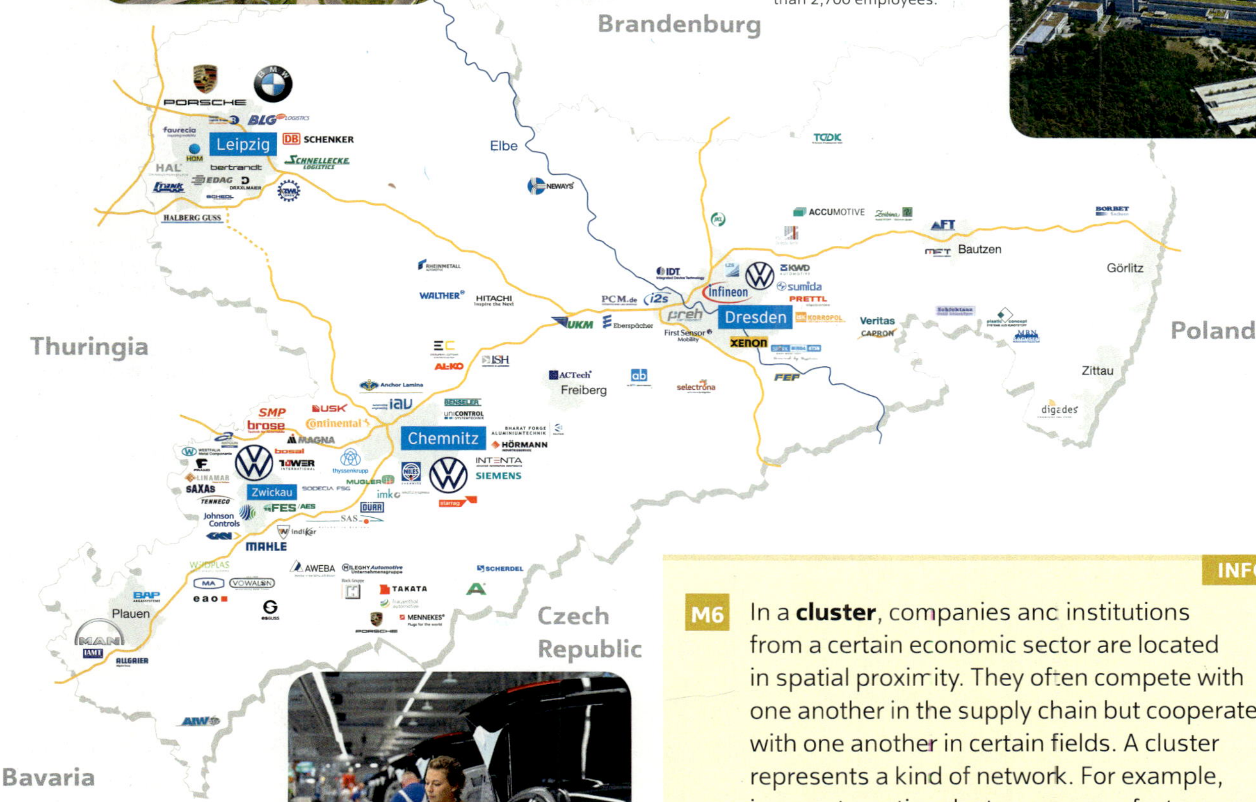

Volkswagen plant in Zwickau with some 8,000 employees (photo: employees on the production line for electric cars)

M4 The automotive industry – locations in Saxony.

INFO

M6 In a **cluster**, companies and institutions from a certain economic sector are located in spatial proximity. They often compete with one another in the supply chain but cooperate with one another in certain fields. A cluster represents a kind of network. For example, in an automotive cluster, car manufacturers, component suppliers, research institutions and public authorities work closely together. In global competition, clusters help to drive innovation and increase the competitiveness of companies.

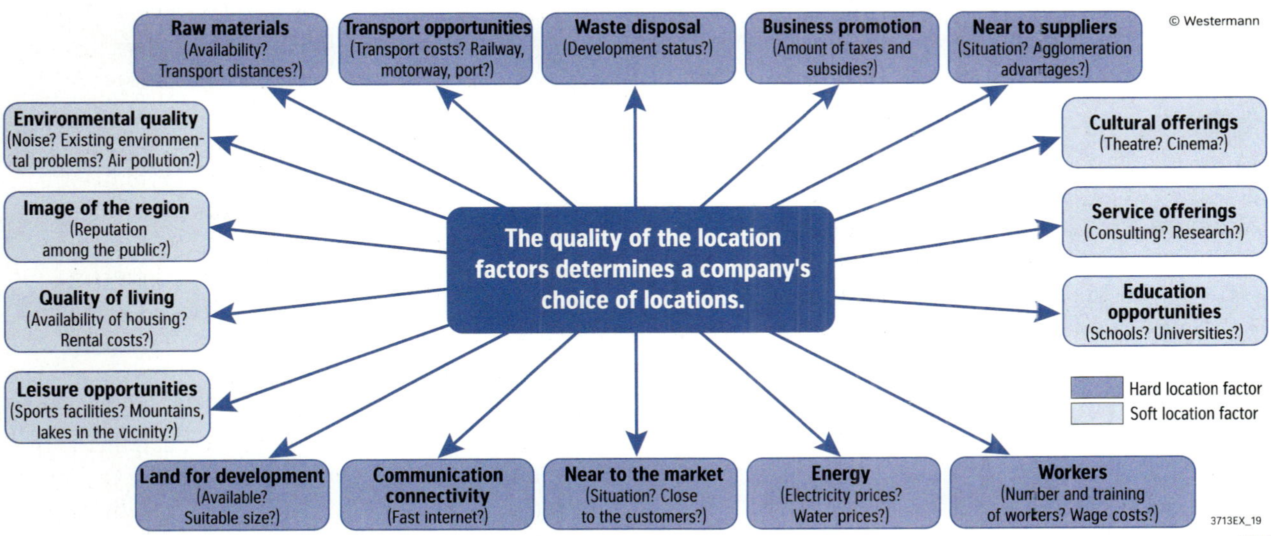

M5 Hard and soft location factors, an overview

Wenn du diese Aufgaben erfolgreich bearbeitet hast, kannst du …

… den Unterschied zwischen harten und weichen Standortfaktoren erklären.

… mithilfe eines Modells erklären, was ein Cluster ist.

*… die Fachbegriffe **location factor** (**Standortfaktor**) und **cluster** (**Cluster**) erklären.*

The automotive industry – worldwide locations

"BMW invests millions in its largest plant in the USA in Spartanburg."

"Volkswagen has 125 production sites around the world."

"The largest Mercedes plant is now in China."

Many companies in the automotive industry have locations all over the world. What are the reasons for it?

1. Describe the location of the production and assembly plants of the global player BMW (M4, M5, atlas).
2. (C) Give reasons why the BMW Group manufactures in
   A China (M2, M4 – M10, internet).
   B Eastern Europe (M2 – M4, internet).
3. Interpret the cartoon M1. 209
4. a) Present the data from M3 in a different form (e.g. only one Y axis, index values). 217
   b) Evaluate whether the presentation in M3 is manipulative. 215
5. "BMW – Future through new production sites?" Comment on this question (M1 – M11).
6. (A) You are a manager at BMW. Name two countries that you think would be suitable for new factories. Give reasons for your choice (atlas). 200, 201

agreement	–	Vertrag
assembly plant	–	Montagewerk
automobile sales	–	Automobilabsatz
economic zone	–	Wirtschaftszone
production site	–	Produktionsstätte
tax advantage	–	Steuervergünstigung

Helpful words and phrases for task 3:

In the cartoon, you can see ...
The cartoon has a caption.
The ... symbolises/stands for ...
The message of the cartoon is ...
From my point of view, ...

M1 „Ja! Ich sehe eine Besserung der unternehmerischen Aussichten."

M2 When European companies relocate part of their production abroad, they look for good location factors.

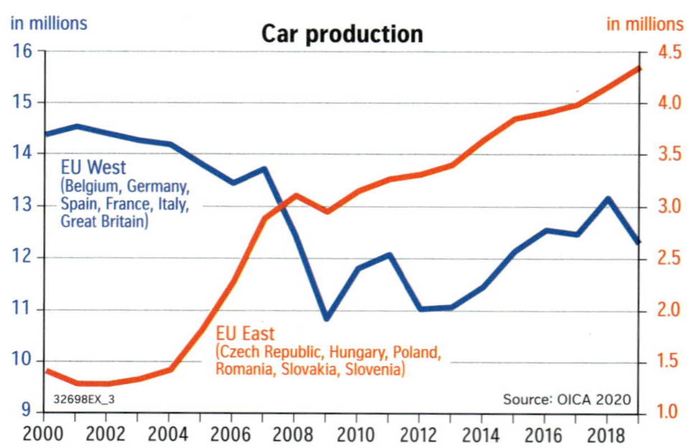

★M3 Important location advantages for the automotive industry are offered by economic zones in Eastern Europe. These include, for example, the European sales market, low labour costs, skilled labour and tax advantages.

INTERNET

M4  The *BMW Group* gives an overview and information about the company's worldwide locations.

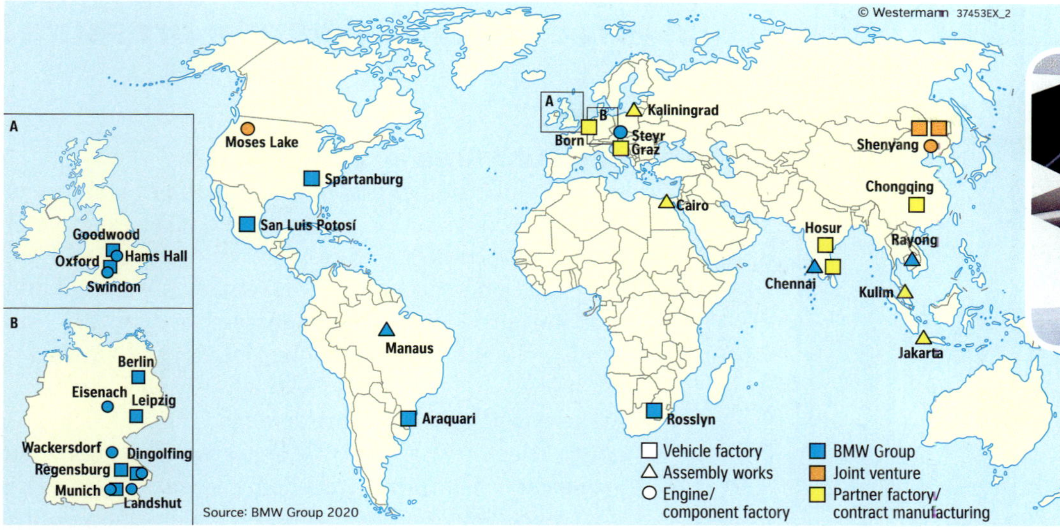

M5 Production sites and assembly plants of the global player BMW

M11 Special models are offered for the Chinese market to meet local tastes (here: 530e in the long Le version)

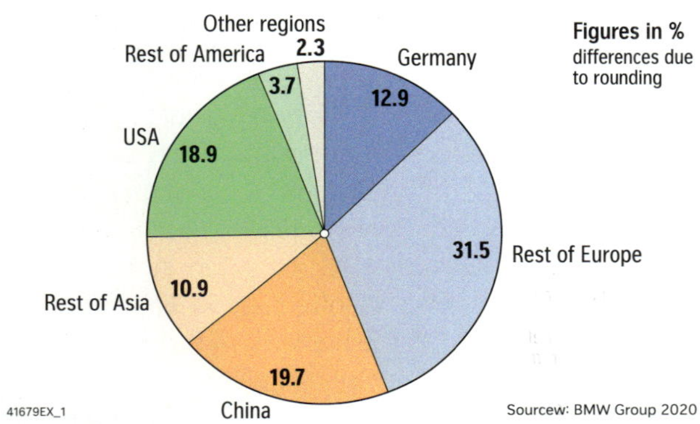

M6 BMW Group's sales in 2019 by region (in million euros)

Year	BMW sales in China* (in 1000)	BMW sales in Europe (in 1000)	BMW sales worldwide (in 1000)
2011	234	858	1669
2013	392	860	1964
2015	466	1003	2258
2017	595	1103	2469
2019	725	1084	2538

Source: BMW Group 2015, 2020 *excluding Hong Kong and Macau

M9 Development of BMW Group's automobile sales (vehicles sold in 1000s)

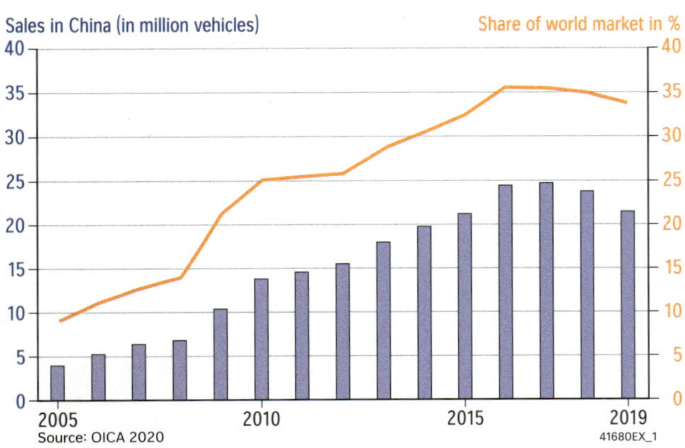

M7 Chinese passenger car market and China's share of the world market

In China, the BMW Group has been working with its Chinese partner Brilliance in a **joint venture** since 2003. The state-controlled joint ventures ensure that many Western models can also be built in China. Through these cooperations, both profits and know-how flow to China. Joint ventures have helped BMW to open up markets in growth regions like China. The joint venture agreement between BMW and Brilliance, which was intended to expire in 2028, was extended by 22 years in 2018. The joint venture obligation for car production in China expires in 2022.

M10 BMW also makes cars in China

BMW hat in der Corona-Krise im zweiten Quartal 25,3 Prozent weniger Autos verkauft als im Vorjahreszeitraum. Einziger Lichtblick war China, wo der Absatz um 17,1 Prozent zunahm.

Source: ZDF: Absatz von BMW bricht um ein Viertel ein. www.zdf.de, Mainz, 07.07.2020, accessed: 12.11.2020

M8 Pressemitteilung aus dem Juli 2020

Wenn du diese Aufgaben erfolgreich bearbeitet hast, kannst du ...

- *... Gründe für weltweite Produktionsstandorte erläutern.*
- *... die Zukunftsperspektiven des Automobilstandortes China erörtern.*
- *... den Fachbegriff **joint venture** (**Joint Venture**) erklären.*

WHAT WE HAVE LEARNED

What is the effect of globalisation pressure? – In Germany

M1 Workers on the production line at a VW plant in Chengdu, China. Volkswagen operates 125 manufacturing plants in 31 countries worldwide (2020).

Industry 4.0 – the digitalisation of work

156 Increasing global competitive pressure forces companies to constantly develop new production concepts and strategies. The goal is to reduce costs and increase productivity. This takes place, for example, with respect to **Industry 4.0**. The term Industry 4.0 summarises different innovations in the course of digitalisation. Digitalisation is changing the world of business and work and is often even considered to be the fourth industrial revolution. At the centre of this is the **smart factory**. In the smart factory, machines coordinate production processes independently, robots cooperate with humans in assembly, and driverless transport vehicles handle logistics orders independently. **Artificial intelligence (AI)** is also used in Industry 4.0. .

In order to strengthen their competitiveness, companies often also outsource parts of their production and leave the manufacturing to other companies, they use **outsourcing**. Individual parts, but also complete components, are delivered **just-in-time** from the supplier companies to the production line. This saves costs for development and storage.

158 The advancing digitalisation has already led to companies bringing production steps back home to their own country in recent years. **3D printing** could play a special role here. With 3D printers, plastics, metals and other basic materials are fused into new objects. Individual parts are then no longer produced by suppliers from abroad but can be manufactured with the help of 3D printing at the place of production.

The Ruhr Area is changing ...

In the course of digitalisation and globalisation, entire economic regions have also changed. One example of this is the Ruhr Area. The economic importance of the Ruhr Area goes back to the rich coal deposits and iron and steel production. With industrialisation in the 19th century, the Ruhr Area developed into an industrial conurbation. 160

At the end of the 1950s, the demand for coal fell. In addition, cheaper coal could be imported. The Ruhr Area increasingly fell into an economic crisis. Mines were closed and workers lost their jobs.. Competition on the world market also grew in the steel industry in the 1970s. In addition, cheap alternatives to steel were introduced to the market in the form of plastics and aluminium. The **coal crisis** was joined by the **steel crisis**.

To overcome the economic crises and the **monostructure** of the Ruhr Area, a far-reaching **structural change** with greater **diversification** was brought about. Mining and the iron and steel industry were replaced by other industries (e.g. the automotive industry) and by service industries. Numerous universities and technical colleges were also established in the Ruhr Area. 162

In addition, the Ruhr Area offers good conditions for **start-ups**. In particular, the network of companies, universities and research institutions as well as a good infrastructure are of central importance for the settlement of young and innovative companies.

Regional and global locations

When choosing the best location, industrial companies take many location factors into account. A distinction is made between **hard** and **soft location factors**. Hard location factors are easy for companies to calculate. Hard location factors include, for example, the infrastructure, the sales market or the available workforce. Soft location factors, such as the image of a region, are more difficult to measure. Due to increasing competitive pressure, companies are looking for locations worldwide. Companies often choose their location in **clusters**. In these networks, companies (e.g. suppliers) and institutions (e.g. research institutes) cooperate closely with each other within a certain economic sector in spatial proximity. 164

Companies search for locations worldwide. In this way, among other things, the companies want to open up new markets. BMW, for example, operates 31 production sites and assembly plants in 14 countries worldwide. In China, BMW works with its Chinese partner Brilliance in a **joint venture**. 166

M2 View of Essen – first half of the 20th century

M5 Essen today – in 2017, Essen was presented as the European Green Capital

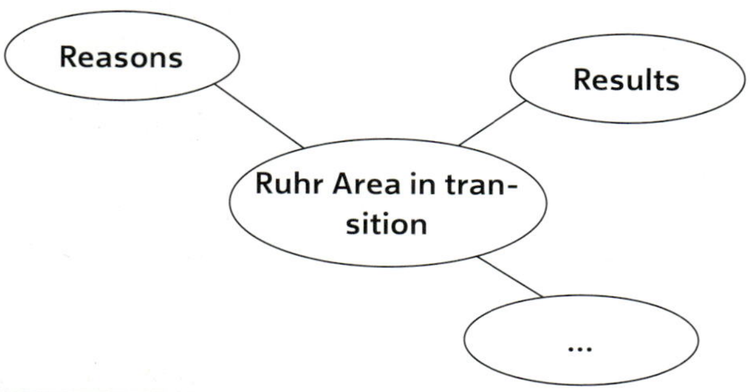

M3 Mind map

1 a) Describe photos M2 and M5.
b) Explain the differences. Use technical terms.

2 Draw a mind map on the subject of "The Ruhr Area in Transition". 199

M4 DER SPIEGEL, No. 36/2016

3 a) Comment on the cover picture of "Der Spiegel".
b) Design a cover picture for a magazine with the topic "The Future of Work".

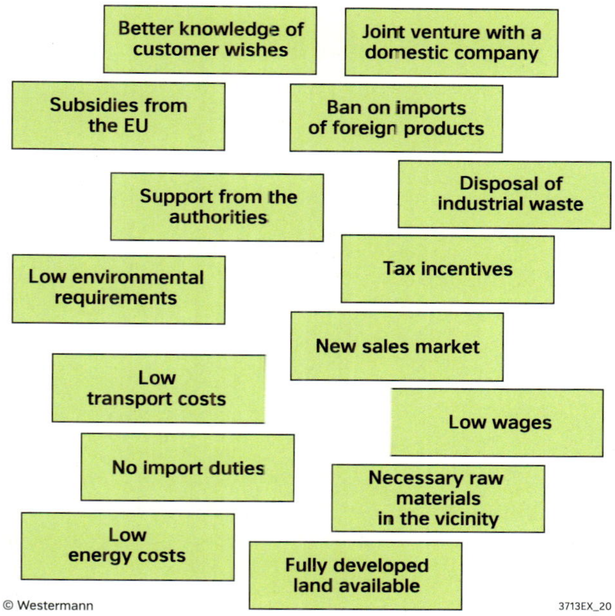

© Westermann 3713EX_20

M6 Location factors for setting up a production facility abroad

4 Name five location factors that are decisive for a car manufacturer to set up a plant in China. Give reasons for your choice (M1, M6).

5 You are an advisor to companies that are searching for suitable locations. Your task is to recommend a suitable location to the companies. Practise your approach:
- Decide on any company of your choice that you would like to advise.
- Make a list of important location factors for this company.
- Research suitable locations.
- Make a well-founded decision about a location for the company. 213
- Give a presentation on the results. 211, 218

The world economy in the process of globalisation

Regions facing the pressure of globalisation, worldwide and in one country – Vietnam

It is said that globalisation and digitalisation are changing the world and everywhere. Is this true? And do they have the same effects everywhere?

1. Describe the different effects of globalisation that have already occured worldwide (M1).
2. Locate two regions that have been globalised to different degrees and illustrate the differences:
   A within America (M1, atlas).
   B within Africa (M1, atlas).
3. Characterise how Vietnam has changed through globalisation since 1995 (M6, M8).
4. Make a podcast (3 minutes) in which you discuss how Vietnam is a country coping with "fragmentation" (M2, M3, M5, M8, M9).
5. Vietnam's economy is booming. Evaluate the role globalisation plays (M2, M6, M8 – 10).

cinnamon	–	Zimt
fragmentation	–	Fragmentierung
poverty line	–	Armutsgrenze
stagnate	–	stagnieren

M2 Boomtown Hanoi – local fragmentation:
Ⓐ International IT experts, web designers and start-ups work in skyscrapers downtown; English is spoken. Next door ...
Ⓑ Overcrowded neighbourhood with poor infrastructure; a modern apartment elsewhere is unaffordable for many locals.

Developments in the economy and society are not the same everywhere. Countries and regions have unequal preconditions. They also compete with one another worldwide. Some locations benefit from globalisation, e.g. through their situation, while others are left behind. Thus, there is fragmentation worldwide (global), between countries (national), within countries (regional) and even within cities (local) into parts or fragments that are globalised to different degrees in economic terms. This is referred to as global, regional or local **fragmentation**. As a result of crises, even strongly globalised subregions can become losers.

M3 Why globalisation does not have the same effect everywhere.

***M1** Global **fragmentation** – economic regions of the world and their significance in the context of globalisation

M4 Cinnamon – tourist attraction (cinnamon festival) and export hit

M5 Life in a village among rice terraces in the mountains

	1995	2019
Inhabitants	75 million.	96 million
GNI per capita	250 US-$	2,540 US-$
Inhabitants below the poverty line	53 %	2 %
Total GDP	21 billion US-$	262 billion US-$
Total imports	8 billion US-$ (47 % raw materials; 27 % machines)	253 billion US-$ (30 % machines, electronics; atomic reactors 9 %)
Total exports	5 billion US-$ (21 % rice, fish; 19 % crude oil; 11 % textiles)	264 billion US-$ (36 % electronics; 14 % textiles, shoes; 7 % rice, fish, coffee)
Employment by sector	I. Sector: 69 %; II. Sector: 13 %; III. Sector: 18 %	I. Sector: 37 %; II. Sector: 28 %; III. Sector: 35 %
Fish production	1.2 million t	7.4 million t
Coffee production	0.25 million t.	1.7 million t
Tourist arrivals	1.4 million	18.0 million.
Foreign direct investments	1.8 billion US-$	16.1 billion US-$

Sources: CEIC, FAO, World Bank

***M6** Data on the development of Vietnam

M7 Aquaculture in the Mekong Delta (here: shrimps) is displacing rice farmers.

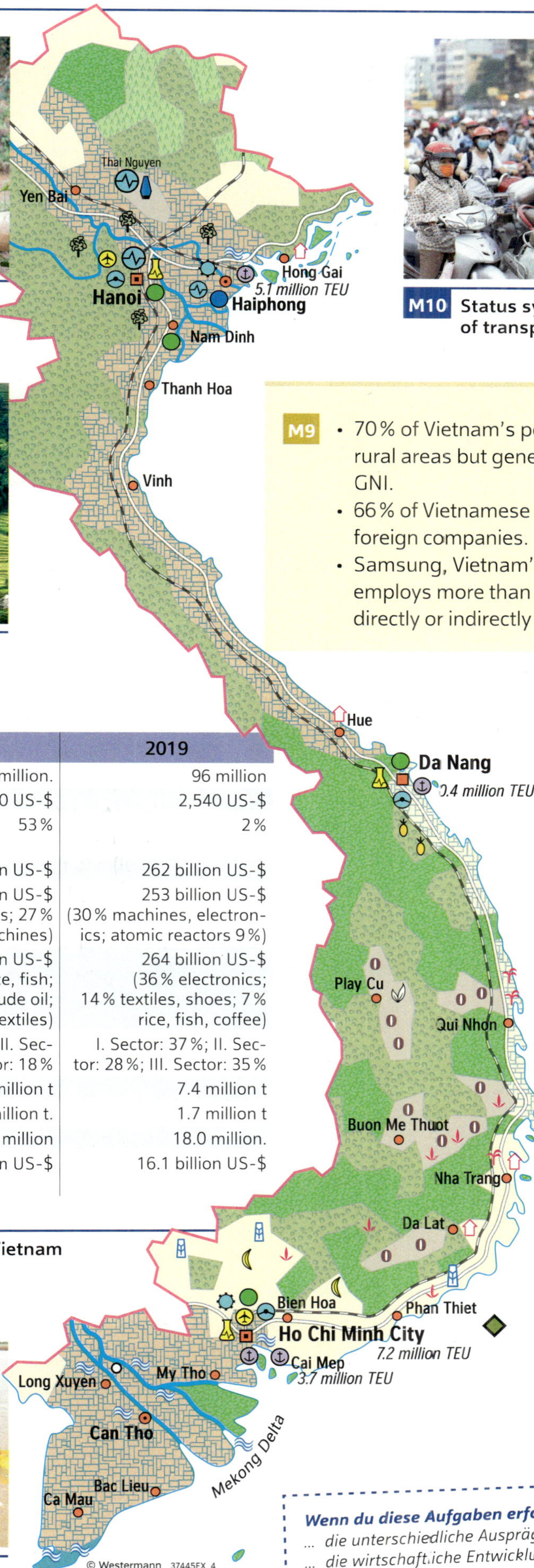

M8 Economic map of Vietnam.

M10 Status symbol and top means of transport

INTERESTING!

M9
- 70 % of Vietnam's population live in rural areas but generate only 20 % of GNI.
- 66 % of Vietnamese exports come from foreign companies.
- Samsung, Vietnam's largest investor, employs more than 160,000 people directly or indirectly in the country.

Wenn du diese Aufgaben erfolgreich bearbeitet hast, kannst du …
… die unterschiedliche Ausprägung der Globalisierung erläutern.
… die wirtschaftliche Entwicklung Vietnams und die Bedeutung der Globalisierung aufzeigen
… die Folgen der Globalisierung kritisch bewerten
*… den Fachbegriff **fragmentation** (Fragmentierung) erklären.*

ORIENTATION

Global cities – the example of Singapore

New York, London, Tokyo, Singapore - they are among the most famous and attractive cities in the world. With globalisation, their influence has continued to grow. What makes such cities stand out as well as important?

1. Describe and explain the distribution of global cities (M4, atlas).

2. A Analyse whether Singapore is an example of a global network of flows of trade, data and people (M2, M4, S. 149).
B Make a chart showing the functions of a global city and the corresponding characteristics of Singapore (M1 – M7).

3. Compare the present-day global cities (M5) with those that have a special future potential concerning location, size, importance (M3, M4, atlas, internet).

4. Research a selected metropolis (M4) and characterise it as a global city (M3, internet). 213

5. The attraction of Singapore has led to a serious shortage of space, so that land is being reclaimed from the sea. Using satellite images (Tuas, Semakau), report where land is being reclaimed and for which purpose (internet). 203, 206

surplus of importance	–	Bedeutungsüberschuss
freight traffic	–	Frachtaufkommen
hub	–	Drehscheibe
innovation capacity	–	Innovationsfähigkeit
passenger traffic	–	Passagieraufkommen
poverty line	–	Armutsgrenze
salary	–	Gehalt
shared accomodation	–	Gemeinschafts-unterkunft
stock exchange	–	Wertpapierbörse
think tank	–	Denkfabrik

Helpful words and phrases for task 3:

... as important/large as ...
Less/more than ...
About/roughly ...
Approximately ...
A majority of ...
A small/high percentage of ...
Most ...

M1 Singapore skyline

- Founded in 1819 as a station for the China to Europe trade route, today a symbol of a rising city through globalisation.
- Youngest global city (5.8 million inhabitants).
- Second largest container port in the world; logistics hub for 25 percent of world trade; Changi Airport has as many passengers as New York (JFK).
- Fourth largest financial centre in the world; diamond and securities exchange.
- Headquarters of 10 of the 100 largest global players; home to many Asian corporate headquarters, computer centres, service providers.
- Representations of international organisations (WTO, ASEAN, IMF, World Bank),
- One of the 10 most visited cities in the world.

M2 Singapore ("Lion City")

Cities are **global cities** if they have an excessive global significance. Some locational factors must be globally outstanding, e.g. innovative capacity or location. Depending on the significance of the functions, they are divided into ranking levels. Global cities in the highest category usually achieve top scores in all four areas:
Economy: global control, e.g. global players via supply chains: which industrial zone in which country, which sales markets, which political contacts or service providers are selected?
Transport and logistics: global transport hub: international shipping and air traffic, trade and goods handling, internet hub.
Finance: international financial centre (stock exchange, bank headquarters, financial services).
Internationality: seat of international political institutions, trade fairs, congresses, world-class culture, unique tourist attractions.

M3 Characteristics of global cities

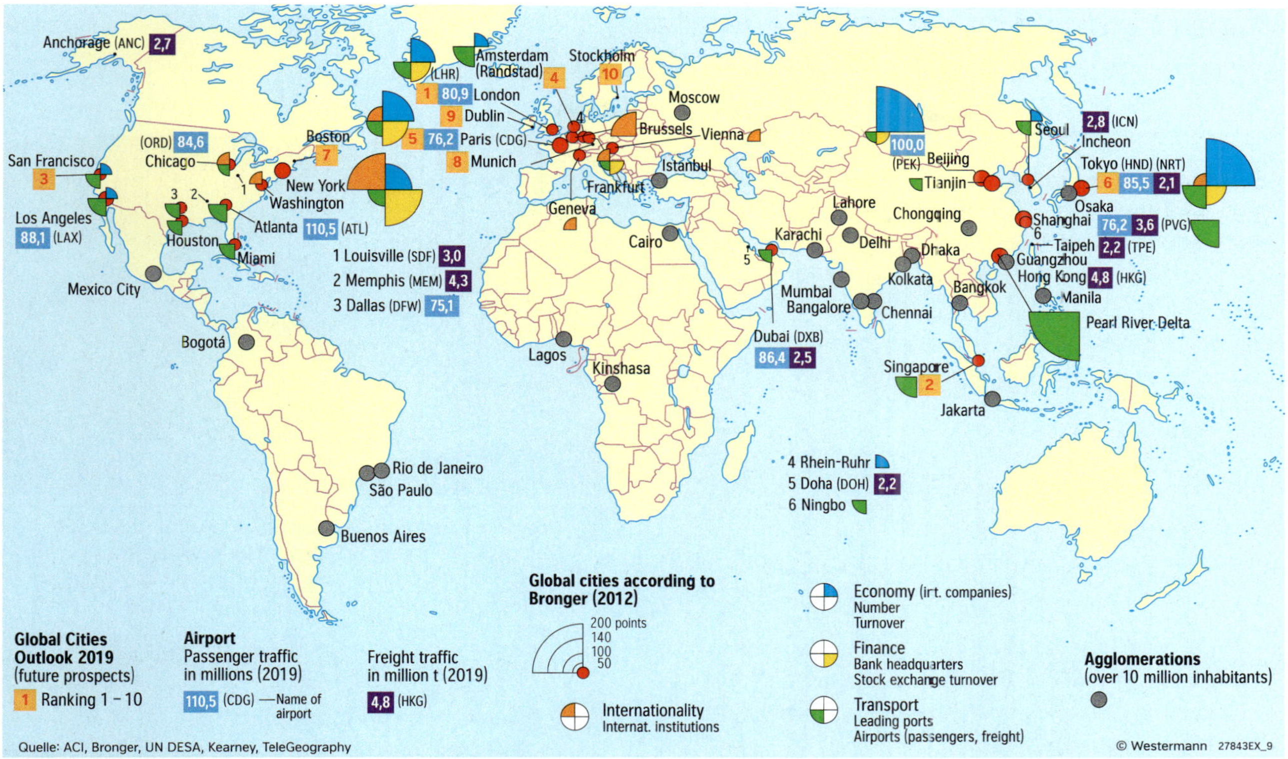

M4 Location of global cities and the largest airports

You have been living in Singapore for 5 years and work for a German company in biotechnology. What is special about Singapore?

Nowhere else do so many employees, 55%, work in knowledge-intensive professions. This is due to international research institutes in biotechnology, AI and measurement technology, as well as the excellent science parks and numerous (elite) universities: Yale, TU Munich, Cambridge, ETH Zurich - all are represented. Singapore is a hub for innovative technologies and a globally sought-after test location.

Do you like living in Singapore?

I like the internationality. In everyday life and in working life. You have to imagine that over 7,000 companies from the EU, the USA and East Asia alone are located here, 1,400 of them from Germany. The quality of life is extremely high. And there are rarely any traffic jams.

A global city without traffic jams?

Yes! The network of bus, underground and transit train routes is an international pioneer. Every place can be reached in record time. There is a smart mobility concept, developed in the think tank Future City, where they are working on sustainable building concepts. The slogan is: Singapore works on solutions for future cities. Clearly, global standards are being set.

That sounds fantastic. But what is the price?! What is the catch?

Only certain sections of the population benefit. Life here is very expensive. We work in research and have high salaries, so we can afford it. But many employees, unqualified support staff, cannot pay that and have been pushed into marginalised areas. 500,000 people live below the poverty line, in some quarters it is more 25 percent.

M5 Interview with Dr. Özlem Sahin, a neuro-informatics expert, on living and working in Singapore

M6 The Merlion (Lion Fish) Fountain

M7 Cheap shared accommodation

Wenn du diese Aufgaben erfolgreich bearbeitet hast, kannst du ...

- ... Global Cities weltweit lokalisieren und ihre Standortqualitäten nach verschiedenen Merkmalen erläutern.
- ... Singapur als Beispiel einer Global City charakterisieren.
- ... den Fachbegriff **Global City** (**Global City**) erklären.

Important places of globalisation: global cities and special economic zones

When we buy a smartphone, a T-shirt or a bicycle, many people in different countries have previously been involved in its production, often in one of the countless special economic zones worldwide.
How do such economic zones operate and what role do they play in the global economy?

1. Describe the global development of special economic zones and their importance for individual countries (M3, M4).
2. List the features of a special economic zone (M1, M2, M5).
3. Ⓒ Write a fictional interview on the following question: "In your opinion, what are the advantages and aims of the Dinh Vu Special Economic Zone in North Vietnam?" with
   A the management of a global player.
   B the Vietnamese Ministry of Trade and Industry (M3, M5, p. 171, M8).
4. Research a selected special economic zone according to various aspects, e.g. position, location factors, cluster focus (M1 – M9). Present your results to the class. 210, 218
5. Special economic zones and global cities are both important locations of globalisation. Note down what they have in common or what makes them different (M6).

cargo turnover	–	Warenumschlag
egress	–	Ausgang
enterprise	–	Unternehmen
executive board	–	Konzernleitung
high wage sector	–	Hochlohnsektor
low wage sector	–	Niedriglohnsektor
networking impulse	–	Vernetzungsimpuls
peel shrimps	–	Garnelen pulen
special economic zone	–	Sonderwirtschaftszone
tax allowences	–	Steuervergünstigungen

Helpful words and phrases for task 3:
to conduct a survey about
Our main aim is ...
What/who/where/why/how ...
Could you please specify your answer.

M1 Working conditions in special economic zones: In Thailand, 14-year-olds from the poor neighbouring country of Myanmar spend 11 hours a day peeling shrimps for the world market. Their wage is 8 euros a day.

INFO

M2 **Special economic zones (SEZs)** are spatially delimited business parks with a special status. They serve to promote the economic development of a country. The state attracts investors with customs and tax allowances, provision of infrastructure, and low environmental regulations and wages. Modern special economic zones specialise in certain functions, e.g. trade and transhipment (often ports) or production purely for export.

Generation of exports in SEZs in %			
Mexico	40	Honduras	60
Guatemala	35	Indonesia	35
Dom. Republic	50	Vietnam	60

Source: UNCTAD

M3 Share of special economic zones in exports (2016)

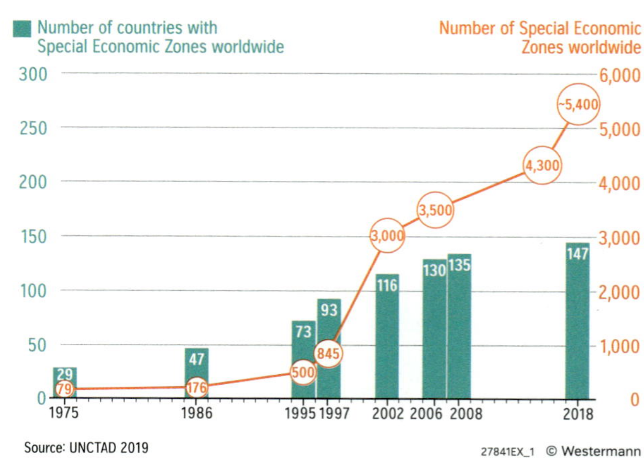

M4 Global importance of special economic zones

Northern spotlight – Dinh Vu Industrial Zone

Hosting over 420 investment projects – representing a cumulative value in excess of US-$ 10 billion – Dihn Vu Industrial Zone (DVIZ) is a well-established option for expansion to northern Vietnam.
In addition to successfully hosting industry champions such as Chevron and Tamada, the completion of Lach Huyen Deep seaport – capable of docking ships up to 100,000 TEU – places DVIZ in direct proximity of northern Vietnam's largest point of sea-based egress. Dinh Vu specializes in petrochemicals and heavy industry. Investors in the park include Petrolimex, Petro Vietnam, Nakashima, and Bridgestone, among others.

Resources for investors

Investors in DVIZ benefit from access to utilities and regulatory assistance that ensure operations function at full capacity. For firms seeking quick alternative sources of production, the ability to procure ready-made factories, and tap into Hai Phong's high skilled labour pools, ensure operations can be started and staffed in a timely manner. Exporters benefit from close proximity to Cat Bi International Airport, the newly constructed Highway 5, and proximity to the Chinese border.

Cost savings instruments

For companies interested in leveraging Vietnam's cost competitiveness to its full potential, unconditional tax holidays present direct cost savings opportunities that can be factored into costs immediately. Within Dihn Vu, all investors are entitled to a four-year tax holiday on Corporate Income Tax, followed by nine years at 50 percent of the going CIT. Furthermore, enterprises exporting 100 percent of their goods to international markets qualify for a 100 percent exemption for import and export duties as well as a free VAT on certain purchases.

Source: Dezan Shira & Associates: Vietnam's Industrial Zones – 3 Locations for Your Business. Vietnam Briefing, www.vietnam-briefing.com, Kowloon, Hong Kong 28.05.2019, accessed: 23.10.2020

M5 **Dinh Vu Special Economic Zone (DVIZ) in Haiphong, North Vietnam**

M7 **The location of Haiphong**

M8 **Handling of goods in the special economic zone in the container port of Haiphong**

INTERNET

M9  You can explore this export region of Vietnam by looking at the websites and satellite images of the special economic zone in Haiphong (Dinh Vu Industrial Zone) and its port.

Special Economic Zones		Global Cities
"Engines" of delivery chains: production locations	**Function for global companies**	Management headquarters, control of goods production
Production location (automation), logistics locations	**Foreign direct investments targeted to**	Administration, research, development
Regional or national impact, means of global networking (participation) within supply chains	**Economic effects**	Global impact, global networking effects
Transport, logistics	**Infrastructure**	Airport, internet hubs
Low wage sector, highly qualified	**Jobs**	High wage sector, executives

M6 **Globalisation locations in comparison – Examples**

Wenn du diese Aufgaben erfolgreich bearbeitet hast, kannst du …
- *… die globale Bedeutung von Sonderwirtschaftszonen und ihr Rolle in der Globalisierung erläutern.*
- *… am Beispiel der Sonderwirtschaftszone Haiphong unterschiedliche Interessen, die mit Sonderwirtschaftszone verbunden sind, ableiten und bewerten.*
- *… den Fachbegriff **special economic zone** (**Sonderwirtschaftszone**) erklären.*

Globalised rainforests – the significance of the palm oil boom

Tropical rainforests represent a unique global ecosystem. They are particularly valuable as the "green lungs" of our planet.
Nevertheless, they are being cleared on a large scale, 70 percent for the production of beef, soy, cocoa and palm oil. Globalisation plays an important role here, as the palm oil boom in Indonesia shows in particular.

Ⓒ **1.** We frequently use products that have their origin in the tropical rainforest.
   A Write them down and sort them by industry as far as possible (M2, M5, M6).
   B Illustrate your personal consumption.

2. Palm oil is a global merchandise. Assign possible producing and consuming countries to the six phases of the supply chain in M2 (e.g. cultivation: Indonesia; M4, M5, p. 147 M7, atlas).

3. a) List the groups of actors who have a direct or indirect economic or ecological interest in the tropical rainforest (M1 – M10).
   b) For each one, add their assumed aims and assess their sustainability.

4. Draw a causal network on the spatial and ecological effects of the global palm oil boom (M3, M6, M9, M10). 199

Ⓐ **5.** Give your personal point of view on the globalisation of the rainforest through palm oil production.

ban – Verbot
be underway – im Gange sein
combustion – Verbrennung
fruit pulp – Fruchtfleisch
merchandise – Handelsware
peaty soil – Torfmoorboden
postpone – verschieben
smallholder – Kleinbauer/-bäuerin

Helpful words and phrases for task 5:
to conduct a survey about
Their main aim is ...
You can't deny that ...
Considering ...

M1 A worker harvesting fruit on a plantation: the palm oil boom has improved his life, because his wages now also pay for his children's education, a moped and a solid roof.

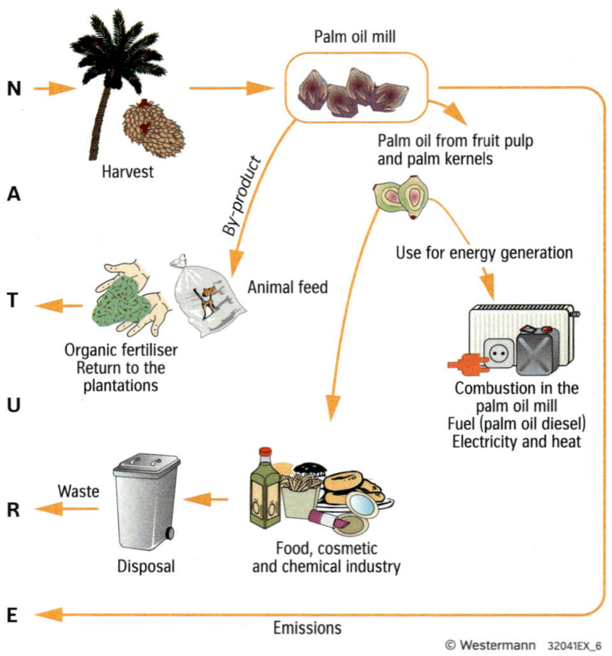

M2 The palm oil production chain: from raw material to consumer

M3 Conservationists rescue orangutans from clearing areas for palm oil plantations. Indonesia and Malaysia are countries with great diversity of species. Many of them count as endangered.

M4 Protest in Sumatra against the EU Parliament's decision to ban palm oil in biofuel from 2021. They fear a loss of sales.

M8 The rainforest ecosystem stores moisture and absorbs CO2 on a global scale. But the area under oil palms has tripled since 1990.

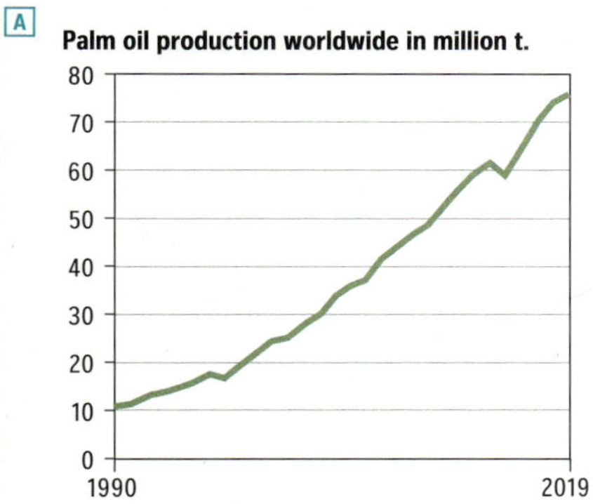

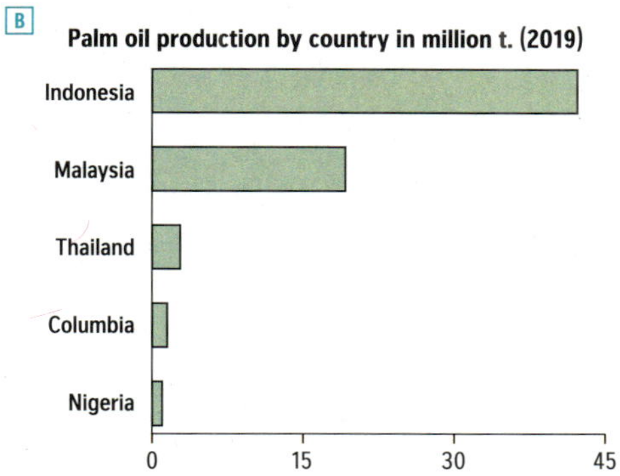

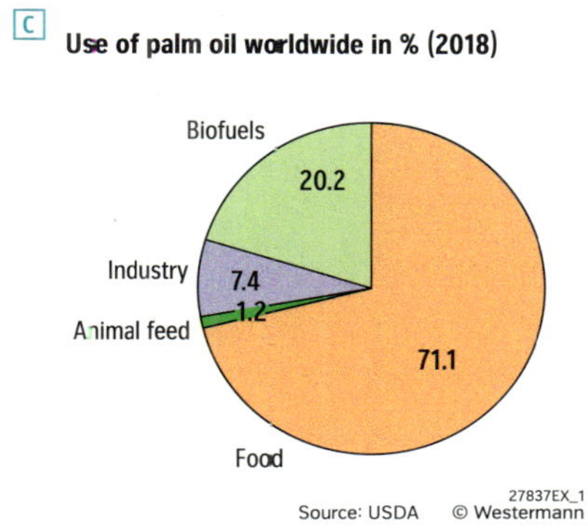

M5 Data on global palm oil production

In the Indonesian province of Papua, a major attack on nature is underway: tens of thousands of hectares of rainforest are to be destroyed for the world's largest palm oil plantation. Palm oil is booming because the world market is hungry for the cheap raw material. [...] The truth is: our consumption leads to forests being cleared for plantations and pastures, rivers being poisoned by mines, landscapes being ploughed up, and indigenous people being robbed of their land. "Globalisation destroys nature" [...].

Source: Rettet den Regenwald e. V., Hamburg: News: Globalisierung – Abholzen und Natur plündern. www.regenwald.org, 27.03.2020, access: 12.11.2020

M6 Protest by the non-governmental organisation Rettet den Regenwald e. V. from Hamburg

INTERNET

M7 Different interests are shown on the websites of the world's largest palm oil company *Wilmar* (Singapore) and *Rettet den Regenwald*.

2002	2005	2010	2015	2016	2017	2018	2019
2.7	3.3	3.3	2.9	6.1	5.0	3.7	3.8

Source: WRI 2020

M9 Worldwide loss of primary rainforest in millions of hectares

INTERESTING!

M10
- Particularly on peaty Indonesian soils, the new establishment of monocultures (slash-and-burn) releases a particularly large amount of carbon, as the soils release up to 6,000 tonnes of CO_2 per hectare.
- Almost 50 percent (2020) of the world's palm oil is produced by smallholders who use it to feed their families. The protest by the smallholders (M6) was successful: the European Union decided to postpone the ban until 2030.

Wenn du diese Aufgaben erfolgreich bearbeitet hast, kannst du ...
... die Bedeutung von Palmöl als globale Handelsware beschreiben.
... erläutern, welche unterschiedlichen Interessen mit der Nutzung des Regenwaldes und der Palmölproduktion verbunden sind.
... verschiedene räumliche und ökologische Auswirkungen des Palmölbooms erläutern und vor dem Hintergrund von Nachhaltigkeit erörtern.

Global networks – only positive?

Globalisation and digitalisation have shaped the development of the world at an unprecedented pace. Have the cross-border flows of people, goods and data always had a positive effect? Are there downsides?

1. Draw a diagram based on M2 and explain it.
2. The downsides are a consequence of the great intensity of global interconnections. Give reasons for
   A a pandemic (e.g. Corona; M1, M2).
   B invasive species (e.g. stink bug; M5 – M7).
   C organised crime (e.g. drugs; M8 – M10).
3. Research and report 213
   A on the Corona pandemic and its global consequences (Internet).
   B on other pandemics (e.g. Spanish flu, Ebola, African swine fever).
   C on another downside (M4, M10).
4. Discuss whether it is a disadvantage to live in countries with limitless flows of data and goods (M6, M10).

abuse of power – Machtmissbrauch
biodiversity – Artenvielfalt
child abuse – Kindesmissbrauch
crate – Kiste
cross-border flow – grenzenlose Ströme
downside – Schattenseite
garlic – Knoblauch
pace – Tempo
roof tile – Dachziegel
marmorated – marmoriert
rape – Vergewaltigung
stowaway – blinde/r Passagier/-in
torture – Folter
trafficking – (illegaler) Handel
stink bug – Baumwanze

Helpful words and phrases for task 4:

On the one hand, ...
On the other hand, ...
In fact, there are ...
advantage/disadvantage
to weigh up different arguments

M1 Consequences of Corona in downtown Los Angeles (2020): tents for homeless people due to job losses

Date	Numbers infected worldwide (totals)	Countries affected
04.01.2019	1	1
22.01.2020	466	6
01.02.2020	12025	25
01.03.2020	87415	65
01.04.2020	873856	177
01.05.2020	3205153	184
01.06.2020	6076390	185
01.10.2020	33892747	186
01.11.2020	46065027	188
01.12.2020	62995994	189

Source: WHO; cases/infections confirmed by the countries themselves

M2 In 2020 the Corona pandemic brought the entire world to a standstill with lockdowns and closures (e.g. curfews, travel and flight bans, border closures, work bans, export freezes at ports).

AMAZING

M3
- Corona pandemic: 2.7 of the world's 3.3 billion workers suffer shutdowns or disruptions to the economy.
- Example of fake news on the internet: "Garlic helps against Corona".

[Economic] globalisation generates spatial inequalities of various kinds, and winners and losers of globalisation can be identified. [In addition], there are the downsides of globalisation, which are connected with issues such as trafficking in human beings, drugs and weapons, the illegal exploitation of raw materials, the abuse of power by transnational corporations, sex tourism, organ trafficking or illegal financial transactions.

Source: Oßenbrügge, Jürgen: Globalisierung. In: PG, 6/2020, p. 7

M4 The limits of limitless interconnections

M5 The brown marmorated stink bug, native to East Asia, is one of the most dangerous invasive species of recent years with no natural enemies in our region.

A shipment of roof tiles in wooden crates - a gift to Zurich from the Chinese twin city of Kunming - brought the first brown marmorated stink bugs to Switzerland as stowaways in 1998. Through the transport of goods, they multiplied explosively from 2000 onwards from Switzerland to other countries in Europe, the USA and Canada. They are not choosy; they can survive on balconies, in boxes, cars, apartment buildings, containers. They lay eggs twice in warm summers. Now, cars are heat-treated before they are transported by ship to protect them from such invasions. The stink bug uses about 200 host plants. Farmers suffer damages costing millions; often entire harvests are destroyed, especially pears or peppers. Control is extremely difficult because the pest infests so many plants and is very robust.

M6 Globalised nature – the brown marmorated stink bug

INFO

M7 Invasive species are usually brought unintentionally by human action from their natural environment into a foreign area and manage to survive there. They are invasive when their massive spread threatens native biodiversity and ecosystems. Example species:
Animals: Spanish slug (Spanische Wegschnecke), Asian ladybeetle (Asiatischer Marienkäfer), brown rat (Wanderratte);
Plants: Ragweed (Ambrosie), waterweed (Wasserpest)

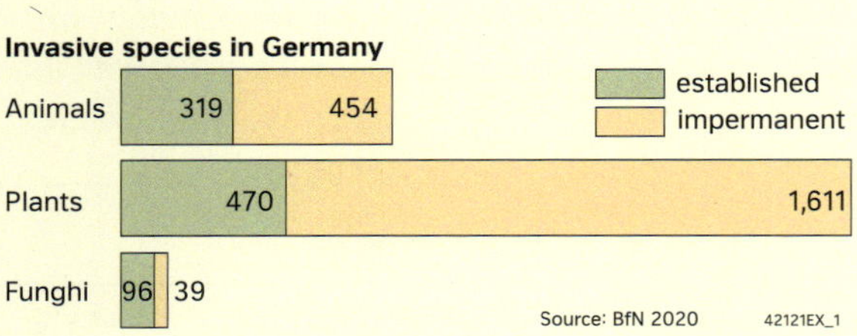

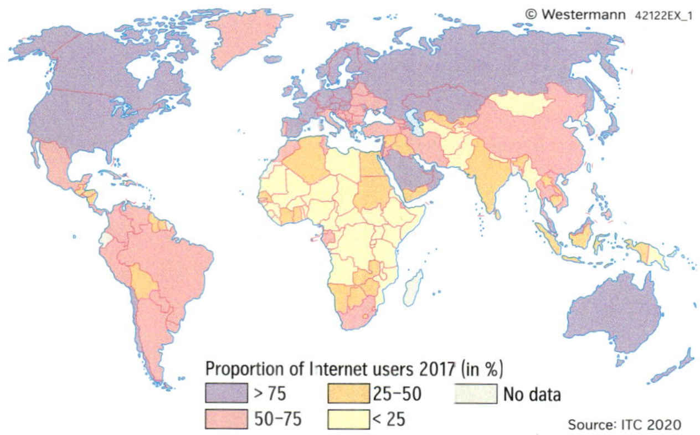

M8 Global internet use: More than half of all people (3.9 billion, 2019) can be influenced by data streams or can (mis)use them without limits.

***M9** Organised crime in a former bunker near Traben-Trarbach: on 5 floors, 25 metres deep, with a powerful computer centre, the site served as a global internet node in the Darknet until 2019. All but one site was illegal. It was used for right-wing extremist agitation, weapon and drug sales, and child pornography. One drug platform alone generated millions in annual sales.

M10 Worldwide, hundreds of thousands of content moderators delete posts, photos and videos around the clock that spread or show torture, murder, rape, and child abuse, for example. There is a focus of activity on the Philippines. More than 1,000 content moderators are working in Germany, e.g. in Essen. The psychological strain is enormous.

Wenn du diese Aufgaben erfolgreich bearbeitet hast, kannst du ...
- *... an Beispielen anführen, wie grenzenlose Vernetzungen über den Transport von Waren, Daten und Menschen negative Auswirkungen für viele Menschen und Länder haben können.*
- *... die Tragweite dieser Vernetzungen und die negativen Seiten der Globalisierung bewerten und eine persönliche Position beziehen.*

Does globalisation make life on earth better? – A debate about winners and losers

The effects of globalisation are a subject of great controversy. Often it is a question of perspective. How are the effects of globalisation to be assessed?

1. A Conduct a debate in your class on the question of whether globalisation makes life on earth better. Use the materials on this and the other pages in the chapter as well as a dictionary and the internet if necessary. Consider the fixed rules of a debate. 220

B Make a chart showing the winners and losers of globalisation. Complete each entry with an argument. Use the materials on this and the other pages in the chapter as well as a dictionary and the internet if necessary. Finally, note down your position on the key question. 218

Many African villages have low-cost Chinese solar power plants. Power supply enables new ways of living and doing business. Charged smartphones promote communication and information, refrigeration technology and digital work create new jobs.

M1 Networking in every corner of the earth (Photo: Solar cells in a village in Zambia)

"We need the investments. Our agriculture is not intensive and productive enough. The investors bring technology and jobs, capital and know-how come into the country," says a state agency that offers land for sale. At the same time, famine has been worsening in East Africa since 2017. After several years of drought, East Africa's biggest locust plague in 25 years and the Corona pandemic further aggravated the situation in 2020. More than 20 million people are threatened by hunger.

M2 Use of agricultural land in Ethiopia (Photo: Workers tend to young plants on a plantation in Ethiopia. They are employed by a big indian company.)

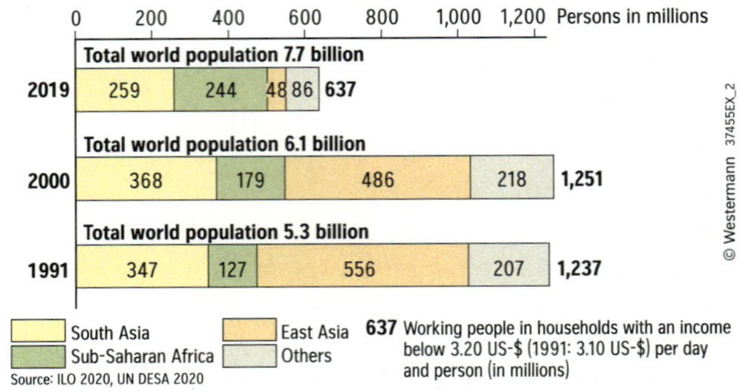

***M3 Working people who have less than 3.20 US dollars a day to live on: in poverty despite working**

DIE ZEIT: *Is humanity better or worse off today than it was two decades ago?*
Winnie Byanyima: *It is better today. Millions of people have escaped poverty, so the world has become better for them.*
DIE ZEIT: *As a critic of globalisation, you concede that globalisation works?*
Winnie Byanyima: *We are not critics of globalisation. We want a different globalisation, because even today one in nine people still goes to sleep beeing-hungry. One in ten has less than two dollars a day to live on, which is the official poverty line. And [...] you should try to live on two dollars a day. That is pretty miserable [...].*

Source: Schieritz, Mark und Pinzler, Petra: „Die Welt wird immer instabiler". www.zeit.de, Hamburg, 02.03.2017, accessed: 28.03.2018.

M4 Interview with Winnie Byanyima from Uganda, politician and head of Oxfam (NGO)

Every night, two trucks from textile factories in Borken (Westphalia) transport loads of linen back and forth to Poland. The company has outsourced the labour-intensive packing of the linen, which used to be done in Borken, to Poland. Wages are much lower in a Polish special economic zone; the headquarters in Borken focuses on administration and design as well as the development of high-quality work processes and the production of special textiles.

M5 Globalised textile industry (example) – effects in Poland and Germany

"Living sustainably in a globalised world without exploiting nature and people: This is the great challenge of the 21st century. Because our prosperity is far too often bought with the misery of others.
It starts in the morning when we take a shower: Almost every shampoo contains palm oil. Rainforests in Indonesia and Malaysia are cleared for the huge plantations - worldwide, the size of a football field every four seconds. Eleven percent of CO_2 *emissions are due to the destruction of forests!*
This continues when we get dressed: 90 percent of our clothes come from Southeast Asia. Often this means 16-hour shifts in stuffy factories, dismissal in case of pregnancy or illness, and starvation wages.
Or at breakfast: children work on many coffee plantations. A kilo of coffee costs ten to twelve euros in Germany. But only 50 cents of that goes to the farmers. The families cannot live on that! That is why the children have to work, 2.3 million in West Africa alone. The list could be continued indefinitely: Our mobile phones, e-bikes and cars contain

Child labour in tanneries in the Hazaribagh district of Dhaka, Bangladesh. The children often work with toxic chemicals without protective clothing.

cobalt. 60 percent of it comes from Congo, where forced labour and environmental destruction in illegal mines are commonplace. We would rightly never accept such conditions in Europe. Why elsewhere? That is why I fight for sustainable supply chains. [...] Now we are called upon to enforce minimum social and ecological standards worldwide, especially the ban on child labour. Consuming sustainably is an important first step in this."

Source: Müller, Gerd: „Ich kämpfe für nachhaltige Lieferketten". In: Green Lifestyle 2020, Media Planet, Berlin, 12/2019, p. 2, translated by: John Atfield

M6 **Dr. Gerd Müller (Federal Minister for Economic Cooperation and Development from 2013 to 2021) on sustainability in a globalised world**

Alibaba was founded in 1999 by Jack Ma as a digital trading platform. Today, the Chinese technology group operates the largest global trading platform and invests in cloud computing and AI technology. It is building its own logistics centres in Germany. Alibaba employs 117,600 people and is the world's seventh-largest technology company in terms of market capitalisation (market capitalisation June 2020: US-$577 billion).

The technology group *Facebook*, which also owns *WhatsApp*, *Instagram* und *Oculus*, was founded by Mark Zuckerberg in Silicon Valley in 2004. The company's services were used by over three billion people and more than 180 million companies worldwide in 2020. The company employs 52,534 people worldwide and operates 17 computer centres of its own. It has a market capitalisation of US-$629 billion as of June 2020. The company made a turnover of 70.7 billion US dollars in 2019 (in comparison the GDP of Luxembourg is 44.1 billion US dollars).

M7 **Importance of technology companies (global players) in globalisation**

The diagnosis made by the Munich sociologist Stephan Lessenich is quite radical:
The citizens of the North live at the expense of the people in the South, he says. They simply outsource the dirty parts of production to them. For us, "resources are extracted, toxins released, waste stored, areas of land devastated, social spaces destroyed, people killed," he accuses consumers, "our abundance robs others of their livelihoods."

Source: Jung, Alexander: Selbstbetrug mit System. In: Der Spiegel, Hamburg, 27/2017, p. 16.

M8 **Globalisation in the northern and southern hemispheres: Professor Stephan Lessenich (inequality researcher)**

Wenn du diese Aufgaben erfolgreich bearbeitet hast, kannst du ...

- *... auf globaler und persönlicher Ebene sowie aus verschiedenen Blickwinkeln die Auswirkungen der Globalisierung beschreiben, erörtern und als Gewinn oder Verlust bewerten.*
- *... ein persönliches Urteil zum Erfolg der Globalisierung begründet und differenziert abgeben.*

Europe – unity is strength

The European Union (EU) is one of the biggest players on the world market, along with the USA and China. What challenges does globalisation pose for the EU? How will these be mastered?

1. Describe the importance of the EU in the global trade in goods (M2, M3, M10, atlas).
2. a) Using the example of Airbus, describe the advantages of a joint project in the EU (M8, M9).
   b) Discuss which disadvantages are also associated with joint projects (atlas, map: "Europäischer Produktionsverbund A380"). 201
3. Explain how companies in the EU are protected from unfair competition (M4).
4. Show the benefits that the EU's common trade policy brings for companies in the Member States with the help of examples (M2, M5, internet). 214
5. Not only can the EU impose punitive tariffs, but its trading partners can do so as well. Discuss the impact of "trade wars" on the EU.
6. (A) "Europe is an economic giant, a political dwarf, and a military worm." Comment on this statement by the firmer Belgian Foreign Minister Mark Eysken.

competitive – wettbewerbsfähig
consumer protection rule – Verbraucherschutzvorschrift
dwarf – Zwerg
competition – Wettbewerb
joint project – Gemeinschaftsprojekt
punitive tariff – Strafzoll
single internal market – gemeinsamer Binnenmarkt
subsidised – subventioniert
tariff – Zoll / Abgabe
trade barrier – Handelsbarriere
trade policy – Handelspolitik
wide-body aircraft – Großraumflugzeug

Helpful words and phrases for task 2:

The figures indicate a/an/that …
It becomes clear that …
My findings show that …
Because of …
That is why …

M1 Session of the European Parliament in Strasbourg

The central engine of the EU is the single internal market, which is one of the largest interconnected economic areas in the world. Goods within the EU are duty-free.
As a trading partner, the EU can act more effectively than any individual state could, because it has greater influence in negotiations and a more significant position in the **World Trade Organisation (WTO)**.
European companies benefit from the EU's trade policy. Trade barriers such as tariffs and quotas could be removed, giving companies better opportunities to compete globally. Conversely, the EU can impose tariffs on goods from other countries to protect the single internal market. This is one way to prevent dumping, for example. Dumping is a form of unfair competition. Products are offered at prices below their actual value. European companies would not be able to compete with this.

M2 The single internal market of the EU

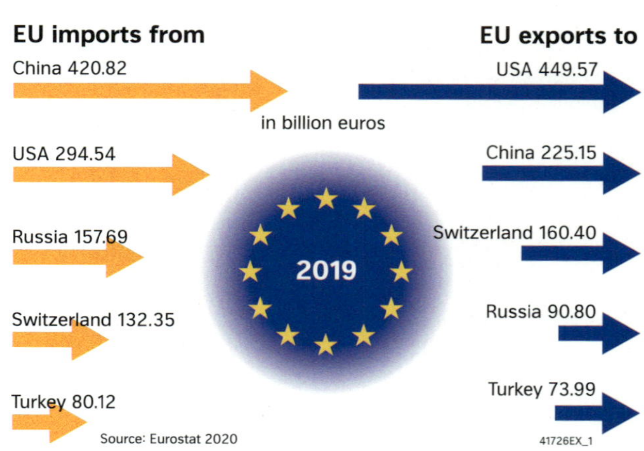

M3 The EU's largest trading partners in goods (2018)

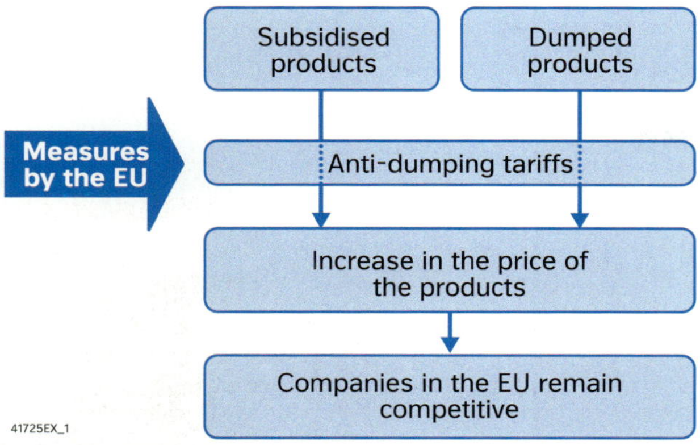

M4 **Options for the EU to ensure fair competition. Dumped products in 2016 included aluminium, bicycles, cement, chemicals, ceramics, glass, paper, solar panels and steel from China.**

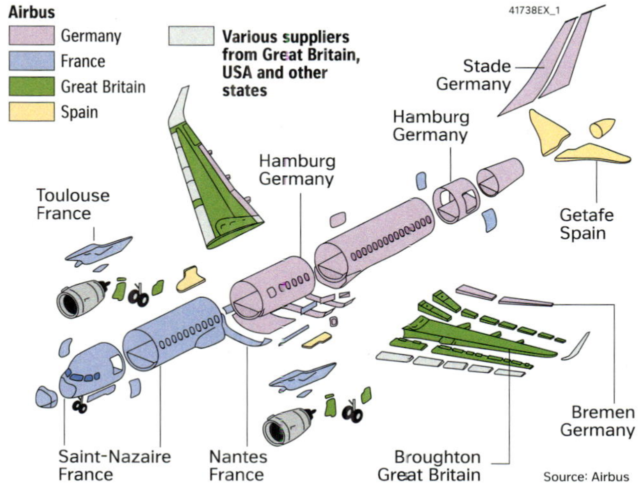

M8 **The Airbus – a joint project**

The EU's trade policy is essential for the European economy as it affects growth and employment. More than 36 million jobs in the EU depend on exports beyond its borders. On average, every €1 billion exports to non-EU countries supports more than 13,000 EU jobs. The EU's trade policy also protects Europeans by ensuring imports respecting consumer protection rules.

Source: European Parliament (ed.): Making the most of globalisation: EU trade policy explained. www.europarl.europa.de, Berlin, 03.06.2019, accessed: 25.11.2020

M5 **Strong together – in foreign trade**

INTERNET

M6  You can find examples of the trade policy of the EU here:
→ Handelspolitik der EU

The Airbus company was founded in 1970. The aim was to build a European wide-body aircraft and thus become a competitor to the American aircraft manufacturer Boeing. The planning, development and production of a wide-body aircraft swallows up enormous sums of money. As a joint project, several countries shared the costs and the risk. Sixteen factories were built in Germany, France, Great Britain and Spain. The individual parts are manufactured here. In addition, parts come from 6,000 subcontractors. Final assembly takes place at several locations, depending on the type of aircraft. Airbus initially won the competition with Boeing. However, both companies had to struggle with the effects of the Corona crisis. Aircraft were ordered, but then their delivery was not accepted.

M9 **The Airbus – a joint project**

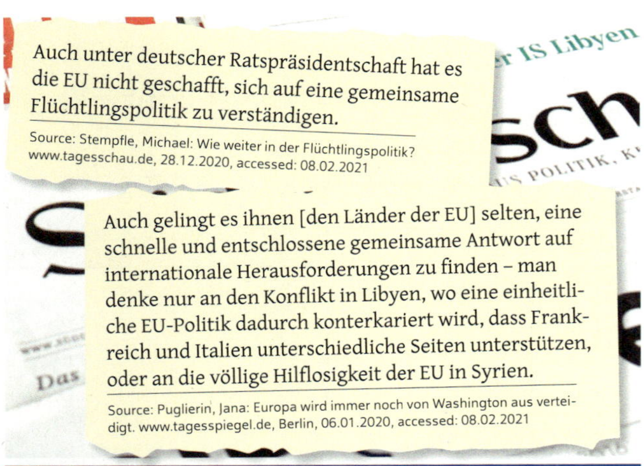
Auch unter deutscher Ratspräsidentschaft hat es die EU nicht geschafft, sich auf eine gemeinsame Flüchtlingspolitik zu verständigen.

Source: Stempfle, Michael: Wie weiter in der Flüchtlingspolitik? www.tagesschau.de, 28.12.2020, accessed: 08.02.2021

Auch gelingt es ihnen [den Länder der EU] selten, eine schnelle und entschlossene gemeinsame Antwort auf internationale Herausforderungen zu finden – man denke nur an den Konflikt in Libyen, wo eine einheitliche EU-Politik dadurch konterkariert wird, dass Frankreich und Italien unterschiedliche Seiten unterstützen, oder an die völlige Hilflosigkeit der EU in Syrien.

Source: Puglierin, Jana: Europa wird immer noch von Washington aus verteidigt. www.tagesspiegel.de, Berlin, 06.01.2020, accessed: 08.02.2021

M7 **Gemeinsamkeit – ohne Probleme?**

INFO

M10 The **single internal market** of the **European Union (EU)** removes all barriers for people, goods, services and capital at country borders. The EU single market is comparable to a single large country in which trade can take place unhindered.

Wenn du diese Aufgaben erfolgreich bearbeitet hast, kannst du …

- *… die Bedeutung der EU im weltweiten Handel erläutern.*
- *… die Vorteile einer gemeinsamen Handelspolitik erklären.*
- *… positive und negative Auswirkungen der Globalisierung auf die EU erörtern.*
- *… die Fachbegriffe **common market** (**gemeinsamer Binnenmarkt**), **European Union (EU)** (**Europäische Union**) und **World Trade Organisation (WTO)** (**Welthandelsorganisation**) erklären.*

Changes due to globalisation and digitalisation – Dresden

Digitalisation is changing the demands on space, its use and thus also the landscape. This can be shown using the example of Dresden. How does digitalisation affect the region around Dresden? Which framework conditions and infrastructures were decisive in turning Dresden into a business location of global importance?

1. List location factors for international companies in the high-tech industry in Dresden (M1, M5, M8, atlas).
2. To what extent is the Dresden area a high-tech cluster? Explain (M1, M8).
3. "Dresden is booming" (M3). Assess this statement with the help of M6 and M7.
4. (C) Evaluate Dresden's importance as a business location, also with regard to future development.
   A Prepare a presentation in the role of an employee of the Office for Economic Development. 211
   B Write a report for a newspaper.
5. Prepare a presentation on the development of Saxony using the data and presentation possibilities of Eurostat (M4, M8, internet). 212
6. Using the example of Dresden, discuss the effects of globalisation and digitalisation on locations, companies and employees.

construction	–	Baugewerbe
economic development	–	Wirtschaftsförderung
educational facility	–	Bildungseinrichtung
manufacturing industry	–	verarbeitendes Gewerbe
promotion	–	Förderung
provision of services	–	Erbringung von Dienstleistungen
scientific facility	–	Wissenschaftseinrichtung
warehousing	–	Lagerei

Helpful words and phrases for task 6:

Concerning ...
As far as ... are/is concerned, ...
From ... point of view/perspective, ...
... might say that ...

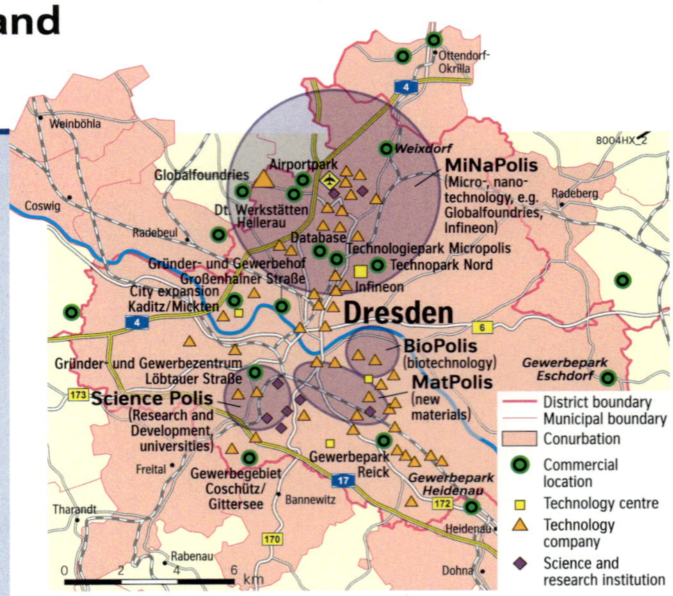

M1 Dresden: development of clusters

"Orient globally - act locally" is the motto of the Office for Economic Development of the state capital Dresden. It is working specifically towards the internationally competitive and sustainable development of business, education and research in Dresden. The modernisation of infrastructure and the promotion of start-up and technology centres are also part of this.

M2 Goals of economic development in Dresden

Dresden is booming. [...] A future vision for sustainable living in an open city is currently being created under the "Open City Dresden – Shared responsibility for sustainable urban development" agenda, which focuses on social and economic development as well as the issues of mobility, energy efficiency and climate protection. Future generations will also benefit from smart solutions like the optimally networked "Smart City". By investing intensively in urban infrastructure and a start-up friendly environment, Dresden is creating the best conditions for ensuring new, innovative businesses keep establishing themselves in the city to encourage and contribute to the structural changes.

Source: Landeshauptstadt Dresden: Dresden. The best decision. Dresden 2017, p. 18

M3 Dresden also promotes its business location internationally.

INTERNET

M4  *Eurostat* provides extensive data on Saxony and Dresden.

M8 Locations of many high-tech companies in Saxony.

- Technology centres, technoparks
- Incubators for start-ups
- Cooperation with universities and research institutions
- Cooperation between the University of Applied Sciences Zittau/Görlitz and the Society for Knowledge and Technology Transfer in Dresden (GWT-TUD)
- DRESDEN-concept (networking of 24 Dresden institutions, including the TU Dresden, the Fraunhofer and Max Planck Institutes, the institutes of the Leibniz Association and the Helmholtz Centre Dresden-Rossendorf)
- Silicon Saxony e. V.
- Location of all six high-tech industries (M2). Throughout Europe, only in Dresden are of these key technologies are represented.

M5 High-tech clusters in Dresden

	Inhabitants	Employees	Inward commuters	Outward commuters	Unemployment rate in %
2000	477807	215610	73845	33793	–
2005	495181	205905	78563	35165	16,4
2010	523058	224364	85169	43398	12,0
2015	543825	246809	89227	51744	7,6
2019	556780	273544	96812	56466	5,3

Source: Landeshauptstadt Dresden, various years

M6 Data about Dresden

M9 Office and commercial building in Dresden's city centre, one of 90 world trade centres recognised by the World Trade Centers Association, including educational and scientific facilities.

Economic sector	Companies	Employees	Development of employment 2013–2018 in %
Manufacturing industry	1123	32212	+4,4
Construction	2371	11190	+4,4
Trade, transport, warehousing	4415	40498	+10,4 (incl. hotels)
Information and communication	1311	13755	+37,1
Provision of services	6760	50666	+14,6

Source: Statistisches Landesamt Sachsen, Kommunale Statistikstelle, Bundesagentur für Arbeit

M7 Companies and employees by economic sector (2018)

Wenn du diese Aufgaben erfolgreich bearbeitet hast, kannst du ...

... den Wirtschaftsstandort Dresden beschreiben.
... die Entwicklung des Wirtschaftsstandorts zum Hightech-Cluster auch in Bezug auf die zukünftige Entwicklung beurteilen.
... die Auswirkungen von Globalisierung und Digitalisierung auf den Wirtschaftsstandort Dresden erörtern.

Spatial Disparities in the EU – Catalonia and Extremadura

The EU is a unified economic area, but there are great disparities – both between the member states and within the individual states.
Where are the structurally weak and the economically strong regions? How do they differ from each other?

1. Describe Barcelona in Catalonia and the village in Extremadura (M2, M3).
2. Compare Catalonia with Extremadura (M1 – M3, atlas). 217
3. Note down possible reasons why Extremadura can be described as a structurally weak region (M1, M3, atlas).
4. (C) Compare the disparities between the sub-regions in Spain with those in other countries
   A using the atlas.
   B using Eurostat data. 212
   C using Diercke WebGIS. 204
5. Relate Catalonia and Extramadura to the Blue Banana model (M4).

deficiency	–	Defizit
economic performance	–	Wirtschaftsleistung
structurally strong	–	strukturstark
structurally weak	–	strukturschwach

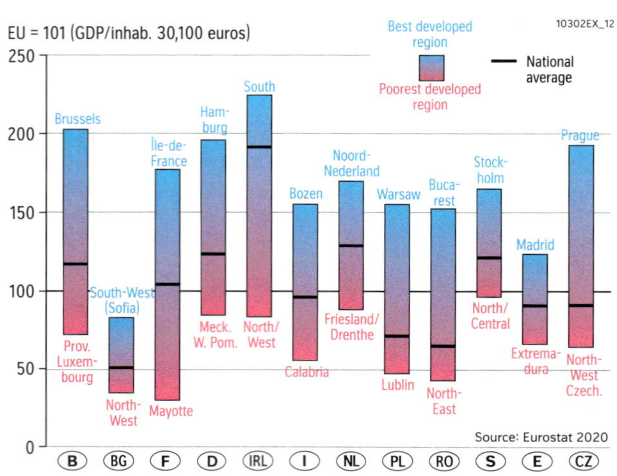

***M1 Regions with differing levels of development in selected EU countries (2018)** 217

- Population (01.01.2019): 7,672,699 inhabitants
- Population density: 238.97 inh./km² (16.32 %)*
- Contribution to national GDP: 20 %
- GDP per capita (2018): 30,769 euros
- Share of economic sectors in GDP (2018): services 74.18 %, industry 19.72%, construction 5.17 %, agriculture 0.93 %
- Share of small enterprises with less than ten employees: 38.12 %
- Unemployment rate (2019): 10.87 %
- Employees by economic sector (2019): services 75.00 %, industry 17.90 %, construction 5.52 %, agriculture 1.58 %
- Main industries: automotive industry, metal and mechanical engineering, chemical industry

*Share of total Spanish population
Source: EURES 2020, Statistical Institute of Catalonia 2020

M2 Catalonia, a structurally strong region with the economic centre of Barcelona

- Population (01.01.2019): 1,066,998 inhabitants
- Population density: 25.63 inhabitants/km²
- Contribution to national GDP (2018): 1.6 %
- GDP/capita (2018): 18,174 euros (25,854 euros)*
- Share of economic sectors in GDP (2018): services 69.05 %, industry 14.50 %, construction 7.86 %, agriculture 8.59 %
- Share of small enterprises with less than ten employees (2018): 96.57 %
- Unemployment rate (2019): 19.68 % (13.92 %)*
- Employees by economic sector (2019): services 69.89 %, agriculture 12.85 %, industry 11.13 %, construction 6.14 %
- Main industry: energy sector

*Average for Spain Source: EURES 2020

M3 Extremadura, a structurally weak region

ORIENTATION

Blue Banana

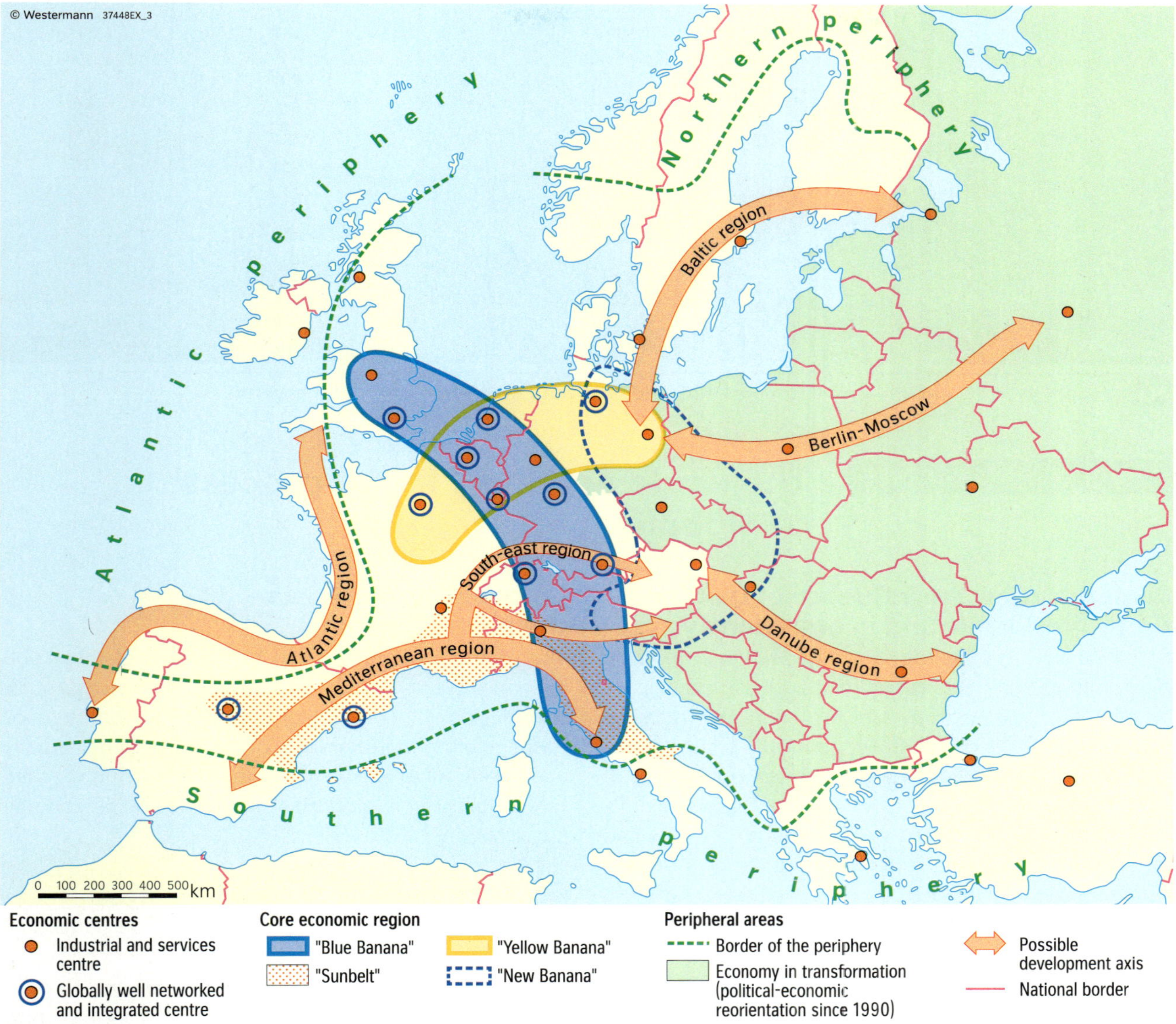

***M4** **Model of spatial disparities in Europe** 210

In 1989, the French scientist Roger Brunet mapped his research results on **spatial disparities** in Europe in order to point out deficiencies in **regional policy**. He used a model to illustrate which regions of Europe have high or low economic power, and what future developments can be assumed from this. He depicted the core economic zone of Europe as a blue banana. The most economically **"active" spaces** with important industrial enterprises, a high share of the total EU population, good infrastructure, a wide range of educational opportunities and a good supply of skilled labour on the labour markets, were concentrated in this zone. The regions with low economic performance were often **"passive" spaces** in rural areas and **structurally weak regions**. His model was further developed by other researchers and adapted to more recent developments. To counteract disparities, the EU regional policy pursues the concept of polycentric development. This means that the EU's funding measures are targeted at many centres.

M5 **The "Blue Banana" spatial model** 210

Wenn du diese Aufgaben erfolgreich bearbeitet hast, kannst du ...

- ... die Lage von strukturstarken und strukturschwachen Ländern und Regionen in der EU beschreiben.
- ... Entwicklungsunterschiede erklären.
- ... ein Raummodell fragegeleitet auswerten.
- ... die Fachbegriffe **spatial disparities** (**räumliche Disparitäten**), **regional policy** (**Regionalpolitik**), **active space – structurally strong region** (**Aktivraum**), **passive space** (**Passivraum**), **structurally weak region** (**strukturschwache Region**) erklären.

M1 Global City (of) London – world economic hub and financial centre of Europe and the world

WHAT WE HAVE LEARNED

The world economy in the process of globalisation

Regions under the influence of globalisation – worldwide and in one country

170 Globalisation shapes the world to varying degrees. Not all areas have the same impact or can participate in the same way. Yet they are often right next to each other. They have different economic and social preconditions. This leads to **fragmentation**, globally (e.g. USA – Africa), regionally (e.g. Vietnam: Hanoi region – mountainous region in the north) or locally (e.g. within Hanoi: banking district – slums next door). With the simultaneous coexistence of spaces that are of different significance for globalisation, the contrasts intensify.

Global cities – the example of Singapore

172 **Global cities**, such as Singapore, are future-oriented world cities and globally outstanding locations. They control the world economy, e.g. global players with their supply chains. But they are also global hubs for data, financial flows, logistics and transport networks (shipping and air traffic). Many top-ranking cultural and tourist facilities add to their dynamics and attractiveness.

Special economic zones as important locations for globalisation

174 In countless **special economic zones** worldwide, sometimes called workbenches of the world, foreign investors produce for the world market at attractive special conditions (low wages, tax advantages, low environmental standards). In Vietnam, for example, there are global players like Samsung, Siemens, Toyota. Economic sectors in many countries have become dependent on foreign investors.

Globalised rainforests – the palm oil boom

The global economy is also taking hold of the global rainforest ecosystem, which as a "green lung" binds carbon from CO_2 and releases oxygen. Global demand for palm oil, with its versatile uses, is rising enormously, as is palm oil production - at the price of further rainforest destruction, global warming, and socio-economic sustainability. 176

Global networks – only positive?

The rapid progress of networking – global and digital – has also produced its downsides. Flows of people across borders can spread viruses explosively through travel, for example, so that pandemics (e.g. Corona) can bring the networked world economy to a standstill and claim many lives. The intensive exchange of goods repeatedly causes economic and ecological damage through invasive animal species, and with unlimited data traffic, parallel worlds on the internet (Darknet) generate organised crime. 178

Does globalisation make life on earth better?

Whether a region or a person is a winner or loser of globalisation depends on one's perspective. Globally, the proportion of people in poverty has decreased since 1990, but the contrasts between rich and poor, the powerful and the weak, have increased. Technological companies like Facebook are seen as the big winners, while low-wage workers are often seen as the losers. 180

Europe – unity is strength

182 For the member states of the EU, the **common single internal market** is an advantage. Goods are duty-free within the EU, and as a trading partner the EU carries more weight than a single state. Together, trade barriers can be removed, and dumping can be resisted. The EU also has an important position in the **World Trade Organisation (WTO)**.

Changes due to globalisation and digitalisation – Dresden

184 Dresden has developed into an important business location, the high-tech cluster of Silicon Saxony. This has been achieved through the networking of global players, technology centres, incubators for start-ups, universities, research institutes and business development institutions.

Spatial disparities in the EU – Catalonia and Extremadura

186 In the EU, there are economically **active spaces** and **passive spaces**, which are depicted, for example, in the Blue Banana spatial model. In Spain, Extremadura is an example of a **structurally weak region**, while Catalonia is an example of a **structurally strong region**.

1 a) Name characteristics of a global city.
b) Choose three characteristics and check whether they apply to London (M1, Internet).

2 Interpret the map (M4). Use expressions like "borderless global networking", "winners and losers". 201

3 a) Draw a graph similar to the one in M3. Evaluate example [A] in M2. For this exercise, enter the evaluation from the perspective of Vietnam and from the perspective of German companies on each ray between plus and minus (use two colours).
b) Give your final evaluation of the situation (M2).
c) Analyse example [B] in M2. Draw a graph for each of the different interest groups, e.g. the Nestlé Group, the country of Malaysia.

[A] Vietnam's textile sector is growing, export figures to Europe are increasing. In order to produce more cheaply and efficiently, Vietnam imports more and more textile machines and robots. Thus, Vietnam has grown to become the fourth most important textile machinery importer in the world. Important suppliers come from Europe, including Germany and Italy. German exports here rose by 41 percent in 2018 (to 160 million US dollars), while Italian exports rose by 18 percent (to 100 million US dollars).
[B] The Nestlé Group is a buyer of large quantities of palm oil. With 1,735 oil mills in 24 countries around the equator, Nestlé uses 455,000 tonnes of palm oil (2019). In the process, about 150,000 hectares of rainforest are cleared annually for new plantations. Europe is an important sales market.

M2 Europe as an important global trading partner – two examples

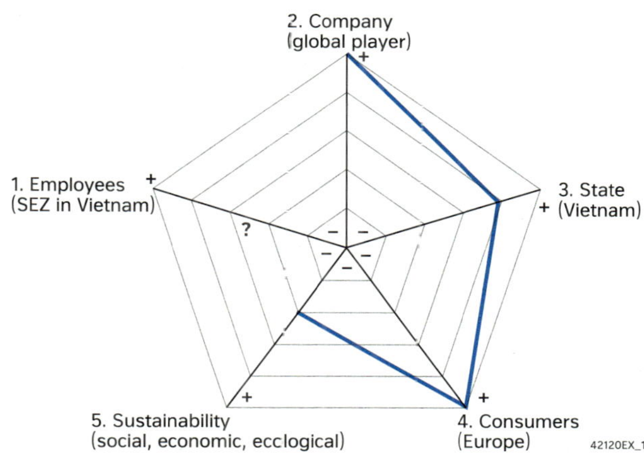

Note: to assess other questions, the parameters may be changed or exchanged, e.g. country of Italy, workers in India.

M3 Evaluation from different perspectives – example: Vietnam's textile exports are increasing (value network – diagram).

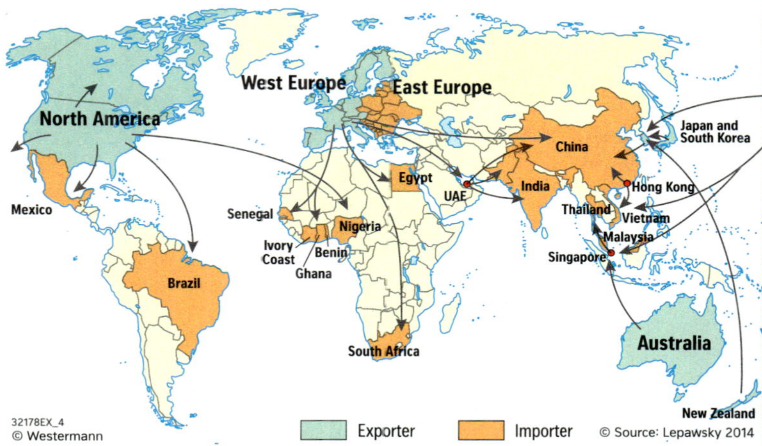

M4 Flows of illegally shipped electronic waste by region worldwide

Apply and practise

PROJEKT VOR ORT

Der Globalisierung auf der Spur – eine Erkundung

An vielen Stellen bestimmen Globalisierung und Digitalisierung jeden Tag unser persönliches Leben und Wirtschaften. Das meiste ist uns unbewusst, auch wenn es unmittelbar sichtbar ist – in unserer Heimatstadt, am Schulort oder zu Hause.

1. Erkundet doch einmal, wie sich heute Globalisierung in eurer nächsten Umgebung und Heimatstadt erkennen lässt und welche Bedeutung sie erlangt hat. Projektideen findet ihr hier. Tipp: Arbeitet in Kleingruppen. Recherchiert möglichst viel konkret vor Ort. 221, 222

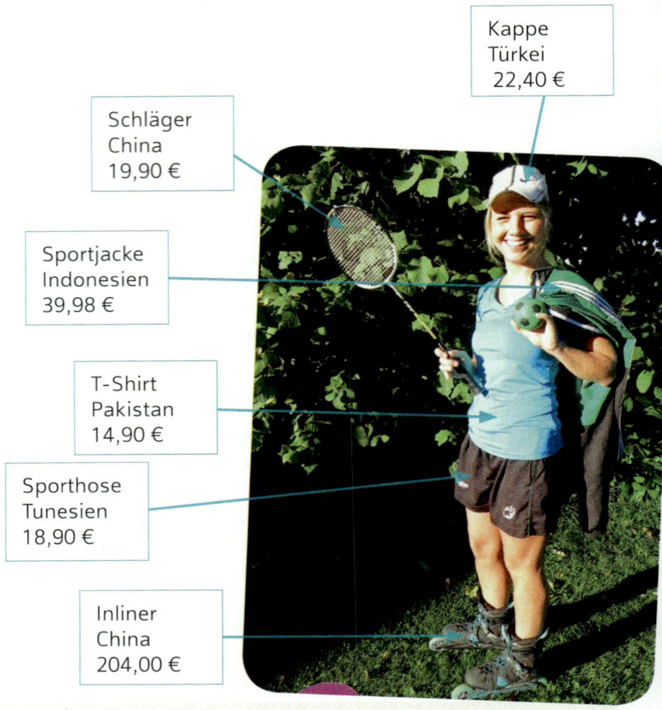

A Aus welchen Ländern stammen die Gegenstände bei uns im Haushalt, die Elektrogeräte und Möbel?
Welche Global Player haben sie produziert? Ist die Produktion nachhaltig?

B Woher stammen die Nahrungsmittel bei uns zuhause? Auf welchen Transportwegen? Bewerte deren Nachhaltigkeit. Welche Bedeutung haben Fastfoodketten? Wie beeinflussen sie die Ernährung in anderen Ländern der Welt?

C Globalisierung hautnah: Woher stammen unsere Textilien?
Ist die Lieferkette transparent? Welche Folgen hat der (niedrige) Preis woanders auf der Welt? Ist die Produktion nachhaltig? Wie wirtschaftet der Global Player? Welche Folgen hat der globale Absatz auf den Kleidungsstil in den Ländern?

Tipps
- Die Herkunftsländer möglichst vieler Produkte zu Hause (und einige Lieferketten) zusammenstellen und damit eine große thematische Karte anfertigen und interpretieren.
- Eine Fotoaussstellung oder Pro- und Kontra-Diskussionen zu Vor- und Nachteilen der Globalisierung ausrichten.
- Infos zur kulturellen Globalisierung gibt es in der Mediathek der Bundesanstalt für politische Bildung (bpb).

M1 Projekt 1: Globalisierung bei uns zu Hause

Informationsquellen:
- Etiketten an Kleidung, Nahrungsmitteln und Haushaltswaren
- Internetauftritt der Heimatstadt mit Fachbereichen (konkrete Anfragen bei Kontaktadressen)
- Homepage der Global Player oder Betriebe vor Ort
- Betriebserkundung vor Ort nach Voranmeldung

Vorgehen:
- Vor der Erkundung und Recherche genau die Untersuchungsfragen überlegen. Auch an wen oder welche Adresse man sie richtet. Eine gute Planung hilft, Zeit zu sparen und gute Ergebnisse zu erzielen.
- Beachte die Methodenseiten zur Erkundung und Präsentation. 218

M2 Infos und Hinweise

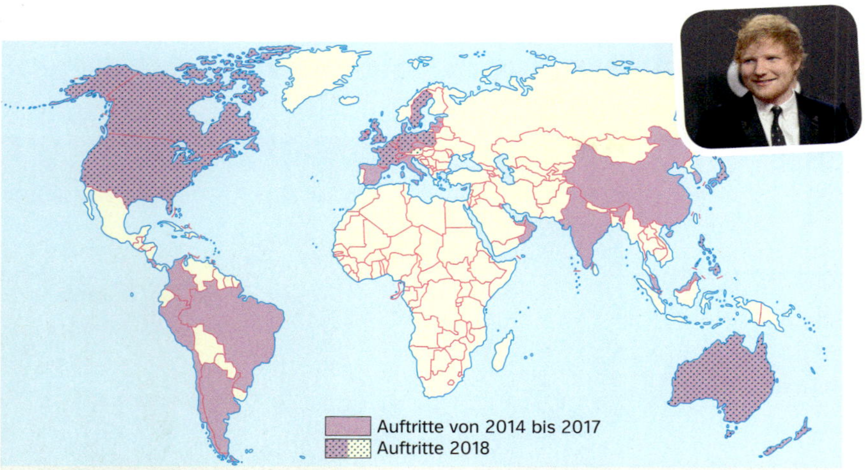

A Welche Künstlerinnen und Künstler sind in deiner Klasse oder deinem Freundeskreis besonders beliebt?
Aus welchen Ländern stammen sie? Wo geben sie Konzerte, wo nicht? Wie werden sie vermarktet? Über welche Portale oder Kanäle werden sie verbreitet? Wo werden sie produziert? Gleicht sich die Musik durch die Globalisierung weltweit immer mehr an? Kommt es zu Anpassung oder Verlust von Konsumgewohnheiten? Kommt es zu einem Verlust an kultureller Vielfalt?

B Globalisierung durch Nachrichten:
Welche Nachrichtenmedien sind die wichtigsten bei dir zu Hause? Welches Medium ist für dich selbst am wichtigsten? Wie groß ist der Anteil globaler Berichterstattung? Über welche Länder erfährt man etwas? Über welche erfährt man nichts? Wie sind die Informationen zu beurteilen? Welche Rolle spielen Fake News?

M3 Projekt 2: Globalisierung über unsere Medien

Welche international tätigen Firmen (Global Player, Hidden Champions) haben einen Standort in deiner Stadt?
An welchem Standort? Wie kann man das Unternehmen im Steckbrief vorstellen? In welcher Weise ist es global verflochten? Welche Bedeutung hat es für die Stadt? Welche Rolle spielt die Digitalisierung?

Tipp: Das Projekt mit BIPARCOURS vorstellen: mit einem Gruppenselfie starten, Stationen mit Infos oder Karten zum Standort oder auch Fotos, Texte, Audiodateien einbauen. Zum Abschluss die Stationen anlegen und auf der App veröffentlichen (siehe Seite 51).

M4 Projekt 3: Globalisierung in unserer Stadt

Wann? (Tag)	Wer?	Wie? (Präsentation)	Was? (Projekt)
Montag, 9.50 Uhr	Gruppe 1	Reportage	Nachrichten global: SPON (Spiegel Online)
Montag, 10.15 Uhr	Gruppe 2	Pro- und Kontra, Fotoausstellung	Globalisierung hautnah: Kleidung in Leverkusen – Kenia
Montag, 10.45 Uhr	Gruppe 3	BIPARCOURS	Global Player: Bayer
…	…	…	…

M5 Klassenaushang zur Planung der Projektpräsentation

IN FOCUS

USA – the world's leading economic power

C Aspect 1: What is the global significance of agriculture?

The USA generates about one fifth of the world's annual income. American companies do business all over the world and American products are known throughout the world.

Key question: What is the importance of the US economy and what changes have taken place in the different sectors of the economy? 223

1. Present your results in an impactful presentation. 219
   Tips on structure:
   - Firstly, describe the areal conditions for agricultural use in the USA: climate, surface forms, soils (atlas, internet).
   - Describe the location of areas with cultivation and areas with livestock (atlas).
   - Report on the position of the US agriculture in world trade.
   - Then give reasons why the US agriculture has been able to achieve this position.
   - Outline opportunities and problems possible arising from the changes in agriculture.
   - Finally, evaluate the possible future developments of the US agricultural sector.

   Add further information by researching other sources, e.g. atlas, maps: Vereinigte Staaten von Amerika – physisch; Nordamerika – Klima; Nordamerika – Landwirtschaft; Great Plains – Landnutzung; Kalifornien – Landwirtschaft.

M2 Fully automated almond harvest in California. More than half of the quantity of fruit, vegetables and nuts produced in the USA is produced here, even though farmland accounts for only three percent of the total agricultural land in the USA.

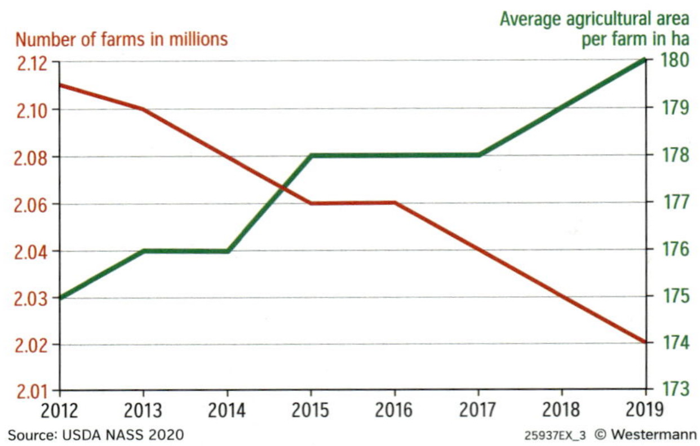

M3 The development of farms in the USA

AMAZING

M1 Number 1 in the USA: John Malone owns the largest agricultural area with a total of 890,000 ha (2019).
Number 1 worldwide: The USA exported agricultural products worth over US-$ 150 billion in 2019. The Netherlands was in a distant second place (US-$ 95 billion).

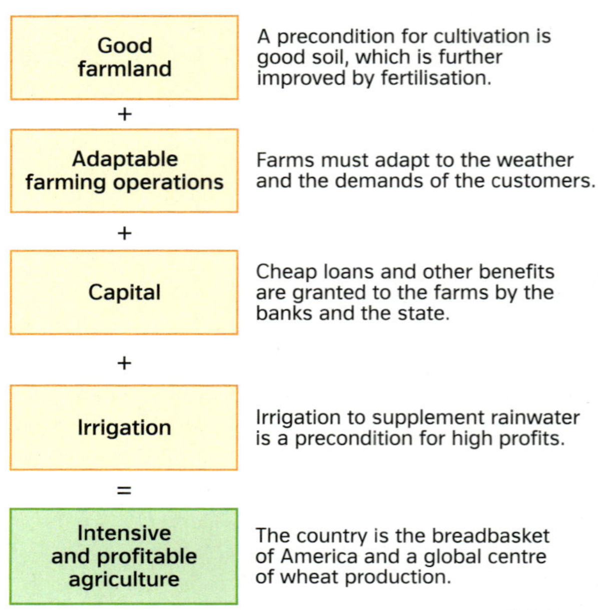

M4 Preconditions for productive grain production in the Great Plains

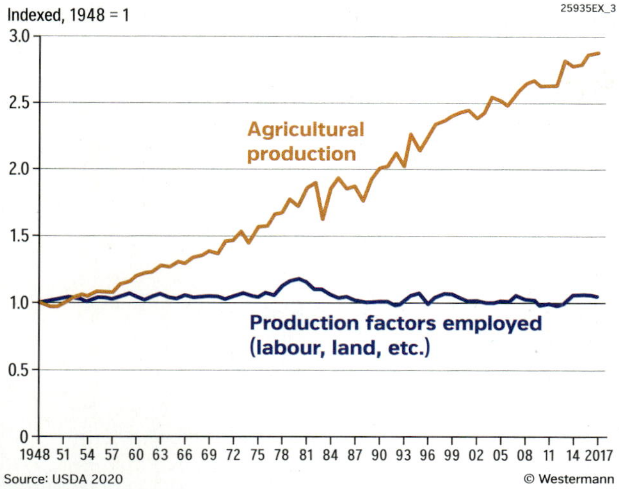

M5 Development of agricultural production

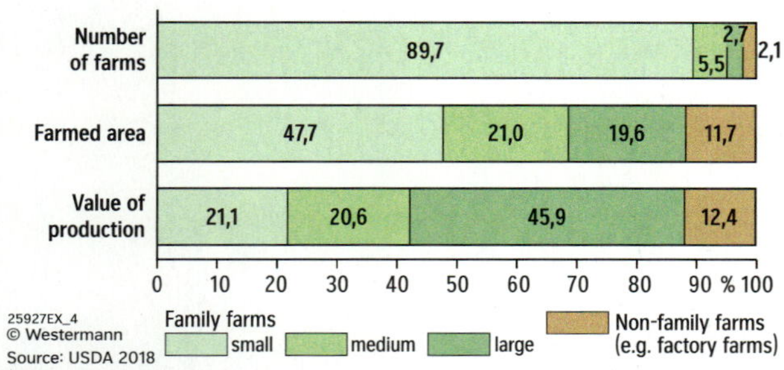

M6 Comparison of family farms of different sizes and factory farms (agribusiness companies)

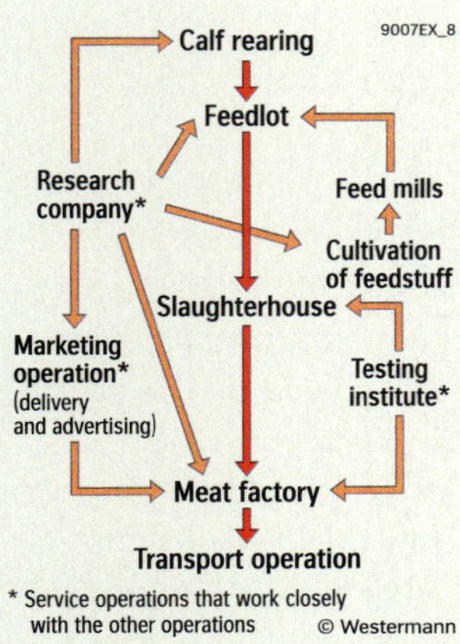

Factory farms that produce more than one million US-$ per year are agribusinesses. They are examples of a concentration process in American agriculture that has been highly successful despite some disadvantages. Among other things, it has made American agriculture by far the most efficient in the world nowadays. The American agroindustry is also very important. With almost 17 million people, it is the largest industry in the world. It includes the many industrially active agricultural processing plants. These, in turn, are usually closely linked to numerous different service companies, such as marketing, transport and research companies.

M7 Agribusiness and agroindustry in the USA

M8 Irrigation in Oklahoma using centre pivots. These machines consist of a pipe that rotates around a central point and irrigates a circular area of almost 50 hectares.

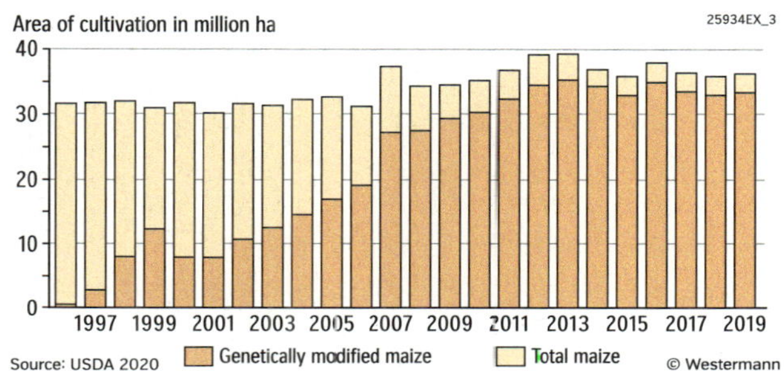

M9 Development of the area under maize and the proportion of genetically modified maize in the USA

In the USA, genetically modified varieties have now become established. The seeds are more expensive, but the costs for pest control are lower. Hardly any food in the USA is free of genetic engineering. [...] Since 1996, genetically modified ingredients have been an integral part of the US food industry. 70 to 80 percent of all products sold today contain ingredients whose DNA has been altered. [...] Today, it is almost impossible for farmers to even find corn or soybeans that have not been genetically modified in some way. But slowly, resistance is growing even in the genetically oriented USA. In 26 states now, consumer advocates and critics [...] have introduced bills that would require the labelling of foods that contain genetically modified ingredients.

Source: ©Thorsten Schröder for ZEIT ONLINE (www.zeit.de) of 17.02.2014 „Gentechnik in den USA: Die Gen-Nation", Zeit Online GmbH, Hamburg, accessed: 26.10.2020

M10 Genetic engineering in the USA

IN FOCUS

USA – the world's leading economic power

C Aspect 2: Transition and global significance of industry

The USA generates about one fifth of the world's annual income. American companies do business all over the world and American products are known throughout the world.

Key question: What is the importance of the US economy and what changes have taken place in the different sectors of the economy? 223

1. Present your results in an impactful presentation. 219
   Tips on structure:
   - Firstly, describe the location of the industrial regions in the USA and their economic characteristics.
   - Report on the importance of the US industry.
   - Give reasons why the US industry has been able to achieve this position.
   - Outline opportunities and problems arising from the transition in industry.
   - Finally, evaluate the possible future developments of the US industrial sector.

   Add further information by researching other sources, e.g. atlas, maps: Vereinigte Staaten von Amerika – physisch; Vereinigte Staaten von Amerika – Wirtschaft; Silicon Valley – Computerindustrie; USA – Entwicklung der Automobilindustrie

M2 "Tin Lizzy", the car produced in assembly line production at Ford's Detroit plants from 1908, and automated body construction at a Ford plant in Chicago today.

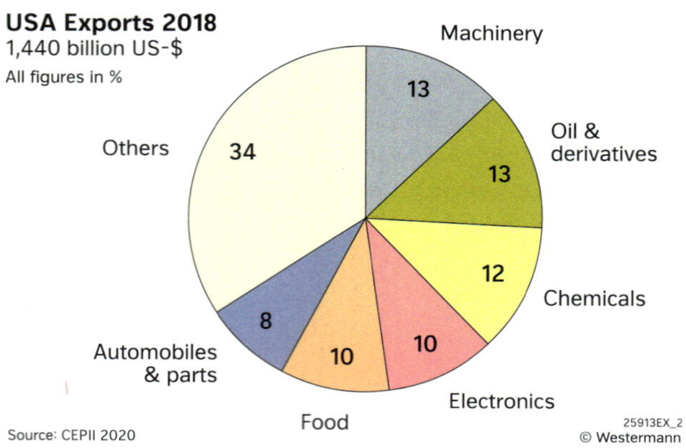

M1 USA: Exports

The Manufacturing Belt in the northeast of the USA is a very old industrial region. The basis for the development of this industrial region was the mineral resources of iron ore and coal. Iron and steel works were built. Detroit became the centre of the automotive industry. The economic rise of the Manufacturing Belt suffered a setback in the 1960s. The demand for steel declined. Factories were closed which resulted in mass unemployment. Structural change was needed in order to create new jobs. Tax advantages and favourable loans for companies in the high-tech industry and for service companies made it possible to attract new businesses.
The Sunbelt benefited from the crisis in the Manufacturing Belt. Here, there were well-trained workers, research facilities, tax advantages and government aid. But the oil and gas deposits in the Gulf of Mexico were also the basis for economic growth in the South.

M3 Development of the Manufacturing Belt and the Sunbelt

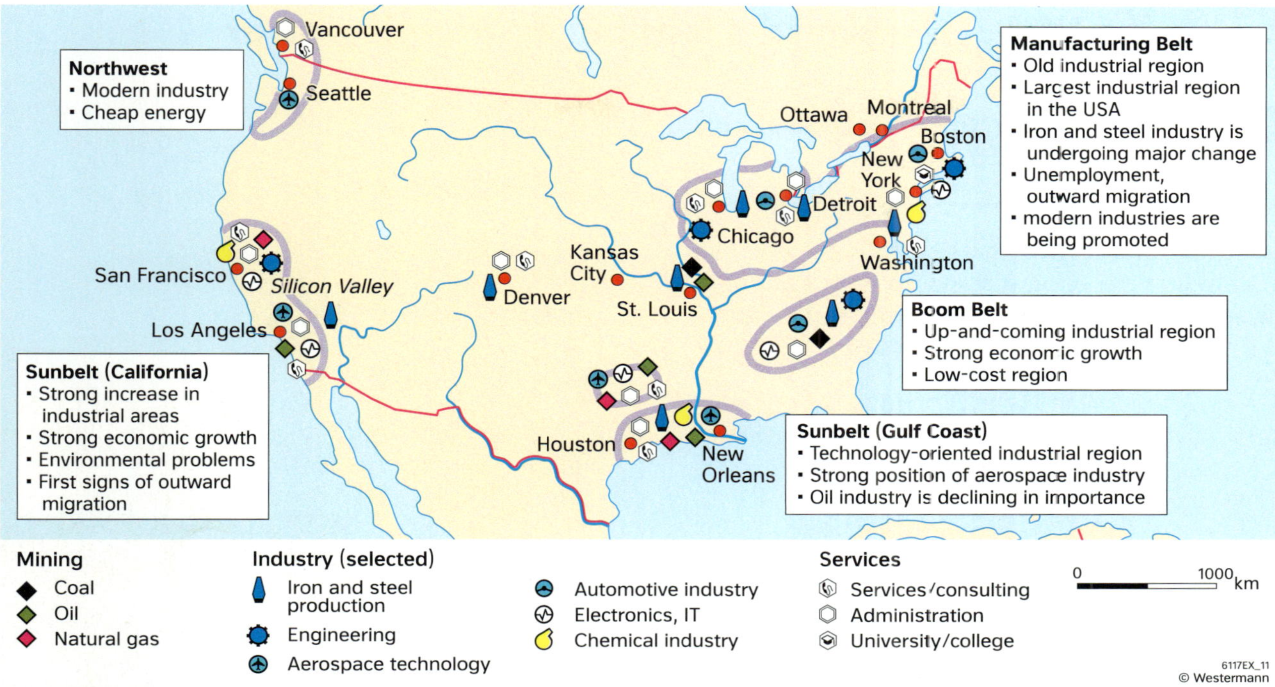

M4 Important economic regions in the USA

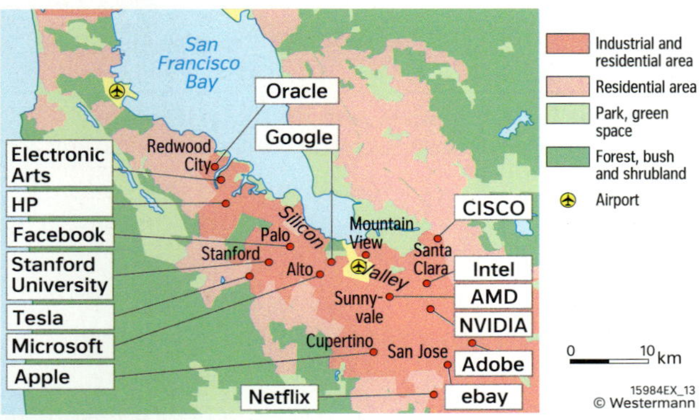

M5 Silicon Valley is the most successful high-tech region in the world. A cluster has developed.

The secret of economic success is not only based on abundant natural resources. The spirit of enterprise also plays an important role. Workers quickly adapt to new demands, and they are willing to change their place of residence if a change of job requires it. Banks invest, even if the risk is high. Entrepreneurs work innovatively. Henry Ford, for example, introduced assembly line production in his car factory at the beginning of the 20th century.
Another factor is a good research and education infrastructure. The development of the Silicon Valley would not have been possible without the impulses from Stanford University. The global company Hewlett-Packard, for example, began with two students working in their garage.

M6 Basis of success

The President has stopped the issuing of new visas for foreign professionals - initially until the end of the year. In view of the 30 million job seekers in the country and the Corona crisis, no additional competition is wanted for the country's own workforce. [...]
The latest measure mainly concerns so-called H-1B visas. These are supposed recruit highly qualified workers that are needed urgently. [...]
That may be true for fields of work like artificial intelligence. But the vast majority of H-1B workers do not work in research labs on the latest cutting-edge technology. Instead, they are employed in IT departments as computer systems analysts. [...]
H-1B visas also make it easier for companies to outsource; with their help, for example, they can quite easily outsource parts of their own IT department or the complete department. Often, US employees must train H-1B colleagues who are then expected – back home – to take over their jobs later. [...]
The current practice is not fair to the H-1B employees either. They are totally dependent on their US employers. When they are dismissed, they also lose their residence permits. It is not uncommon for them and their families to have to leave the country within days. Thus, the H-1B practice weakens the position of all workers.

Source: © Buchter, Heike for ZEIT ONLINE (www.zeit.de) on 29.06.2020: „Mit diesem Einwanderungsstopp hat Trump sogar recht", Zeit Online GmbH, Hamburg, accessed: 12.11.2020

M7 Visas for workers from abroad (2020)

IN FOCUS USA – the world's leading economic power

C Aspect 3: New York – a global city

The USA generates about one fifth of the world's annual income. American companies do business all over the world and American products are known throughout the world.

Key question: What is the importance of New York for the US economy as well as globally? 223

1. Present your results in an impactful presentation. 219

   Tips on structure:
   - Name the characteristics of a global city (p. 172/173).
   - Describe the importance of New York as a global city.
   - Point out the special significance of Manhattan.
   - Then outline opportunities and problems arising from the importance and development of the city.
   - Finally, evaluate the possible future developments of the US economic sector.

   Add further information by researching other sources, e.g. atlas, maps: New York – Manhattan; New York – kulturgeprägte Wohngebiete; New York – Bevölkerungsdichte.

M2 Of the 500 companies with the highest turnover worldwide, 16 have their headquarters in Manhattan

Because of the worldwide networking of the media, global cities are ideal locations for terrorist attacks, because terrorists can attack and destroy the "control centres of the world economy" observed by the entire world. Terrorist attacks like the one on the 11th of September 2001 have shown how vulnerable global cities are.

M3 Global cities are vulnerable

US cities are involved in global competition, with New York undoubtedly [the most important]. [...] The number of headquarters of transnational companies is an important characteristic of the significance of a global city, since the most important decisions are made here and the global company networks are coordinated. With 20 headquarters of the 100 companies with the highest turnover in the USA, New York is the undisputed leader. These are concentrated in just a few blocks south of Central Park and at the southern tip of Manhattan. [...] New York developed into the country's leading press and publishing location early on. [...]

The largest American banks, brokers and insurance companies as well as the most important foreign banks have their headquarters here. [...] With the headquarters of the United Nations located on the East River, New York is also a centre of global political decisions. [...]
New York's museums are as important as those of London or Paris, and its private art market attracts collectors from all over the world. When it comes to drama, musicals and music, New York impresses with a wide range of offerings..

Source: Hahn, Barbara: Die US-amerikanische Stadt im Wandel. Berlin, Heidelberg: Springer 2014, p. 59 – 62.

M1 Expert article on the global city of New York

M4 UN headquarters

M6 Museum of Modern Art

M9 Wall Street in the banking and stock exchange district

From Manhattan, the city's business and cultural centre, any social and ethnic diversity that has evolved is increasingly disappearing. According to the new survey, 57 percent of Manhattan residents have a university education, the majority of whom are white. "Manhattan has become an exclusive zone, an island of the highly educated and well-off professionals," says Andrew Beveridge, a demographer at the City University of New York.
Manhattan has always been a magnet for the ambitious and the smart, but until now, there was also room on the island for other lifestyles [...]. But in the past five years, living in Manhattan has become so expensive that lower income groups can no longer afford to live in the heart of New York.
To be able to maintain its standard of living, an American family that lives on 60,000 dollars a year elsewhere would have to earn 146,000 dollars in Manhattan – 137.9 percent more than the national average. 100,000 dollars would be spent for housing alone. "The people who keep this city alive, the waiters and caretakers, can no longer live in it," says Pedro A. Noguera, a sociologist at New York University.
The new New York has become a fortress of the rich, into which the servants are admitted during the day, but which they have to leave again in the evening for their quarters on the outskirts of the city. [...]
An increasingly globalised labour market has flooded the city with both a global elite and poor fortune-seekers - the city's social gap is becoming more and more like that of the Third World.

Source: „Die neue Festung der Reichen", SPIEGEL.de, Sebastian Moll, 29.8.2006, https://www.spiegel.de/panorama/gesellschaft/new-york-die-neue-festung-der-reichen-a-433262.html, accessed: 20.03.2018

M5 The other side of the coin

Number of headquarters of the 500 companies with highest turnover worldwide, 2020[1]	
Peking	55
Tokyo	37
New York	16
Paris	16
Largest stock exchanges in the world by market capitalisation, 06/2020 (in US-$ trillion)[2]	
NYSE, New York	21.0
Nasdaq, New York	14.6
Japan Exchange Group, Tokyo	5.7
Shanghai Stock Exchange, Shanghai	5.3
Most visited art museums in the world, 2019 (visitors in millions)[3]	
Louvre, Paris	9.6
National Museum of China, Beijing	7.4
Vatican Museums, Vatican	6.9
Metropolitan Museum of Modern Art, New York	6.5

Sources: [1] Forbes 2020, [2] World Federation of Exchanges 2020, [3] The Art Newspaper 2020

M7 Comparison of cities

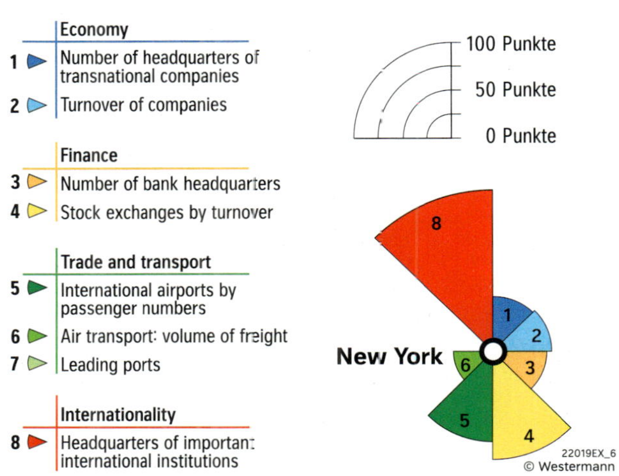

M8 New York, assessed according to Bronger's global city model 210

Vernetzt denken – typisch Geographie!

Ende 2019: In Wuhan (China) bricht eine Viruserkrankung aus – Anfang 2020: Es entsteht eine Pandemie – weltweit brechen Gesundheitssysteme zusammen – in vielen Industriebetrieben steht die Produktion still – Lockdown – die Arbeitslosenzahlen steigen – es kommt auf allen Ebenen zu Wirtschaftskrisen – Maßnahmen zur Bekämpfung der Pandemie schränken die Menschen ein – Tausende demonstrieren – in den Tourismusregionen gehen die Buchungen dramatisch zurück – während sich dort die Natur erholt, treten in den Naherholungsgebieten vermehrt Schäden auf – die Luftbelastungen durch Emissionen aus Industrie und Verkehr sinken ...

Wir leben in einem globalen System, in dem viele Komponenten aufeinander einwirken, voneinander abhängig sind. Im Methodenkapitel lernst du, einzelne Informationen zu beschaffen, auszuwerten und zu präsentieren. Wesentlich ist jedoch, diese Informationen immer wieder mit anderen (z. B. mit dem, was du schon weißt) zu verknüpfen und die sich weiter ergebenden Auswirkungen zu betrachten. So können sich negative Wirkungszusammenhänge ergeben, aber auch positive, chancenreiche. – So vernetzt zu denken und zu arbeiten, das ist typisch für das Fach Erdkunde, für die Wissenschaft Geographie.

M2 **In Arizona versuchte der Milliardär Edward Bass 1991 das System Erde im Kleinen zu simulieren. Unter sehr großen, hermetisch abgedichteten Glaskuppen lebten zwei Jahre lang acht Menschen und 3800 Tier- und Pflanzenarten von der Außenwelt völlig abgeschnitten. Das Projekt scheiterte, die Wechselbeziehungen im Ökosystem Erde waren zu komplex, um sie tatsächlich simulieren zu können.**

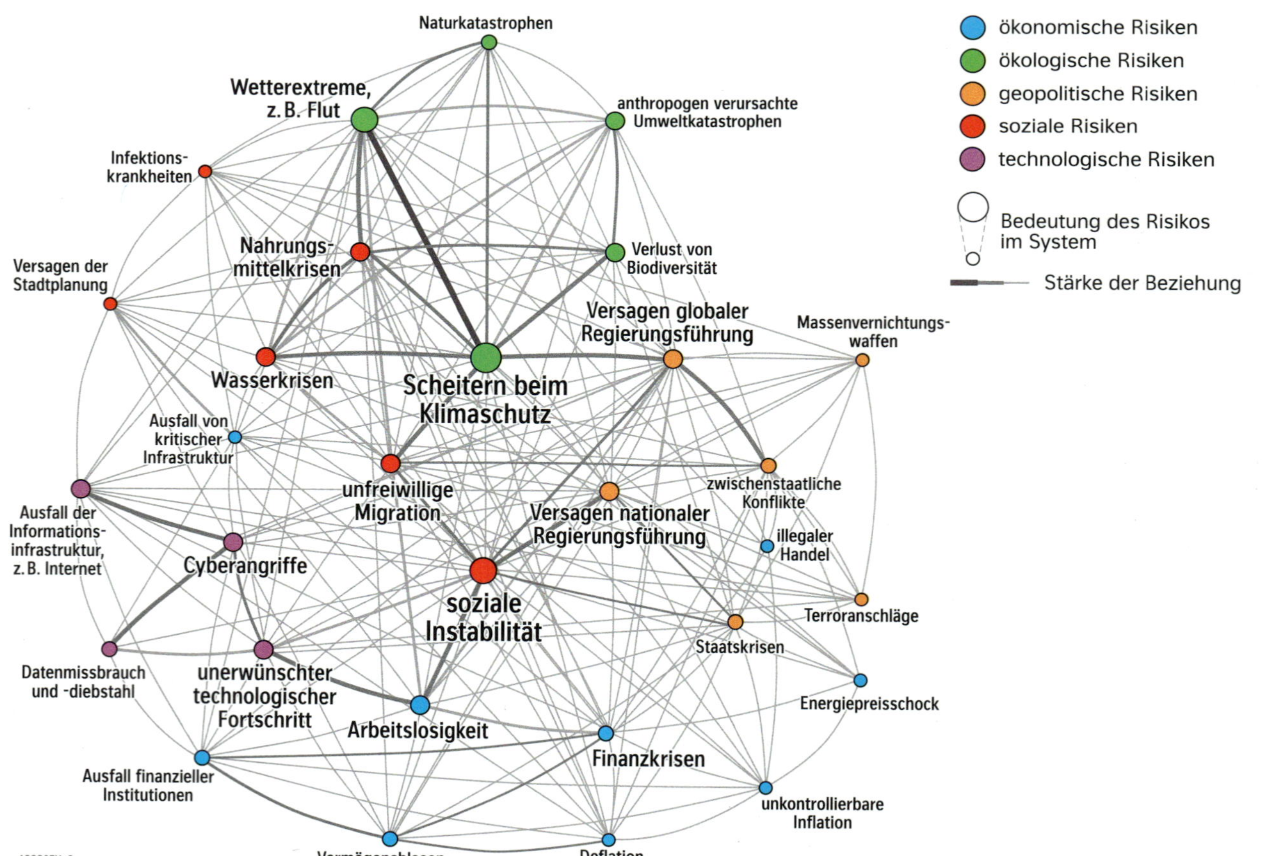

Quelle: World Economic Forum: Global Risks Report 2020

M1 **Alljährlich werden 1000 international anerkannte Expertinnen und Experten befragt, welche Risiken und Herausforderungen sie im kommenden Jahr für besonders bedeutend halten. Dabei wird auch bewertet, welche Auswirkungen diese auf Wirtschaft, Gesellschaft und Umwelt haben könnten.**

Wie ordne ich meine Gedanken? – Mind map, causal chain, causal network and concept map

Es gibt die unterschiedlichsten Möglichkeiten, um Themen grafisch darzustellen.
Grundsätzlich werden die Schlüsselbegriffe grafisch miteinander durch Linien oder Pfeile verbunden. Daraus lassen sich dann die Gliederung (die Struktur) eines Themas und innere Zusammenhänge ablesen.

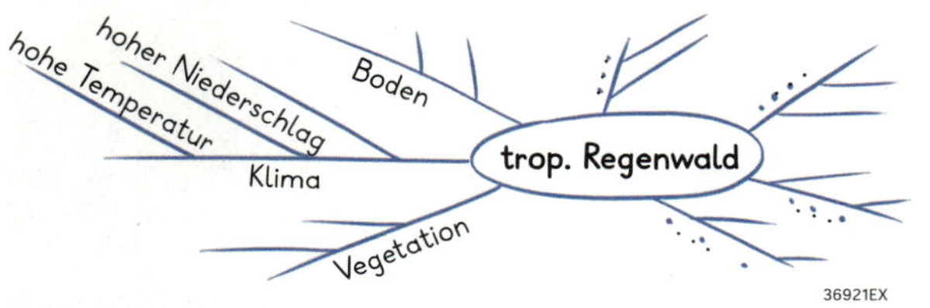

M1 Mind map

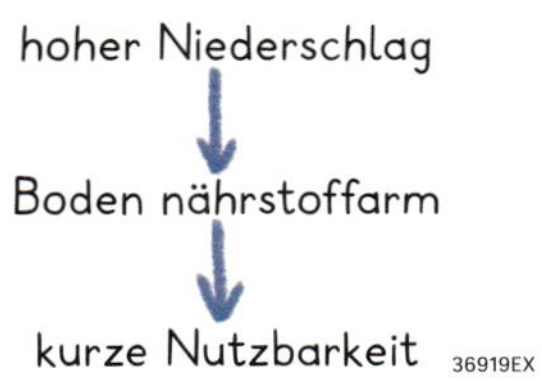

M2 Causal chain (Kausalkette)

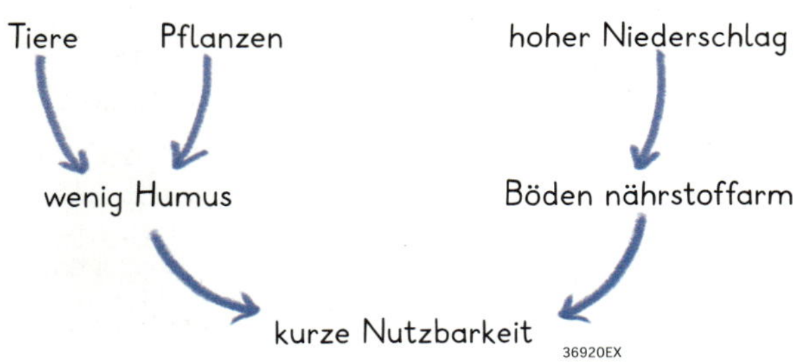

M3 causal network (Wirkungsgefüge)

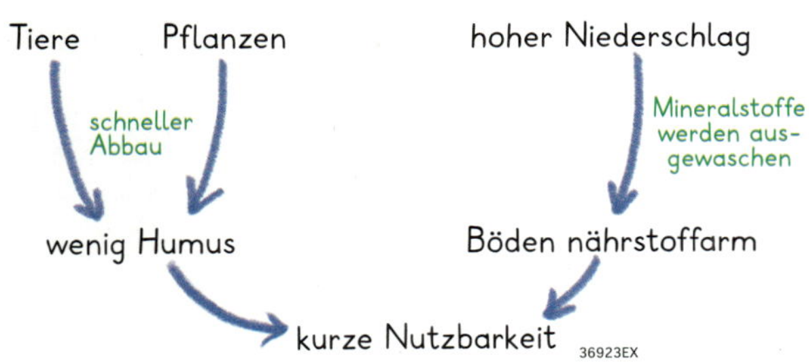

M4 Concept map

INFO

M5 Die **mind map** („Gedankenlandkarte“) hat Haupt- und Nebenäste, die von einem zentralen Thema ausgehen. Sie ordnen die Unterthemen und Begriffe nach ihrer Bedeutung.

INFO

M6 Die **causal chain** (**Kausalkette**) ordnet Begriffe nach Ursache und Wirkung und verbindet sie in einer Linie mit Pfeilen. Nur selten besteht ein Sachverhalt nur aus einer Ursache und einer Wirkung. Daher ist es meist sinnvoller, das Thema als Wirkungsgefüge oder Conceptmap darzustellen.

INFO

M7 Auch im **causal network** (**Wirkungsgefüge**) sind Begriffe nach dem Prinzip von Ursache und Wirkung angeordnet. Man kann damit aber auch kompliziertere Zusammenhänge darstellen z. B.
- mehrere Ursachen und Folgen (mehrere Begriffe auf gleicher Ebene),
- geringere und stärkere Folgen/Auswirkungen (dicke, dünne Pfeile),
- Folgen unterschiedlicher Art (Farbe),
- Wechselwirkungen (Doppelpfeile).

INFO

M8 In einer **concept map** werden die Linien und Pfeile des Wirkungsgefüges noch beschriftet.

Eine concept map erstellen

1. Erstelle eine Liste mit den Schlüsselbegriffen (Schlüsselwörtern) des Themas.

2. Ordne die Begriffe nach dem Prinzip von Ursache und Wirkung. Achte darauf, dass gleich wichtige oder gleichzeitige Gründe oder Folgen in derselben Ebene stehen.

3. a) Verbinde die Begriffe mit Pfeilen. Sie haben die Bedeutung „bewirkt“ oder „hat zur Folge“. Unterschiedliche Stärke der Pfeile kann unterschiedlich starke Wirkung ausdrücken.
b) Färbe Begriffe und Pfeile mit gleicher Bedeutung gleich ein.
c) Beschrifte einzelne Pfeile.

➡ Übungsmöglichkeiten zu Mind map, causal chain, causal network and concept map: Seite 26, 34, 42, 50, 66. 86, 92, 160, 176

Wie orientiere ich mich mit Atlas und (digitalen) Karten?

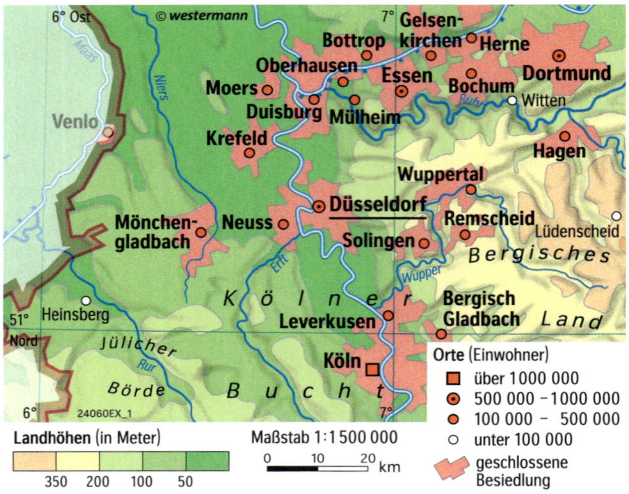

M1 Großer Maßstab – kleiner Maßstab. Ist die Zahl hinter dem Doppelpunkt klein (z. B. 1 : 100), dann spricht man von einem großen Maßstab, ist die Zahl sehr groß (z. B. 1 : 16 000 000) dann spricht man von einem kleinen Maßstab.

Die Karten im **Atlas** sind in **Planquadrate** eingeteilt, die mit einer Kombination aus Buchstabe und Zahl (z. B. A2) bezeichnet werden können. So kann man leicht die im Register angegebenen Orte finden.
Um jeden Punkt der Erde genau bestimmen zu können, wird ein **Gradnetz** verwendet. Es besteht aus **Breiten-** und **Längengraden**. Vom Äquator zu den Polen gibt es jeweils 90 Breitengrade. Vom nullten Längengrad aus (dem Nullmeridian) zählt man nach Osten (East) und nach Westen (West) jeweils 180 Längengrade.
Ein Ort wird mit der Angabe des Breiten- und des Längengrades bezeichnet, so ergeben sich die **geographischen Koordinaten**.

M2 Suchen und Finden in Karten

INFO

M3 **Geographische Koordinaten** mit **Minuten** und **Sekunden**: Jeden Breiten- und Längengrad kann man noch einmal in jeweils 60 Minuten und jede Minute noch einmal in 60 Sekunden unterteilen z. B. 50° 59′ 32″ N / 7° 7′ 43″ O. Also: 50 Grad, 59 Minuten, 32 Sekunden nördlicher Breite ...

Helpful words and phrases:

... large scale (großer Maßstab) – one to one-hundred
... small scale (kleiner Maßstab)– one to sixteen million
... the legend (Legende) contains the symbols (Signaturen) and their explanations.
... coordinates: fifty degrees, fifty-nine minutes, thirty-two seconds north. And seven degrees, seven minutes and forty-three minutes east.

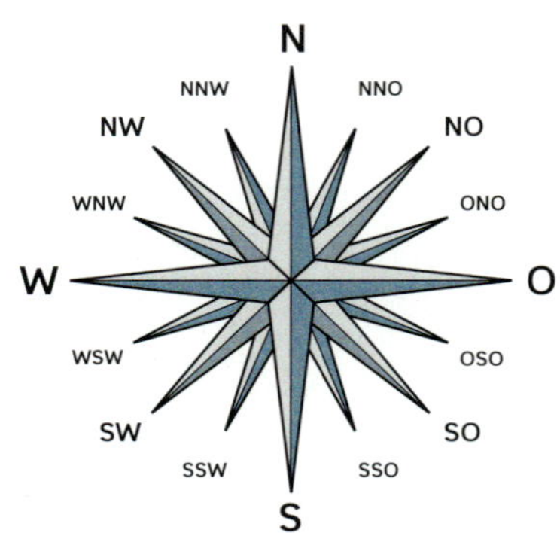

M4 Die Himmelsrichtungen. In englischer Sprache wird Ost zu East (E). Karten sind in der Regel nach Norden ausgerichtet. Sonst gibt der Nordpfeil (rechts) die Ausrichtung an.

Auch **digitale Karten** bieten Informationen: Beispielsweise kannst du in der Anwendung *Google Maps* über das Menü Straßenkarten, Satellitenbilder und Geländekarten (mit der Darstellung von Gebirgen und Tälern) abrufen. Auch diese **Geographischen Informationssysteme (GIS)** arbeiten mit geographischen Koordinaten.
Klicke einen Punkt auf der digitalen Karte an, drücke die rechte Maustaste und wähle im Kasten *„Was ist hier?"* Es werden dir die geographischen Koordinaten angegeben (in Grad, Minuten und Sekunden und als Dezimalangabe).
Einen Ort auf der Erde findet man, indem man den Namen oder die geographischen Koordinaten in das Suchfeld eingibt.

M5 Digitale Karten

➡ Übungsmöglichkeiten: Seite 74

Wie beschreibe und interpretiere ich eine thematische Karte?

Du hast nun genügend Hintergrundwissen, um viele Materialien nicht nur beschreiben, sondern auch interpretieren zu können.

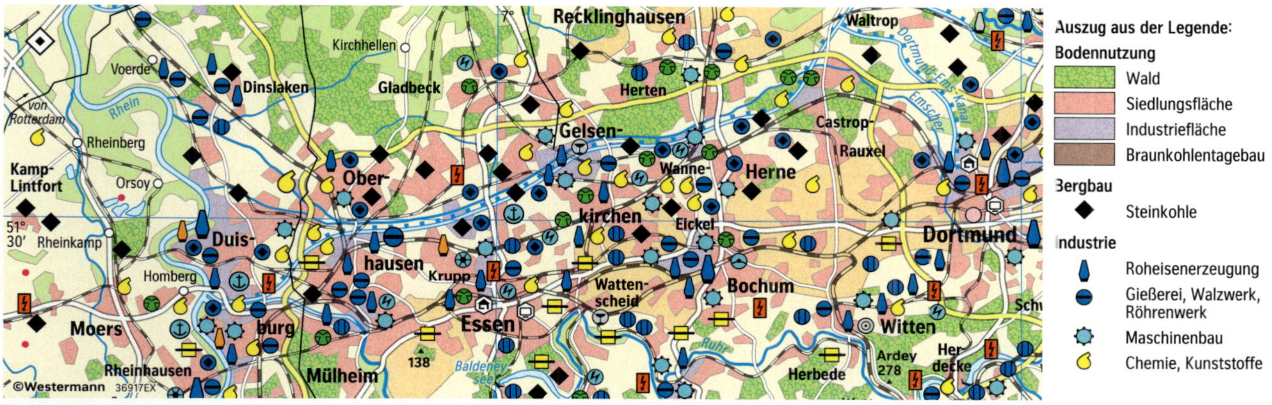

M6 Die Wirtschaft des Ruhrgebietes Mitte des zwanzigsten Jahrhunderts (Ausschnitt Diercke Weltatlas S. 40)

Eine thematische Karte beschreiben und interpretieren

Beschreibung

① Wo? Wann? Was? – das Thema der Karte
Lage und Größe des Raums, Thema der Karte, evtl. Zeitraum, für den die Angaben gelten

② Was sind die genauen Inhalte?
a) Legende (→ Signaturen, Farben)
b) Verteilung der Signaturen: Wo gibt es besonders viele, wo besonders wenige (Extremwerte)?
c) Gliederung der Auswertung z. B. nach Himmelsrichtungen

③ Was ist die Kernaussage der Karte?
Zusammenfassung der Aussage der Karte in wenigen Sätzen

NEU: Interpretation

④ Wie lässt sich die Aussage der Karte erklären? Welche weitere Entwicklung könnte sich in dem Raum ergeben?
Gibt es einen Zusammenhang zwischen den einzelnen Elementen der Karte? (z. B. dort, wo viele Signaturen A sind, sind auch viele Signaturen B – dort, wo viele Bodenschätze sind, gibt es auch viel Industrie)
Warum ist dieser Zusammenhang so? (z. B. die Industrie verarbeitet ...)
Welche Probleme könnte es in dem Raum geben? (z. B. durch die vielen Industriebetriebe ist das Verkehrsnetz sehr gut ausgebaut, es wird aber auch sehr belastet)
Wie könnte sich der Raum weiterentwickeln? Welche Zukunftsperspektiven gibt es? (z. B. der Raum wird wahrscheinlich auch in Zukunft wirtschaftlich stark sein ...)
Wenn nötig oder sinnvoll, recherchiere nach Zusatzinformationen. 213

INFO

M7 Beschreibung und Interpretation
Beschreiben heißt, etwas genau zu betrachten und den Inhalt wieder geben zu können.
Interpretieren heißt, erklären und auf mögliche Auswirkungen hinweisen zu können. Damit das gelingt, benötigst du Hintergrundwissen. Dieses musst du möglicherweise eigenständig suchen und bewerten → recherchieren 213.

Helpful words and phrases for the interpretation (4.):

... the map shows / the topic of the map is...
... it is a small- / large-scale map of ...
... the area shown is ...
... we can see a pattern in the distribution / location of...
... many... are located close to... because...
... there is a concentration of... because...
... from ... we can conclude that...
... but ... can lead to...
... there is a risk that ...
... this offers opportunities / chances ... because...
... can be useful for the area
... can have good / bad consequences for the region: ...

➡ Übungsmöglichkeiten zur thematischen Karte: Seite 38, 48, 50, 123

Wie nutze ich einen digitalen Atlas? – Diercke Atlas | Die App

Der digitale **Diercke Weltatlas** umfasst alle Karten deines gedruckten Weltatlas. Egal ob auf dem Smartphone oder auf dem PC: Die digitale Version des Atlas enthält die aktuellsten Karten und ist einfach zu benutzen. Aber sie kann noch mehr als der gedruckte Atlas: Zoome in Karten, miss Entfernungen, verändere die Karte oder verknüpfe sie mit einer anderen. Probiere es aus!

Arbeiten in und mit der Karte
In der Menüleiste links kannst du verschiedene Werkzeuge benutzen, um deine Karte zu verändern. Du kannst sogar selbst Signaturen einfügen und deine eigene thematische Karte erstellen:

- Umkreise wichtige Aspekte oder markiere farblich.
- Miss Entfernungen zwischen Punkten.
- Schreibe direkt in die Karte.
- Füge neue Signaturen in dein Kartenbild.
- Lösche deine Bearbeitungen.
- Speichere einen Screenshot.
- Schneide einen Teil der Karte aus und kopiere ihn.
- Verbirg die Karte.

M1 Kartenwerkzeuge

Veränderung der Kartenebenen
In den Karten sind Signaturen gleichen Themas zu einer Ebene zusammengefasst. In dem Legendenfenster kannst du auf „Ebenen" klicken. Jetzt kannst du einzelne oder mehrere Ebenen mit deren Signaturen in deiner Karte durch einen Klick auf das Auge neben den Ebenen weg- oder zuschalten. Dies hilft dir, eine Fragestellung genauer zu untersuchen.

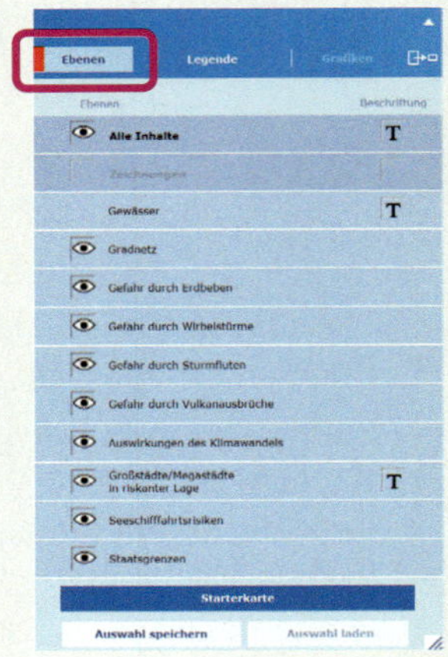

M3 Umgang mit der Legende

Möchtest du zwei unterschiedliche Karten des selben Raumausschnittes miteinander vergleichen, so kannst du sie problemlos nebeneinandersetzen. Nutze die obere Menüleiste. Klicke auf ▣ und du kannst eine zweite Karte neben die erste setzen. Mit einem Klick auf ▣ verknüpfst du die Karten. Bewegst du jetzt die eine, bewegt sich die andere mit. Auch das Zoomen funktioniert bei beiden Karten dann gemeinsam. Ein Klick auf ▣ leitet dich zum **Diercke Globus Online**. Die Karte wird in den Globus integriert und du kannst weiter mit ihr arbeiten (siehe unten). Zusatzinformationen zu der ausgewählten Karte sind mit einem Klick auf ▣ abrufbar.

M2 Verknüpfen von Karten auch mit dem Diercke Globus

Wie nutze ich einen digitalen Globus? – Diercke Globus Online

Möchtest du von deinem Zuhause über den Globus nach New York fliegen? Oder selbst eine thematische Karte erstellen? Oder möchtest du wissen, wie eine dargestellte Karte im Satellitenbild aussieht? Dann nutze den Diercke Globus Online. Auf *schueler.diercke.de* kannst du das kleine Zusatzprogramm herunterladen. Als Kennwort gibst du den Online-Schlüssel ein, der vorne in deinem Diercke Weltatlas eingedruckt ist. Los gehts!

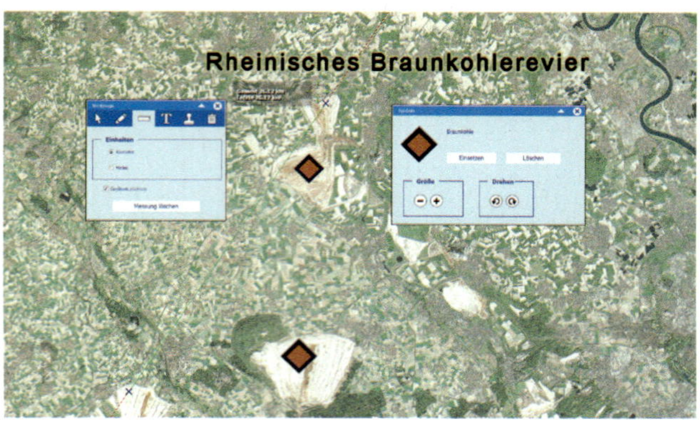

Unter dem Button „Menü" kannst du „Ebenen" auswählen und damit einzelne Signaturen zuschalten. Aber du kannst auch Karten integrieren und diese in ihrer Transparenz verändern. Über den Werkzeuge-Button kannst du Texte, Signaturen und Markierungen einfügen und so deine eigene thematische Karte erstellen. Entfernungen kannst du mithilfe des Lineals abmessen.

M4 Erstellung von Karten

Wie nutze ich Google Earth als geographische Informationsquelle?

Digitale Globen wie Google Earth können nicht nur genutzt werden, um Satellitenbilder anzuzeigen. Sie dienen auch als geographische Informationsquelle. Mit Werkzeugen in Google Earth kannst du Entfernungen und Umfänge messen, Räume im zeitlichen Verlauf betrachten und sogar thematische Karten erstellen.

Google Earth nutzen

① Was will ich wissen?
Überlege, welche Fragestellung du untersuchen möchtest. Handelt es sich dabei um etwas, das man als Karte darstellen kann?

② Welches Google Earth Werkzeug kann dir bei der Beantwortung der Frage helfen?
Wenn du Entfernungen oder Umfänge messen möchtest, nutze das Lineal (M5). Zeitliche Veränderungen kannst du mit dem Zeitschieberegler herausarbeiten (M7). Mit Streetview gehst du auf eine virtuelle Exkursion (M6).

③ Was ist die Erkenntnis?
Fasse zusammen, welche Informationen du mithilfe von Google Earth gewonnen hast. Beantworte deine Fragestellung.

Luftlinie: Messen der direkten Entfernung zwischen zwei Punkten
- Klicke in der Menüleiste auf das Lineal.
- Wähle die Registerkarte „Linie" und das geeignete Längenmaß (z. B. Meter) aus.
- Klicke auf den Startpunkt und dann auf den Endpunkt deiner zu messenden Strecke.

Wegstrecke: Ermitteln nicht gerader Strecken (z. B. Straßen) oder von Umfängen (z. B. Tagebau)
- Klicke in der Menüleiste auf das Lineal.
- Wähle die Registerkarte „Pfad" und das geeignete Längenmaß (z. B. Kilometer) aus.
- Klicke auf den Startpunkt und bei jeder Änderung des Verlaufs.

M5 Nutzung des Werkzeugs ‚Lineal'

Gehe auf eine virtuelle Exkursion und finde heraus, wie dein Untersuchungsraum aussieht.
- Ziehe mit der Maus das kleine Männchen im rechten Bildrand in die Karte. Die Straßen, bei denen Streetview verfügbar ist, färben sich blau ein.
- Lasse das Männchen an der gewünschten Stelle los.
- Bewege dich mit den Cursortasten durch die Straßen und untersuche deinen Raum.

M6 Gehe auf virtuelle Exkursion mit Streetview

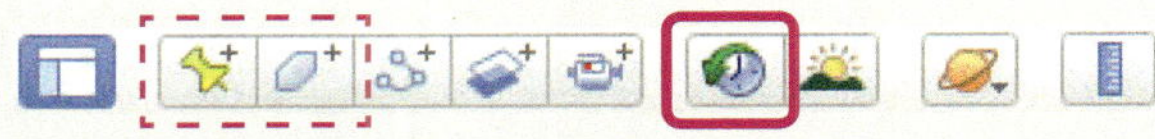

Zeitschieberegler: Untersuchung eines Raumes im zeitlichen Verlauf
- Klicke in der Menüleiste auf den Zeitschieberegler.
- Innerhalb der erscheinenden Leiste kannst du den Knopf verschieben. Das jeweilige Datum der Aufnahme erscheint über der Leiste.

Für Google Earth Profis:
Durch das Einfügen von Polygonen und Markierungen kannst du selbst thematische Karten erstellen. Zum Beispiel kannst du die Ausdehnung eines Stadtteils heute und in der Vergangenheit je mit einem Polygon kartieren.
Tipp: Bei der Registerkarte „Stil – Farbe" kannst du die Transparenz herabsetzen. Teste den Effekt.

M7 Ein Blick zurück mit dem Zeitschieberegler

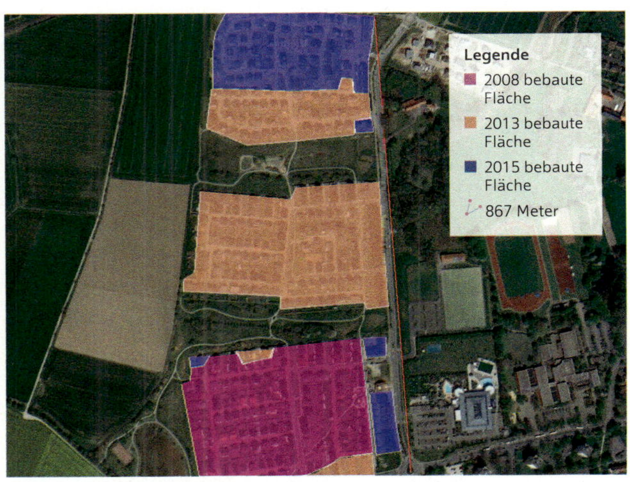

M8 Schülerprodukt: Wie hat sich unser Wohnort verändert? Das Beispiel Soest.

Wie zeichne ich eine Kartenskizze?

Eine Kartenskizze zeichnen

1. Inhalte auswählen.
   Was willst du zeichnen? (z. B: Gebirge, Flüsse)

2. Nach **einfachen Formen** suchen.
   Auf welche Formen lassen sich die Umrisse reduzieren? (eine Kombination z. B. von Dreieck, Trapez, Halbkreis)

3. Die Umrisse zeichnen, die Inhalte einzeichnen.
   Am besten ist es, frei **aus der Hand** zu zeichnen. Hab keine Angst, stark zu **vereinfachen**!

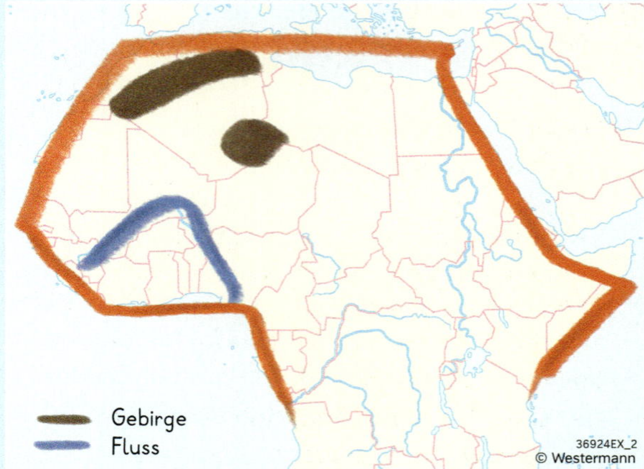

Wie nutze ich ein WebGIS?

Als **Geographisches Informationssystem (GIS)** wird eine Software bezeichnet, mit deren Hilfe man Daten über einen Raum

- mit einer Dateneingabe erfassen,
- in einer Datenbank verwalten,
- über eine Datenauswertung analysieren und
- mit einer Datenausgabe präsentieren kann.

Man unterscheidet dabei das WebGIS und das auf Computern fest installierte GIS-Programm (Desktop-GIS).

INFO

M1 In einem **WebGIS** (Online-GIS) werden aus einer Datenbank Daten vorgegeben. Diese werden dann zu digitalen Karten verarbeitet, welche über das Internet abrufbar sind. Bekannt sind Online-Kartendienste wie *Google Maps*, *HERE*, *Bing Maps* oder *OpenStreetMap*.
Durch eine einfache Aktion können hier von überall zu jedem Ort aktuelle Daten abgerufen werden (z. B. Straßensperrung, Stau).

INFO

M2 **GIS-Software** wie sie von Behörden (z. B. Stadtverwaltungen), Firmen (z.B. Energieversorger, Netzbetreiber, Logistikunternehmen) oder Universitäten genutzt wird, ist auf lokalen Rechnern und Netzwerken installiert. Sie bieten die Möglichkeit, eigene Daten einzugeben und auch eigene Klassifizierungen zu erstellen. So können immer sehr schnell aktuellste Karten für alle Einsatzzwecke zur Verfügung gestellt werden.

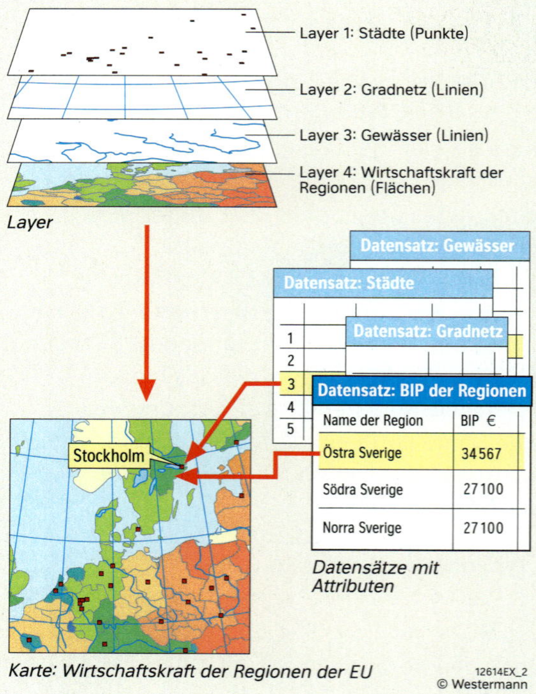

Eine Karte besteht aus verschiedenen **Layern**. Jeder Layer kann entweder einzelne Punkte, Linien oder Flächen beinhalten.
Jeder dieser Layer kann einzeln sichtbar und aktiv oder aber auch unsichtbar geschaltet werden. So könnte man die Inhalte einer Karte nur auf das Gewässernetz reduzieren oder nur die Städte und die Verkehrswege darstellen.
Durch das Zu- und Wegschalten der Layer ist es einfach möglich, Karten zu vergleichen, neu zu erstellen und Abfragen von Daten durchzuführen.

M3 Das Prinzip der Layer

Wie werte ich einen Fachtext aus?

Wie werte ich einen Fachtext aus?

1 Was will ich wissen?
Die Fragestellung an den Text

2 Lesen und Unbekanntes klären.
Unbekannte Begriffe markieren und die Bedeutung recherchieren. 213

3 Lesen und unterstreichen
Genaues Durcharbeiten, Unterstreichen der Schlüsselwörter/Schlüsselaussagen

4 Das Wichtigste notieren
Das Wichtigste (Schlüsselwörter, Schlüsselaussagen) herausschreiben (der Fachbegriff dazu lautet: „exzerpieren“, man fertigt ein „Exzerpt“, einen Auszug, an). Dabei auf eine sinnvolle Gliederung (z. B. nach Unterthemen) achten!

5 Was ist die Kernaussage? Was sind die Antworten auf meine Fragen?
Zusammenfassung der Kernaussage

6 NEU: Hat der Text eine bestimmte Aussageabsicht (Intention)?
Ist der Text eher objektiv oder eher subjektiv? Wer ist die Autorin oder der Autor 213? Will der Text objektiv informieren (z. B. Lexikon, Fachbuch)? Oder stellt der Text das Thema aus einer bestimmten Sichtweise dar (z. B. Zeitungskommentar, Werbung, Karikatur 209)? Wird die Aussageabsicht offen deutlich oder ist sie versteckt?

Helpful words and phrases:
- *... the text is taken from an encyclopaedia / a newspaper / a scientific magazine*
- *... the author gives facts about ...*
- *... the authorgives his or her opinion on...*
- *... is shown in a very positive / negative way / objectively*

Wie beschreibe und interpretiere ich ein Foto?

Ein Foto beschreiben und interpretieren

Beschreibung

1 Was? Wo? Wann?
Welchen Ort zeigt das Bild? Wo und wann wurde es aufgenommen (Jahreszeit)? – Oft sagt die Bildunterschrift dazu schon etwas. Wo liegt der Ort? → Atlas

2 Welche Einzelheiten kannst du erkennen?
Gliedere nach Vordergrund, Mitte, Hintergrund. Was ist besonders auffallend? Gibt es z. B. Siedlungen und große Gebäude? Menschen, die etwas tun? Besonders auffällige Naturerscheinungen?

3 Was ist die Aussage des Bildes?
Zusammenfassung aller Informationen zu einer Kernaussage

NEU: Interpretation

4 Wie kann man das Dargestellte erklären, welche Folgen könnten sich ergeben?
Warum ist das so, was du im Bild siehst? Welche Ursachen gibt es für das, was du im Bild siehst?
Überlege mögliche positive oder negative Folgen oder Entwicklungsmöglichkeiten für die Zukunft.
Wenn nötig oder sinnvoll, recherchiere nach Zusatzinformationen. 213

Helpful words and phrases for the interpretation (4.):
- *... the foto / picture shows ...*
- *... it was taken in summer / autumn / winter / spring*
- *... the foto shows a rural / urban / industrial / ... area*
- *... in the foreground / background / middle you can see ...*
- *... we can explain ... by looking at ...*
- *... is caused by ...*

→ Übungsmöglichkeiten zur Fotointerpretation:
Seite 44, 53, 58, 60, 64, 73, 123, 142

Wie beschreibe und interpretiere ich Satellitenbilder?

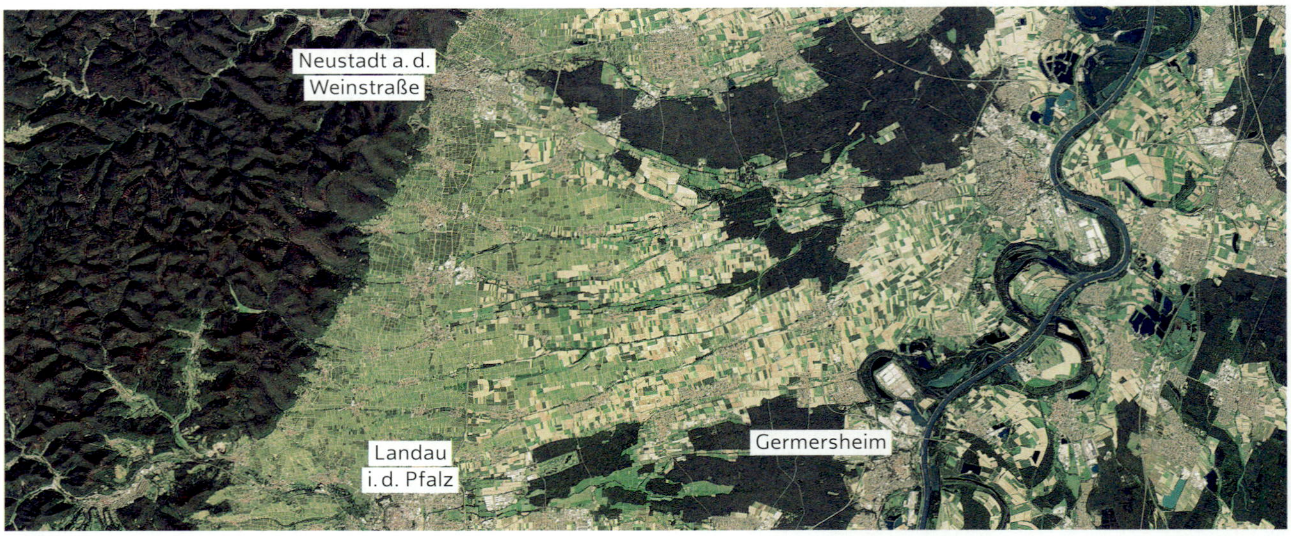

M1 Das Rheintal südlich von Speyer, Höhe ca. 90 km

Ein Satellitenbild beschreiben und interpretieren

Beschreibung

1 Wo? Wann? Was? – das Thema des Satellitenbildes
Welcher Raum wird gezeigt? Wie groß sind die Entfernungen? Wo ist Norden im Bild? Wann wurde das Bild aufgenommen?

2 Welche Einzelheiten kannst du erkennen?
Gliedere den Bildausschnitt in Teilräume. Beschreibe einzelne besonders auffällige Objekte (Städte, Flughäfen, Seen usw.).

3 Was ist die Aussage des Bildes?
Zusammenfassung aller Informationen zu einer Kernaussage

NEU: Interpretation

4 Wie kann man das Dargestellte erklären?
Erkläre, was du beschrieben hast. Das geht umso besser, je mehr du über den Raum weißt. Du solltest den Atlas zur Erklärung hinzunehmen.Es ist auch sinnvoll, sich zusätzliche Informationen zu verschaffen. 213

Helpful words and phrases for the description (1.):

- *... the satellite image shows (landscape / mountains / coast / city)*
- *... the image shows an area of 45km x 20km (say: "45 by 20 kilometres") ...*
- *... north is on the top / right / left ... side of the image ...*
- *... the image was taken in summer / winter ...*

➡ Übungsmöglichkeiten zum Satellitenbild: Seite 146, 160, 172, hinterer Buchdeckel Innenseite

INTERNET

M2 Google Maps bietet nicht nur viele Karten und Zusatzinformationen, man kann auch das Satellitenbild einschalten. Willst du dir Fotos von einem bestimmten Ort anschauen, dann klicke auf das gelbe Männchen unten rechts. Dann kannst du dir in den blau gekennzeichnetet Straßen Fotos anschauen.

Mit Google Earth Timelapse kannst du die Veränderung von Räumen auf Satellitenbildern im Zeitraffer beobachten (S. 90/91).

schueler.diercke.de bietet dir einen Zugang zum Diercke Globus 202. Damit kannst du die gesamte Welt in Satellitenbildern (Flughöhe 15 bis 12 800 km) und Karten in 3D und mit einem Flugsimulator bereisen.

Helpful words and phrases for the description (2.):

- *... we can see different areas / sections of the image ...*
- *... some features are striking (auffällig) ...*
- *... the airport / the lake / the river / the coastline / the city centre ...*
- *... the landscape is clearly structured ...*
- *... a large, dark green area in the west ... probably forest ...*
- *... east of the dark green area, we can see light green areas ... probably fields...*
- *... further east are light green areas with a rectangular pattern ...*

Helpful words and phrases for the interpretation (4.):

- *... woods / fields / settlements are located ...*
- *... different fields near the mountains ...*
- *... vineyards (Weinberge) are located there, because...*
- *... in the valley (Tal) / on the plain (Ebene) there are fields, because...*
- *... we can see a regulated river course (begradigter Flusslauf) / the course of the river has been shortened...*
- *... straight / regular riverbed ...*
- *... in the mountainous area we can see dense forest because ...*

M

Wie beschreibe und interpretiere ich Tabellen und Diagramme?

Eine Tabelle beschreiben und interpretieren

Beschreibung

1 Wo? Wann? Was? – das Thema der Tabelle

2 Was sind die Inhalte?
a) Informationen aus Tabellenkopf und Vorspalte entnehmen, die Maßeinheiten beachten!
b) Welches sind die Extremwerte?
c) Wird eine zeitliche Entwicklung dargestellt? Wie entwickeln sich die Werte?
d) Ist es sinnvoll, Werte zu vergleichen?

3 Was ist die Kernaussage der Tabelle?
Fasse die Aussage der Tabelle in ein bis zwei Sätzen zusammen.

NEU: Interpretation

4 Wie lässt sich die Aussage erklären, welche Folgen könnten sich ergeben?
Recherchiere gegebenenfalls nach Zusatzinformationen. 213

Ein Diagramm beschreiben und interpretieren

Beschreibung

1 Wo? Wann? Was? – das Thema

2 Was sind die Inhalte?
a) Informationen in der Beschriftung der Achsen: Maßeinheiten beachten (z. B. Jahre)!
b) Welches sind die Extremwerte?
c) Wird eine zeitliche Entwicklung dargestellt? Wie entwickeln sich die Werte?
d) Ist es sinnvoll, Werte zu vergleichen?

3 Was ist die Kernaussage des Diagramms?

NEU: Interpretation

4 Wie lässt sich die Aussage erklären, welche Folgen könnten sich ergeben?
Recherchiere gegebenenfalls nach Zusatzinformationen. 213

➡ Übungsmöglichkeiten zum Erstellen:
Seite 10, 12, 30, 46, 48, 56, 74, 158

Wie erstelle ich selbst Tabellen und Diagramme?

Welche Form der Darstellung ist für dein Vorhaben am Sinnvollsten: **Tabelle** oder die Darstellung in einem **Balken-**, **Säulen-**, **Linien-** oder **Kreisdiagramm**.

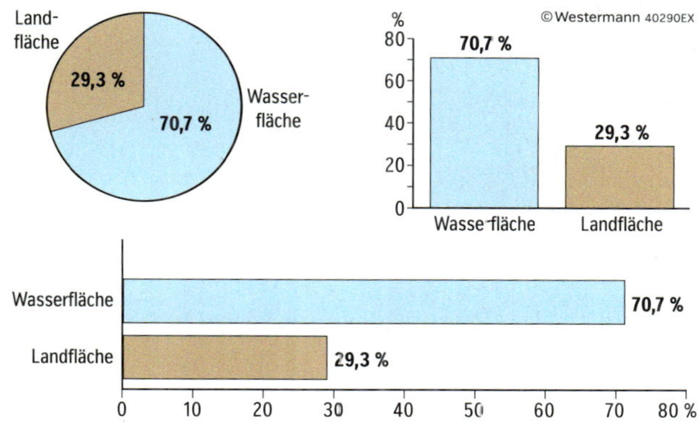

M3 Das Kreisdiagramm (auch: „Tortendiagramm" oder pie chart) macht besser als ein Balken- oder Säulendiagramm Anteile an einem Ganzen deutlich.

Tabellen und Diagramme selbst erstellen

1 Die Achsen zeichnen
Was sind die höchsten Werte? Wie lang müssen demnach die beiden Achsen sein?

2 Die Achsen beschriften
Thema, Zahlenwerte
Achtung: gleiche Abstände zwischen den Teilstrichen (z. B. je Jahr ein Zentimeter).

3 Die Säulen, Balken oder Linien zeichnen
Die Säulen/Balken einzeichnen
Die Linien zeichnen. Zuerst die Werte zwischen den beiden Achsen markieren (z. B. Punkte). Dann die Markierungen verbinden.

4 Den Titel formulieren
Tabellen und Diagramme kannst du auch mithilfe von Computerprogrammen erstellen.

Helpful words and phrases:
... the chart (Tabelle) / bar chart (Säulendiagramm) / line graph (Liniendiagramm) / pie chart (Kreisdiagramm) ...
... shows ... / gives information on ... /
... numbers / values / percentage ...
... the number of / the ratio (Verhältnis) of ... and ...
... the development of ...
... between ... and ... rises / falls / stays stable ...
... the maximum / minimum of ... is ...
... the most important information is that ...

Wie beschreibe und interpretiere ich ein Klimadiagramm?

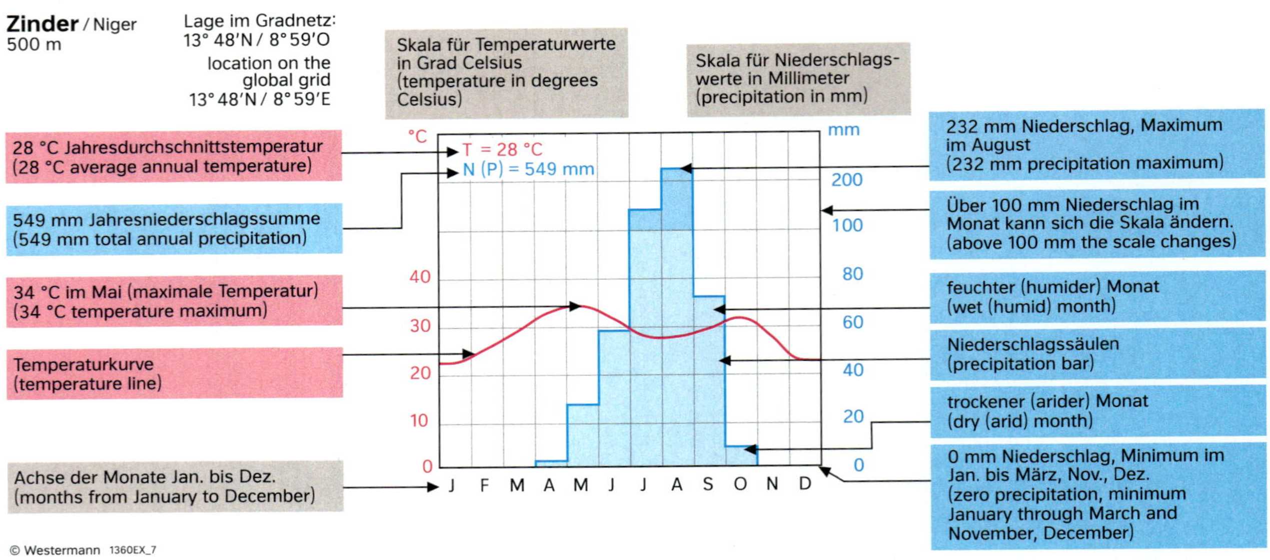

M1 Klimadiagramm der Station Zinder (Niger)

Ein Klimadiagramm beschreiben und interpretieren

Beschreibung

1 Wo? Wann? Was?

2 Was sind die Inhalte?

a) Verlauf der Durchschnittstemperatur?
b) Wie hoch sind die Niederschlagssummen?
c) Wie entwickeln sich die Werte? Welches sind die Extremwerte?

3 Was ist die Kernaussage des Klimadiagramms zum Klima an diesem Ort?

NEU: Interpretation

4 Wie lassen sich die Werte des Klimadiagramms interpretieren?

- Wie lässt sich der Jahresgang von Temperatur und Niederschlag erklären?
- Wie kommt es zu den Extremwerten? Gibt es einen Zusammenhang zwischen den Niederschlag und Temperatur?
- Welche Auswirkungen hat dieses Klima auf die Natur und die Menschen?
- Wie sind Anbaumöglichkeiten für Pflanzen oder die Voraussetzungen für Tourismus?
- Recherchiere gegebenenfalls nach Zusatzinformationen. 213

Helpful words and phrases for the description:

... the graph shows the climate of ...
... its location in the grid system is ...°N / S and ...°E / W.
... the metereological station is located ... m above sea level
... the average annual temperature
... the total annual precipitation is ...
... the maximum / minimum temperature is...
... the maximum / minimum monthly precipitation is ...
... the hottest / coldest / wettest / driest month is ...
... the rainy / dry season is from ... to ...
... there are ... arid / humid months ...

Helpful words and phrases for the interpretation:

... the temperature course results from the location in the northern / southern hemisphere (Nord- / Südhalbkugel)
... plants growth is poor because of hot temperatures and little rain

Wie zeichne ich ein Klimadiagramm?

Tipps zum Zeichnen eines Klimadiagramms:

Du solltest darauf achten,

- *dass 10 °C auf der Temperaturachse jeweils 20 mm Niederschlag entsprechen,*
- *dass es möglicherweise sinnvoll ist, bei hohen Niederschlagswerten den Maßstab zu verändern (bis 100 mm: 1cm je 10 mm; über 100 mm: 1cm je 100 mm. Siehe auch M1),*

Sehr schöne Klimadiagramme kannst du mit Diercke Klimagraph erstellen: diercke.westermann.de/Klimagraph

➡ Übungsmöglichkeiten zum Klimadiagramm: Seite 74

Wie werte ich eine Bevölkerungspyramide aus?

Eine Bevölkerungspyramide interpretieren

Beschreibung

1 Die Form der Bevölkerungspyramide
Beschreibe die Umrisse der Pyramide.
Welchem Modell entspricht sie am ehesten?
Nenne die auffälligsten Werte/Extremwerte.

Interpretation

2 Wie lässt sich die Form der Pyramide erklären?
- Hat die Pyramide eine breite Basis, d. h. gibt es viele Geburten?
- Verjüngt sich die Pyramide nach oben stark, das heißt sterben viele Menschen früh?
- Ist die Lebenserwartung hoch?

3 Abweichungen vom Modell
- Beschreibe, wo sich starke Abweichungen vom Modell ergeben (z. B. viele Frauen in einer Altersgruppe).
- Erkläre diese Abweichungen – dazu musst du dir Zusatzinformationen beschaffen.

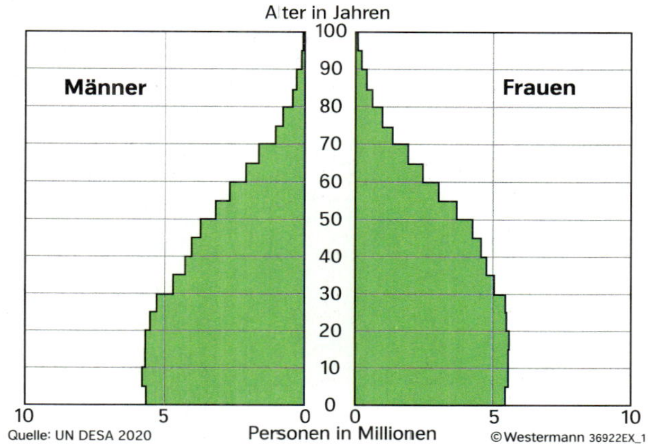

M3 Bevölkerungspyramide von Mexiko 2020

4 Folgerungen und Probleme
Stelle die Folgerungen und Probleme dar, die aus der Bevölkerungspyramide ablesbar sind, zum Beispiel: die Bedeutung der Bevölkerungsentwicklung für die Nahrungsmittelversorgung, die Bildung, den Arbeitsmarkt, die Versorgung der Senioren.

Formulierungshilfen zur Interpretation 201, 207

Wie beschreibe und interpretiere ich eine Karikatur?

Eine Karikatur beschreiben und interpretieren

1 Beschreibung
Wer und was ist dargestellt (z. B. Personen, Orte, Probleme)?
Wie ist es dargestellt? Wieso wirkt die Karikatur **übertrieben** und **witzig**? (Was ist anders, z. B. größer/kleiner dargestellt als in Wirklichkeit?)
Welche Bedeutung hat die **Bildunterschrift**?

2 Interpretation
Auf welche Probleme oder Sachverhalte soll aufmerksam gemacht werden? Welche Meinung wird vertreten? Was soll mit der Karikatur bei Betrachtern erreicht werden?

3 Eigene Stellungnahme
Was sagst du selbst zu dem Dargestellten?

Helpful words and phrases for the interpretation:
In the caricature "Die Kreuzfahrer" a cruise ship is depicted moving at full speed …
It is labeled "Klimakonferenz" (Climate Conference) and emits from its chimney instead of smoke very many …
In the foreground clings … and calls in vain …
The ship is intended to show that countless … are being produced at the climate conference, but that in the end they do not help the Earth, but …
The Earth clings to a palm tree…
The Earth seems desperate because …
The artist wants to draw attention to this…
I agree / disagree with the the cartoonist …
In my opinion, we should…

Bildelement	Darstellung	Interpretation
Wasser	sehr im Vordergrund	überflutet alles (Klimawandel)
Schiff „Klimakonferenz“	groß, fährt weg	Klimakonferenz groß, unbeweglich, kümmert sich nicht …
Blätter	…	…

M2 Hilfe zu Beschreibung und Interpretation

Wie arbeite ich mit Modellen?

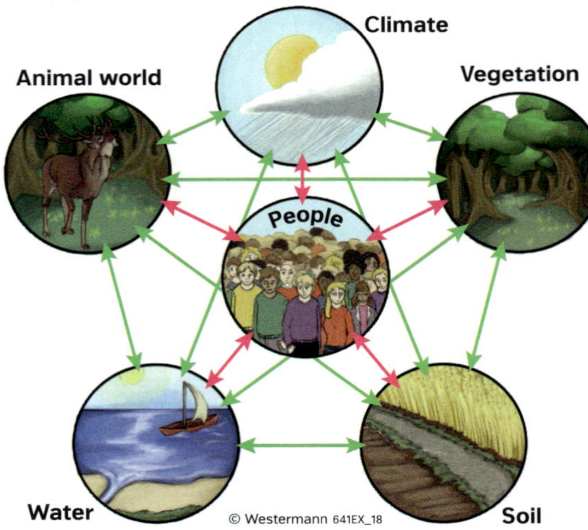

M1 Das Modell eines Ökosystems kann man auf alle Ökosysteme anwenden: Auf die unterschiedlichen Landschaftszonen (z. B. die gemäßigte Zone oder die Zone des tropischen Regenwaldes). Man kann auch die Erde als großes Ökosystem ansehen und das Modell darauf anwenden – oder auch auf den heimischen Garten; auch diesen kann man als Ökosystem betrachten.

➡ Übungsmöglichkeiten zum Modell: Seite 28, 48, 112, 114, 132, 134, 174, 187

M5 Ein **Modell** ist ein vereinfachtes Abbild der Wirklichkeit. Es reduziert einen Sachverhalt auf das Wesentliche (z. B. Pendlermodell). Mithilfe von Modellen kann man viele komplizierte Sachverhalte übersichtlich darstellen, sich gut merken und einfach erklären.

1 2 3 Mit Modellen arbeiten

1 Das Thema des Modells
Hinweise dazu gibt die Bezeichnung des Modells.

2 Was ist der genaue Inhalt?
a) Sieh dir die Beschriftungen genau an, die Farben, die Verbindungen zwischen einzelnen Begriffen.
b) Was wird hier verallgemeinert? Was ist das Typische, das hier dargestellt werden soll?
c) Zu welchen Einzelbeispielen aus der Wirklichkeit passt das Modell?

3 Was ist die Kernaussage des Modells?
Fasse die Aussage des Modells in ein bis zwei Sätzen zusammen.

Wie verknüpfe ich Informationen aus verschiedenen Materialien?

1 2 3 4 5 Informationen aus verschiedenen Materialien verknüpfen

1 Was will ich wissen?
Welche Fragen möchte ich mithilfe der Materialien beantworten?
Welche Informationen benötige ich?

2 Die Materialien ordnen

3 Die Auswertung
Notiere von jedem Material die Kernaussage und die Schlüsselbegriffe/Schlüsselwörter.

4 Die Zusammenfassung
Unterstreiche in deinen Notizen mit jeweils einer Farbe alle Informationen, die zu einem Thema oder Unterthema gehören. Schreibe diese auf ein frisches Blatt in sinnvoller Reihenfolge. (Es ist sinnvoll, dahinter immer einen kurzen Verweis zu schreiben, aus welchem Material du den Begriff hast.)

5 Die Kernaussage
Fasse deine Auswertung in wenigen Sätzen zu einer Kernaussage zusammen.
Bei komplizierteren Sachverhalten ist es sinnvoll, die Zusammenhänge in einer Mindmap, einem Wirkungsgefüge oder einer Conceptmap darzustellen. 199
Beantworte die Fragen, die du dir am Anfang gestellt hast.

➡ Übungsmöglichkeiten zur Materialverknüpfung: Seite 11, 24, 26, 48, 56, 58, 60, 88

Wie verfasse ich ein Referat?

Ein Referat verfassen

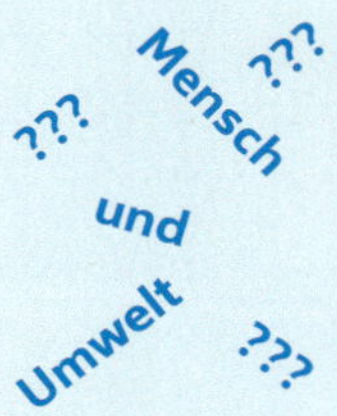

1 Schritt:
Festlegung des Themas/der Leitfrage

2 Schritt:
a) Suche nach Informationen/Recherche
b) Bearbeitung des Themas
Ich beachte immer die Leitfrage.

3 Schritt:
a) Ausarbeitung des Referats
b) Ausarbeitung der Präsentation

4 Schritt:
Vortragen der erstellten Präsentation

Ein gutes Referat (Begriff angelehnt an lat. referre: mitteilen, berichten) ist von verschiedenen Bedingungen abhängig.
Je umfassender du den Inhalt recherchierst, das heißt nachschlägst und aufarbeitest, desto besser wird dein Referat. Dein Auftreten bei der Präsentation wird dann auch viel sicherer sein. Der Inhalt steht bei einem Referat allerdings im Vordergrund.

INFO

M3 Das schriftliche Referat

- **Titelblatt** (der Titel des Referates und Name der Verfasserin oder des Verfassers – dein Name. Man kann das Titelblatt auch bunt mit einem oder mehreren Bildern gestalten.)
- **Gliederung**
- **Einleitung**
  Was ist das Besondere an dem Thema? Warum sollte man sich mit dem Thema beschäftigen?
- **Hauptteil**
  Das Thema wird in einzelne Abschnitte/Kapitel, in sinnvoller, logischer Abfolge unterteilt.
- **Schluss**
  Zusammenfassung des Themas, das Wichtigste
  Eigene Stellungnahme: Wie ist deine persönliche Meinung zu dem Thema?
- **Verwendete Quellen**

Wenn man ein Referat vorträgt, teilt man häufig an die Zuhörerinnen und Zuhörer eine kurze Zusammenfassung aus. Dieses Handout enthält die wichtigsten Inhalte, wichtige Fachbegriffe und vielleicht auch wichtige Materialien (z. B. Tabellen).

Zunächst musst du dein Thema genau abgrenzen. Formuliere dazu eine Leitfrage oder Problemstellung. An ihr orientierst du dich bei der Recherche, das heißt bei der Erkundung und Untersuchung der Materialien.
Für eine gute Materialsammlung muss das Thema deines Referats eng gefasst sein, da du sonst zu viel Material erhältst. Ein gutes Thema wäre z. B. „Der Kokaral-Damm: Eine nachhaltige Lösung für den Aralsee?“, wohingegen „Der Aralsee“ ein schlechtes Thema wäre.
Nun beginnst du, für dein Thema zu recherchieren. 213 Nutze möglichst unterschiedliche Quellen: Bücher, Zeitschriften und, wenn möglich, das Internet. Achte hier vor allem darauf, dass deine Quellen vertrauenswürdig sind! Vorsicht vor Fehlern, Fakes und Fälschungen!

- themenbezogen – gut aufbereitet – verständlich – gut recherchiert – gut gegliedert
- Wie präsentiere ich die Ergebnisse?
- Wie stehe ich vor der Gruppe?

M4 Darauf solltest du achten!

➡ Übungsmöglichkeiten zum Referat:
Seite 54–61, 102–107, 123, 169, 192

Wie nutze ich eine Datenbank?

Datenbanken (databases) dienen dazu, große Mengen an Daten sicher zu speichern, sinnvoll zu strukturieren und damit einfach abrufbar zu machen. Sie können dann problemlos in einem weiterführenden Programm verarbeitet werden. So kann man zum Beispiel für beliebig viele Regionen der Erde die Daten zur Bevölkerungszahl, zur Sterberate usw. speichern. Man kann auch mithilfe einer Abfrage (Query) bestimmte Daten auswählen und dann zur Darstellung in einer Tabelle, einem Diagramm oder einer Karte (z. B. GIS) nutzen.
Datenbanken bestehen aus dem *Datenbestand* (Menge der Daten) und dem *Systemkatalog*, in dem die Ordnungskriterien, die logische Struktur der Datenbank festgelegt ist.

Zur Nutzung der Datenbank startet man eine Abfrage über den Systemkatalog und erhält dann die gewünschten Daten aus dem Datenbestand. Bei vielen Datenbanken wird man durch die Abfrage geführt: Die möglichen Suchkriterien (z. B. Geburtenrate, BNE, Alphabetisierungsrate) werden dann automatisch in der Abfragemaske vorgeschlagen. Man kann auch verschiedene Suchkriterien miteinander verknüpfen.

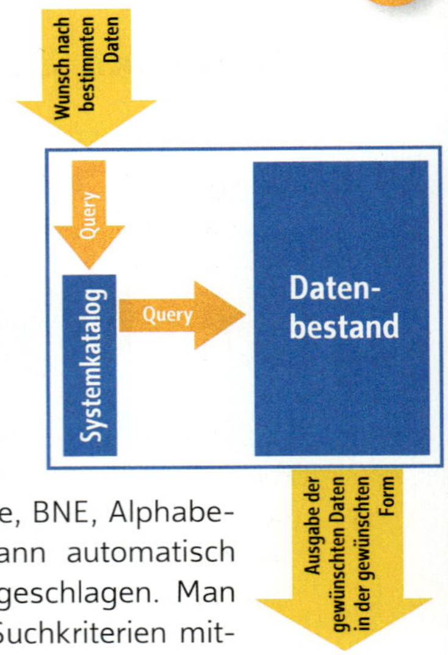

INTERNET

M1 Diese Organisationen geben kostenlosen Zugang zu umfangreichen Datenbanken:
- Food and Agriculture Organisation of the UN, FAO → FAOstat (mehrere Datenbanken mit detaillierten Angaben z. B. zu allen Erzeugnissen des primären Sektors)
- Organisation for Economic Cooperation and Development, OECD → OECD Statistics
- United Nations (z. B. World Population Prospects)
- United Nations Development Programme, UNDP → Human Development Reports, Data
- United Nations Population Fund, UNFPA
- World Bank → open data
- European Union → Eurostat → data
- Destatis (Statistisches Bundesamt)

Ziel: Ich suche die Bevölkerungszahlen aller Staaten der Erde für die Jahre 1950, 2020 und 2100.
Datenbank: UN World Population Prospects (https://population.un.org/wpp/)
Vorgehen:
→ „Interactive Data" → Es erscheint die Abfragemaske „Data Query" → „Total Population by sex (thousands)" → „World" → Es erscheint die Frage: „Do you want to select the sub-locations as well?" → „Yes" → „Next" → Es erscheint eine Liste aller verfügbaren Jahre → „1950" + „2020" + „2100" → „Next" → „Both sexes combined" → „Next" → Es erscheinen die möglichen Ausgabeformen der Tabelle → „By Years/Periods" → „Results" → Es erscheint die Tabelle in der gewünschten Form mit den gewünschten Daten. Wenn man es wünscht → „Export to Excel"

M4 Ein Beispiel

Große Datenbanken werden ständig aktualisiert und immer mal wieder in ihrer Oberfläche verändert. Auch die Möglichkeiten des Datenabrufs und der Datenausgabe werden von Zeit zu Zeit optimiert. Was die Nutzeroberfläche betrifft, gibt es keinen allgemeingültigen Standard. Bei der Nutzung von Datenbanken muss man sich daher immer neu in ihren Aufbau und in ihre Abfragemöglichkeiten eindenken.

M2 Hinweis

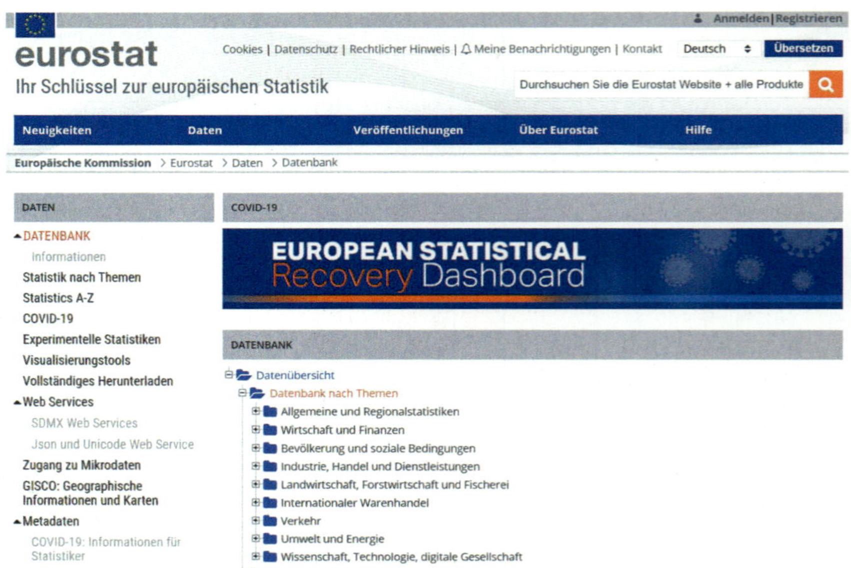

M3 Eurostat: Abruf von Daten zum Bruttoinlandsprodukt (BIP)

➡ Übungsmöglichkeiten: Seite 72, 94, 152, 184, 186

M

Wie gelange ich an geeignete und glaubwürdige Informationen?

INFO

M5 **Recherche** (franz. recherche) bezeichnet die gezielte Suche nach Informationen. Am besten recherchiert man in Bibliotheken und im Internet.

Internetlexika (z. B. Wikipedia): Hier gibt es keine Redaktion, jeder kann an Artikeln mitarbeiten. Deshalb ist der Wahrheitsgehalt manchmal zweifelhaft.
Filmsammlungen: Allein bei YouTube kannst du Zehntausende Lehr- und Erklärvideos abrufen. 219
Webcams liefern Livebilder von der gesamten Erde.

M6 Ergiebige Informationsquellen im Internet

**M9 Größte und bekannteste Suchmaschine: *Google* (globaler Marktanteil 2020: 92 %).
DuckDuckGo verspricht, wenig persönliche Daten von den Nutzern (also von dir) zu sammeln. *ECOSIA* nutzt den Gewinn aus den Werbeeinnahmen, um neue Bäume zu pflanzen (im November 2020 waren es bereits 110 Millionen).**

INFO

M7 **Von Fehlern, Fakes und Fälschungen**
Im Gegensatz zu gedruckten Fachbüchern und Fachzeitschriften sind die Informationen aus dem Internet häufig fehlerhaft oder nicht glaubhaft.
Anzeichen für unseriöse Internetseiten sind:
- Es gibt kein Impressum, der Autor ist unbekannt.
- Die Rechtschreibung ist sehr fehlerhaft.
- Bilder, Tabellen, Diagramme und Texte haben keine Quellenangabe, obwohl sie nicht von demjenigen stammen, der die Seite erstellt hat.
- Die Informationen sind so zusammengestellt, dass sie für etwas werben und nicht mehr objektiv wirken.
- Es gibt keine Angaben zum Erstelldatum.

Richtig recherchieren!

1. **Was will ich wissen?**
Grenze das Thema ab, formuliere Fragen und eine übergeordnete Fragestellung.

2. **Schlüsselwörter finden**
Welches könnten die Schlüsselwörter sein, auf die die Recherche aufbaut?

3. **Suchmaschinen füttern**
Schau dir nun die (ersten zehn) Suchergebnisse an und bewerte sie: Liefern sie genügend Informationen? Sind sie vertrauenswürdig?

4. **Die Suche verändern**
Willst du nach weiteren Quellen suchen?
   - Verwende eine andere Suchmaschine.
   - Gib andere Schlüsselwörter oder dieselben Schlüsselwörter in anderer Reihenfolge in die Suchmaske ein.

Wie verfasse ich Quellenangaben?

Bei Informationen aus Büchern, Zeitschriften und dem Internet ist es sehr wesentlich, jeweils die Quelle anzugeben. Wichtig sind vor allem der Autor und der Titel der Quelle. Für die Form der Quellenangaben gibt es Regeln, die eingehalten werden müssen:
(**Blaue Elemente** verwendest du immer unverändert, die konkreten Inhalte der restlichen Bestandteile ergeben sich aus deiner jeweiligen Quelle.)

- ***bei Büchern (Monografien)***
Nachname des Autors**,** Vorname des Autors**:** Titel des Werkes**.** x**. Auflage.** Verlagsort**:** Verlag Erscheinungsjahr**.**

- ***bei Sammelbänden mit Herausgeber***
Nachname des Autors**,** Vorname des Autors**:** Titel des Aufsatzes**. In:** Titel des Sammelwerkes**. Herausgegeben von** Vor- und Nachname des Herausgebers**.** x**. Auflage.** Verlagsort**:** Verlag Erscheinungsjahr**,** Seitenzahlen**.**

- ***bei Zeitschriften***
Nachname des Autors**,** Vorname des Autors**:** Titel des Beitrags**. In:** Titel der Zeitschrift**,** Heftnummer **vom** Datum**,** Seitenbereich**.**

- ***Seiten im Internet***
Name des Autors**,** Vorname des Autors**:** Titel des Werkes**. In:** Titel der Webseite**.** Erstelldatum**.** URL **[Zugriff:** Datum**].**

M8 Quellenangaben haben eine bestimmte Form

Wie kann ich Materialien kritisch hinterfragen? – Statistiken, Diagramme, Fotos und mehr!

Viele Materialien, die wir auswerten, sind mit einer bestimmten Zielsetzung, mit einer bestimmten Intention verfasst. Die Autoren wollen oft nicht (nur) objektiv informieren, sie wollen eine bestimmte Sichtweise darstellen. Zuweilen wollen sie uns sogar manipulieren.

In den Bildern, Filmen und Texten der Werbung wird das sehr deutlich und ist auch jedem bewusst. Aber auch Informationen, die auf den ersten Blick seriös erscheinen, sind häufig nicht objektiv. Das gilt für Tabellen, Diagramme, Texte, Karten, Fotos und Filme.

Kann man den Statistiken trauen?

Genaue Daten?
Die Bevölkerungszahl der Länder, das BNE, das Durchschnittseinkommen – all diese Werte werden aufwendig berechnet oder auch nur geschätzt – wie auch immer: Wirklich genaue Angaben gibt es nicht häufig. So kennt zum Beispiel niemand die genaue Einwohnerzahl Deutschlands, da Bevölkerungszählungen nur selten durchgeführt werden; die beiden letzten Zählungen fanden 1987 und 2011 statt. Alle aktuelleren Daten werden von Fachleuten fortgeschrieben.
Dennoch haben auch diese „ungenauen Daten" ihren Sinn: Sie vermitteln eine ungefähre Vorstellung von Größenordnungen und ermöglichen Vergleiche mit anderen (genauso ungenauen) Daten.

Zu genaue Daten?
Wenn für Brasilien eine Bevölkerungszahl von 208 362 750 (2017) angegeben wird, dann sollte dich auch das misstrauisch machen. Sehr genaue Daten wirken seriös, sind aber oft auch nur geschätzt. In der Statistik spricht man auch von Schein- oder Pseudogenauigkeit.

„Uralte" Daten?
Nur wenige Daten werden regelmäßig erhoben und veröffentlicht. Das gilt besonders für Entwicklungsländer, aber auch für Industrieländer.

M1

M3 Karikatur

VORSICHT

M4 Vorsicht bei Mittelwerten
Mittelwerte sagen alles – und gar nichts. Wenn die Temperaturen im Durchschnitt immer über 5 °C betragen, kann doch ein strenger Frost die Ernte vernichten. Bei einem Schnitt von 3,1 in der Klassenarbeit kann es sein, dass niemand eine „3" in der Arbeit hat.

Zwei Klimastationen im Vergleich:
Plymouth (England)
Jahresdurchschnitt: 10,8 °C
Abs. Minimum -8,9 °C, abs. Maximum 31,1 °C
Omaha (Nebraska/USA)
Jahresdurchschnitt: 10,8 °C
Abs. Minimum -35,6 °C, abs. Maximum 43,9 °C

M5 Mittelwerte

ERSTAUNLICH I

M2 Genaue Daten wirken überzeugend
Der Polarforscher Robert E. Peary gab am 6. April 1909 folgende Position an: *89° 57′ 11″ nördlicher Breite*. Das entspricht einer Entfernung zum Nordpol von knapp 5 km. Die Angabe von ***11″*** lässt die Ortsbestimmung auf etwa ±30 m genau erscheinen. Damals war aber eine solche Messung noch nicht möglich. Die maximal mögliche Genauigkeit lag bei ±30 km. Für Peary wäre eigentlich also nur die Angabe ***ca. 60′*** möglich gewesen.

ERSTAUNLICH II

M6 Genaue Daten sind häufig sehr ungenau ...
Das renommierte World Factbook des CIA gibt für 2019 die Einwohnerzahl von China mit 1 394 015 977 an und die Einwohnerzahl Deutschlands für 2020 mit 80 159 662.
Die UNO weist 2019 für China 1433,784 Mio. Einwohner aus und für Deutschland 2020 83,748 Mio. Das Statistische Bundesamt der Bundesrepublik Deutschland nennt für 2019 für China 1 397 715 000 und für Deutschland 2020 83 122 889 Einwohner.

Absolute Zahlen und relative Zahlen (Prozent oder eine Umschreibung wie „fast die Hälfte") können sehr unterschiedlich wirken. Zum Beispiel wirkt die Aussage „An der australischen Westküste gab es in diesem Jahr doppelt so viele Haiangriffe wie im Vorjahr!" alarmierender als „Die Zahl der Haiangriffe ist von ein auf zwei gestiegen."

M7 **Absolute oder relative Zahlen?**

Erstaunlich viele ...

Die Zahl der Raucher steigt stark!

Anteil der Raucher sinkt um ein Viertel!

7 % Raucher mehr weltweit!

Alle Aussagen stimmen! Die Ausgangsdaten:

	1990	2015
Weltbevölkerung	5,3 Mrd.	7,3 Mrd.
Raucher	870 Mio.	933 Mio.

M8 **Absolute und relative Zahlen – mit unterschiedlicher Wirkung ...**

Wie manipuliert man mit Karten?

Die Aussage einer Karte hängt wesentlich davon ab, in welche Gruppen (Klassen) man die Daten zusammenfasst und welche Farbabstufungen man wählt. Auch die „Kartengrundlage" (Projektion/Netzentwurf) spielt bei kleinen Maßstäben eine Rolle.

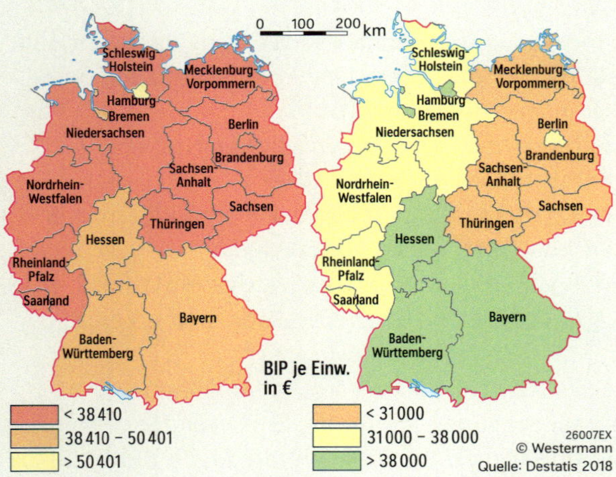

M9

Tipp: Auf den Seiten 72/73 und 202 – 204 kannst du selbst digitale Karten erstellen.

Wie manipuliert man mit Diagrammen?

Durch eine wohlüberlegte Darstellung kann man mit denselben Daten unterschiedliche Wirkung erzielen:

- Wahl der Abmessungen (Höhe, Breite),
- Weglassen nicht gewünschter Zeiträume,
- Verlegen des Ursprungs,
- durch Hinzufügen von „Schätzungen", die einen gewünschten Trend verstärken.

M10

A

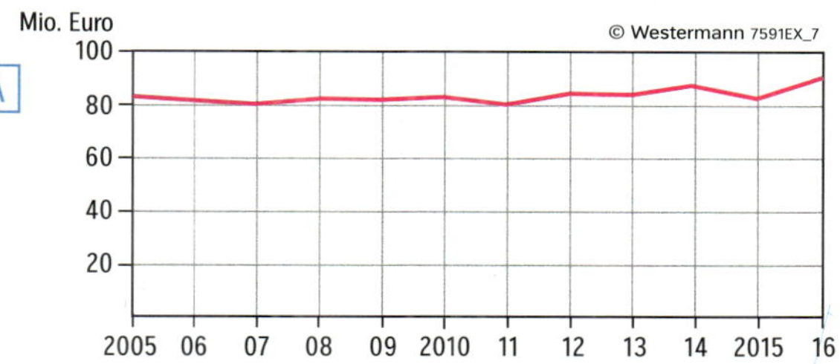

B

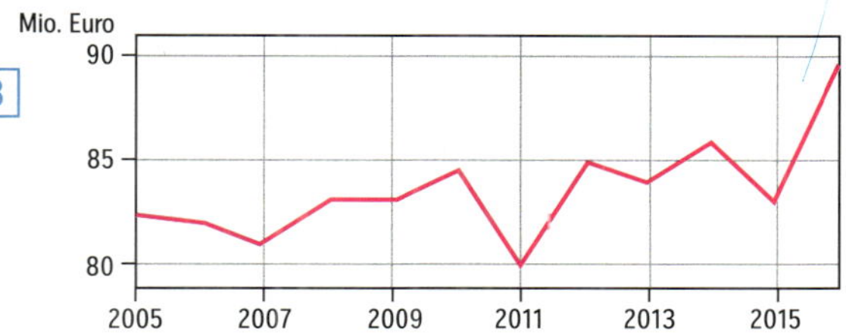

C

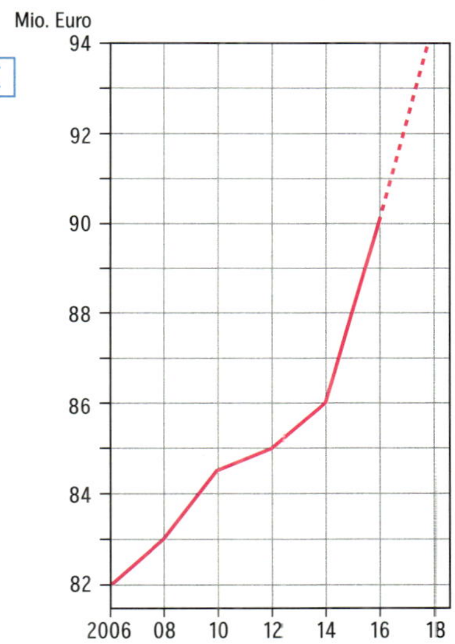

Anmerkung der Geschäftsleitung. „Darstellung im Zweijahresrhythmus zur besseren Übersicht über die Geschäftsentwicklung"

M11 **Welches dieser Diagramme zur Entwicklung der Gewinne würde die Firma Fakecard AG wohl verwenden? A, B und C haben dieselbe Datengrundlage.**

Die richtigen Folgerungen?

Oft erscheinen Schlussfolgerungen sehr logisch, obwohl die Zusammenhänge völlig sinnlos oder zufällig sind: „Die Hälfte aller Todesfälle ereignen sich im Krankenhaus."

Folgerung: Krankenhäuser sind gefährlich ...

M12

M1 Kreative Geographie durch Bildbearbeitung – Köln am Fuße des Vulkans Fuji (Mitte)

Wie manipuliert man mit Fotos?

Fotos können
- gestreckt werden oder gestaucht.
- aus der Froschperspektive etwas mächtig erscheinen lassen, aus der Vogelperspektive winzig
- leicht bearbeitet und gefälscht werden.
- Wichtiges weglassen.

M2 Vorsicht vor Fakes!

M5 Der richtige Ausschnitt machts ...

Wie manipuliert man mit Texten?

Texte sind das beliebteste Medium, um Meinungen darzustellen oder Interessen zu vertreten. Das geschieht auf verschiedenste Weise – offen und versteckt:
- durch Verwendung sprachlicher Mittel (die du aus dem Deutschunterricht kennst), vor allem Übertreibungen,
- durch Überbetonung oder/und Weglassen von Fakten,
- oder gar durch Verfälschung von Fakten – Fakes.

M3

VORSICHT

M4 Bin ich im falschen Film?
Nicht nur Fotos können manipuliert werden, auch Filme! Da wird eifrig geschnitten und retuschiert. Heute besitzt jeder per App die Möglichkeit, Gesichter realistisch gegen andere Gesichter auszutauschen (sogenannte Deepfakes). Du kannst überprüfen, ob du eine Urversion eines Films siehst oder eine veränderte Version. Gib die URL deines YouTube-Films in den YouTube DataViewer ein. Er findet die erste veröffentlichte Version.

Wie gehe ich grundsätzlich mit Informationsmaterialien um?

Was ist die Quelle? Wer ist der Autor? (dazu im Impressum nachschauen. Wenn es das nicht gibt, ist das schon bedenklich. Im Internet auch unter „Kontakt“ nachsehen.)
Für wen haben die Informationen Nutzen?
Für wen haben sie Nachteile?
Nutzt das Material Mittel zur Manipulation (z. B. Farben, sprachliche Mittel, Musik)?
Wie lauten die Kernaussagen des Materials?
Welches Ziel, welche Interessen sind erkennbar?

Bei allen schriftlichen oder mündlichen Präsentationen immer die Herkunft der Materialien, die Quellen (mindestens Autor, Jahr, Titel) angeben! 213

M6

Wie vergleiche ich richtig?

„Man kann nicht Äpfel mit Birnen vergleichen!“ Dieses alte Sprichwort ist im Prinzip richtig. Es gibt jedoch eine Ausnahme: Man muss für beide einen „gemeinsamen Nenner“ finden, einen Aspekt unter dem sich beide vergleichen lassen. Äpfel und Birnen sind zum Beispiel beides Früchte. Beide können reif sein oder unreif.
Man vergleicht verschiedene Dinge immer in Bezug auf ein Thema: Klassenarbeiten in Bezug auf die Note oder die Fehlerzahl, die Ferienwohnung in Bezug auf die Größe oder die Ausstattung und verschiedene Länder in Bezug auf ihre Fläche, die Bevölkerungszahl oder ihr Klima.

INFO

M7 **Vergleichen** heißt, **Gemeinsamkeiten und Unterschiede herausarbeiten** – zum Beispiel aus Tabellen, Diagrammen, Texten, Bildern oder Filmen.

Entwicklungen vergleichbar machen – mit Indexzahlen
Oft möchte man Daten in ihrer zeitlichen Entwicklung miteinander vergleichen (z.B. die Entwicklung der Einwohnerzahl verschiedener Gemeinden). Um diese Entwicklungen vergleichen zu können, nutzt man **Indexzahlen**.Dazu wird ein Wert innerhalb der Zeitreihe (der erste, der letzte oder auch einer in der Mitte) als Basiswert gleich 100 gesetzt. Die übrigen Werte werden dann im Verhältnis dazu umgerechnet (wie bei der Prozentrechnung).Manchmal bietet es sich auch an, den Basiswert auf einen Zeitpunkt zu legen, an dem ein besonderes Ereignis stattgefunden hat, durch das die Zeitreihen beeinflusst werden. Häufig ist es sinnvoll, die Zeitreihen mit den Indexwerten als Diagramm darzustellen.

M8 Indexzahlen

INFO

M9 **Indexzahlen** dienen dazu, unterschiedliche Datenreihen in ihrer zeitlichen Veränderung so aufzubereiten, dass sie vergleichbar werden.

➡ Übungsmöglichkeiten zum Vergleich mithilfe von Indexzahlen findest du auf den Seiten 20, 26, 30, 44, 64, 78, 118, 126, 160, 166, 186

Vergleichen

So gehst du sinnvoll vor:

1. Lege das Thema fest, unter dem der Vergleich stattfinden soll.
2. Überlege, welche Aspekte zum Thema gehören und analysiere die Materialien hinsichtlich dieser Aspekte. (evtl. Anlage einer Mindmap)
3. Suche hier nach **Gemeinsamkeiten** und nach **Unterschieden**.
4. Fasse die wesentlichen Ergebnisse des Vergleichs zusammen.

Die Umrechnung in Indexzahlen
Die Einwohnerzahl von drei Gemeinden hat sich verändert.
Wo fanden die größten Veränderungen statt?
Wie wirkte sich die Wiedervereinigung aus?

Entwicklung der Einwohnerzahlen

	1980	1990	2000	2010
Gem. A	882	817	801	1078
Gem. B	324	318	293	248
Gem. C	223926	235712	247497	294640

Indexzahlen — Basiswert = Anfangswert

	1980	1990	2000	2010
Gem. A	882	817	801	1078
	100	93	91	122

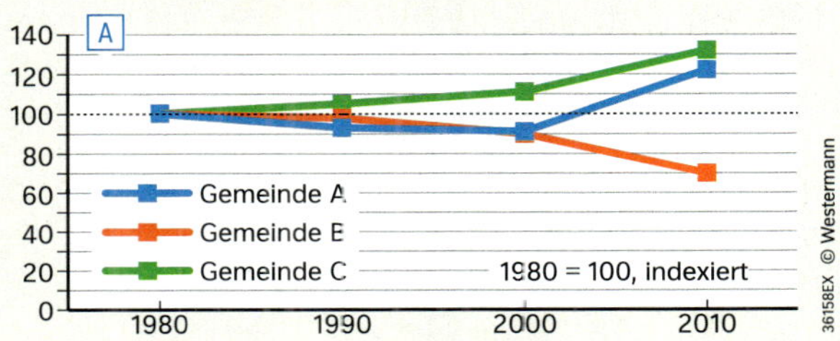

Indexzahlen — Basiswert = 1990 (Wiedervereinigung)

	1980	1990	2000	2010
Gem. A	882	817	801	1078
	108	100	98	132

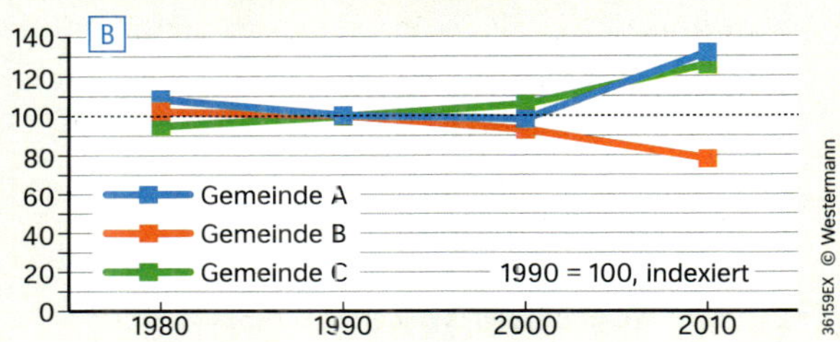

Wie kann ich meine Arbeitsergebnisse präsentieren?

Es ist immer sinnvoll, eine mündlich vorgetragene Präsentation optisch zu unterstützen. Dazu gibt es verschiedene Möglichkeiten: Bilder, Plakate und Schaubilder oder Folien, die an die Wand projiziert werden. Eine Möglichkeit ist es auch, eine Computer-Präsentation zu erstellen. Insgesamt wird so das Gesagte übersichtlicher, anschaulicher und bleibt besser im Gedächtnis haften.

Einen Vortrag präsentieren

1 Vorbereitung:

a) Notiere einen Stichwortzettel oder auf Karteikarten, was du mündlich vortragen willst. Aber Achtung: keine ganzen Sätze (die liest man häufig nur ab), sondern nur Stichworte (vor allem Fachbegriffe) notieren.

b) Bereite die Hilfsmittel vor, mit denen du deinen Vortrag anschaulich und einprägsam gestalten willst (Plakate, Folien, Fotos, Tafelbilder usw.).

2 Durchführung:

a) Nenne das Thema deines Vortrags.

b) Gib eine kurze Übersicht über den Inhalt.

c) Halte dich an die Regeln der 5-A-Technik (M1).

d) Nutze Hilfsmittel, um die Zuhörerinnen und Zuhörer zu fesseln.

e) Plane am Ende Zeit ein, damit die Zuhörerinnen und Zuhörer Fragen stellen können.

Und dann: Das Feedback!
Frag deine Mitschülerinnen und Mitschüler, wie ihnen der Vortrag gefallen hat, was sie behalten haben und was du hättest besser machen können. So kannst du deine Vorträge immer besser gestalten.

Ansehen: *Lies das Stichwort auf dem Stichwortzettel oder der Karteikarte.*
Aufsehen: *Sieh die Zuhörerinnen und Zuhörer an.*
Ansprechen: *Sprich erst dann, wenn du die volle Aufmerksamkeit hast.*
Abwechslung: *Verwende eine abwechslungsreiche Sprache, formuliere frei, lies nicht nur eintönig ab. Setze Hilfsmittel ein, die den Vortrag anschaulich und interessant machen. Sprich langsam und mache Pausen, gib Zeit, dir gedanklich zu folgen.*
Aufrecht stehen: *Steh aufrecht und wende deinen Zuhörerinnen und Zuhörern nicht den Rücken zu.*

M1 Die 5-A-Technik beim Vortragen

Lernplakat: Nur Material aufkleben, das sich gut einprägt – wenige Fotos, besser Diagramme und kurze Texte

- *Übersichtlich und gut lesbar gestalten!*
- *Eine sinnvolle, klare Gliederung formulieren!*
- *Eine kurze Überschrift (fette Schrift) verwenden!*
- *In großer Schrift schreiben, damit die Texte auch aus drei oder fünf Metern Entfernung (in der Klasse) gelesen werden können.*
- *Farben verwenden!*
- *Unterschiedliche Formen verwenden!*
- *Zusammenhänge mit Linien oder Pfeilen deutlich machen (wie in einer Mindmap oder einem Wirkungsgefüge)!* 199

M2 Plakate können noch längere Zeit im Klassenraum hängen bleiben und als Lernhilfe dienen.

- *Die Folien sollen sehr übersichtlich gestaltet sein, denn sie werden ja nur für kurze Zeit projiziert.*
- *Keine langen Texte!*
- *Nur einprägsame Schaubilder!*

M3 Es ist sinnvoll, am Ende der Präsentation die wichtigsten Quellen zu nennen. Dabei kannst du auch noch eine Empfehlung geben, wo sich deine Mitschülerinnen und Mitschüler weiter informieren können.

➡ Übungsmöglichkeiten zur Präsentation: Seite 53, 56 – 60, 102 – 106, 123, 169, 174, 190

Wie präsentiere ich meine Ergebnisse mithilfe eines Computerprogramms?

Computerprogramme, wie zum Beispiel *PowerPoint*, helfen dir, deinen Vortrag zu veranschaulichen oder das Gelernte zu visualisieren. Während du sprichst, kann eine Präsentation das Gesagte unterstützen. Möchtest du ein digitales Plakat erstellen, so kannst du das ebenfalls zum Beispiel mit *PowerPoint* tun (M5).

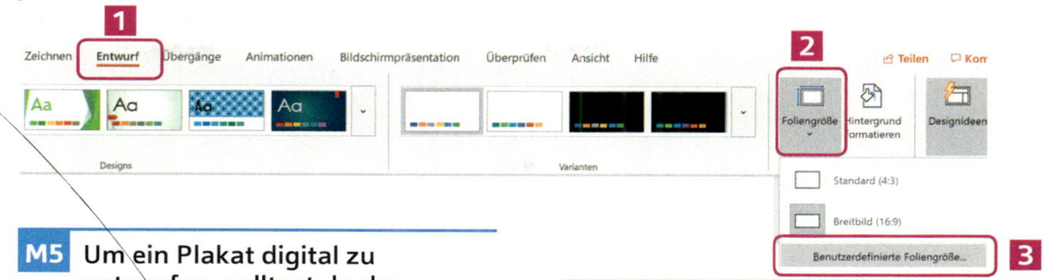

M5 **Um ein Plakat digital zu entwerfen, solltest du das Folienformat anders einstellen. Folge dazu den Schritten 1 – 5. Am besten sehen Plakate im Hochformat aus. Über „Einfügen" kannst du danach viele mögliche Materialien einfügen oder auch Texte und Diagramme selbst erstellen.**

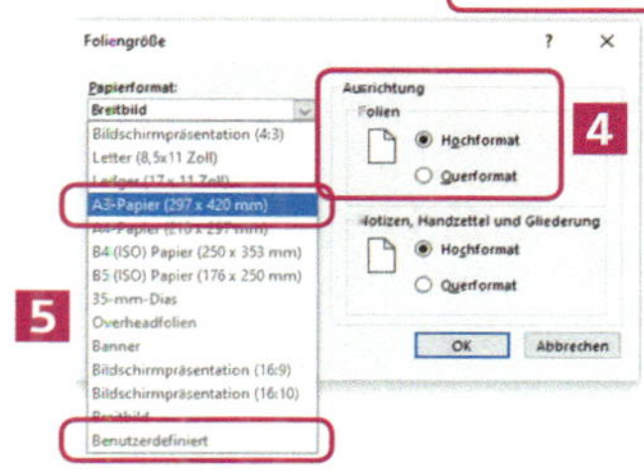

Erstellung einer Präsentation, zum Beispiel mit *PowerPoint*

① **Neue Folie wählen**
Es werden dir unterschiedliche Grundlayouts für die Foliengestaltung angeboten. Möchtest du frei eine Folie entwerfen, wählst du eine leere Folie aus.

② **Einfügen von Bildern/Grafiken/Texten**
Unter „Einfügen" kannst du auswählen, was auf deiner Seite erscheinen soll. Es ist möglich, Textfelder, Diagramme und Tabellen selbst zu erstellen oder Bilder und anderes einzufügen (Quelle nennen!). Du kannst auch Formen oder Piktogramme einfügen, mit denen du Aspekte auf deiner Folie hervorheben kannst.

③ **Benutzerdefinierte Animation**
Was, wann, wie auf deiner Folie erscheint, kannst du hier einstellen. Beachte, dass zu viel Animation vom eigentlichen Inhalt ablenkt.

Wie erstelle ich ein Erklärvideo? 219

Welche Arten von Erklärvideos gibt es?

Legetechnik-Video: ausgeschnittene Figuren und Texte werden vor der Kamera verschoben.
Trickfilm-Video: Puppen (z. B. Playmobil) werden durch zahlreiche Einzelaufnahmen animiert.
Bild-Video: Fotos (Bildrechte!), Zeichnungen usw. werden gefilmt, evtl. gezoomt.
Film-Video: eigene oder kopierte (Bildrechte!) Filmsequenzen werden aneinandergeschnitten.
Dokumentarfilm: dokumentarische Aufnahmen werden aneinandergeschnitten.
Spielfilm: Es werden einstudierte Spielsequenzen aneinandergeschnitten.
Achtung: Text und Film sollten sich ergänzen! (Man sollte nicht nur hören, was man schon sieht!)

M4

Worauf sollte man achten?

Wer ist die **Zielgruppe?**
Danach richten sich die **Sprache** (z. B. die Zahl der Fachbegriffe) und alle filmerischen Mittel:

- Eine klare, einprägsame Sprache mit kurzen Sätzen verwenden. Fachbegriffe benutzen!
- Nicht zu schnell sprechen! Pausen lassen, in denen das Bild wirken kann.

Inhalt:

- Den **Inhalt** auf das Wesentliche reduzieren!
- Klar einem roten Faden folgen!
- **Geräusche** und **Musik** gezielt einsetzen!
- Für **Abwechslung** sorgen!
- **Impressum** und **Quellen** nicht vergessen!

Und: Es muss alles fachlich korrekt sein!

M6

➡ Übungsmöglichkeiten: Seite 156, 160, 192, 194, 196

Wie führe ich eine Diskussion oder Debatte? – Pro- und Kontra

Eine Diskussion führen

1 Vorbereitung
Vor einer Diskussion (lat. „discussio" = Untersuchung) muss man sich genau über den Sachverhalt informieren. Recherchieren! 213 Nur so kann man überzeugende Argumente und anschauliche, konkrete Beispiele zu den Argumenten finden.

2 Durchführung
Während der Diskussion müssen alle die üblichen Gesprächsregeln befolgen. Am besten wird eine Diskussionsleiterin oder ein Diskussionsleiter gewählt.

3 Ergebnis feststellen
Wer hat wen überzeugt? Kann eine gemeinsame Entscheidung getroffen werden?

INFO

M1 Die **Debatte** (franz. débattre: (nieder-) schlagen) ist ein Streitgespräch, das im Gegensatz zu einer Diskussion sehr strengen Regeln folgt. Die Debatte geht von einer Entscheidungsfrage (ja/nein) aus, die im Folgenden debattiert wird. Am Ende steht immer eine Abstimmung für oder gegen den debattierten Sachverhalt. Die Mittel der Debatte sind gute Argumente und eine gute rhetorische Darstellung.

Die **Moderatorin** ist verantwortlich für den geregelten Ablauf der Debatte. Die **Pro-Gruppe** und die **Kontra-Gruppe** vertreten gegensätzliche Meinungen. Die Mitglieder bereiten jeweils gemeinsam die Debatte vor. In der Debatte werden sie durch einen oder mehrere **Sprecher** vertreten. Jede Gruppe wählt einen **Gruppenleiter**. Die **Beobachter** folgen der Debatte und entscheiden durch ihre Stimmabgabe.

***M2** Die Beteiligten

TIPPS

M3 Argumente – Kernstücke von Diskussion und Debatte
Zur Vorbereitung einer Entscheidung sollten alle Argumente pro und kontra genau durchdacht werden. Dazu sollte man den Sachverhalt intensiv recherchieren. 213
Argumente pro und kontra sollte man möglichst prägnant formulieren und nach ihrer Bedeutung ordnen. Wirksam sind Argumente nur dann, wenn man sie mit konkreten Beispielen belegen kann.

Eine Debatte führen

1 Vorbereitung durch die gesamte Gruppe
- Genaue Informationen über das Thema sammeln (Recherchieren 213)
- Die Argumente diskutieren: Welche sind besonders wichtig und überzeugend? Welche konkreten Beispiele können dazu genannt werden?
  Welche Argumente könnte die Gegenseite bringen? Wie sind sie zu entkräften?
- Wählen des Gruppenleiters

2 Die Debatte
- Die Moderatorin eröffnet die Debatte.
- Er führt eine erste Abstimmung des Themas unter den Beobachtern durch.
- Die Gruppensprecher halten ihre Plädoyers pro und kontra (maximal drei Minuten).
- Die Gruppen diskutieren Erwiderungen auf die Argumente der Gegenseite (Gegenargumente).
- Die Sprecher der Gruppen antworten auf die Argumente der Gegenseite.
- Die Beobachter stellen Fragen an die Gruppen.
- Die Gruppensprecher halten ihre Schlussplädoyers pro und kontra (maximal drei Minuten).
- Die Beobachter stimmen ab.

3 Auswertung der Debatte
Reflexion: Wie viele Beobachter haben ihre Meinung geändert und warum? Waren die Argumente und war der Vortrag überzeugend?

➡ Übungsmöglichkeiten findest du auf den Seiten 134 und 180

M

Wie führe ich eine Erkundung durch?

„Geographen verlaufen sich nicht, sie erkunden."
Alte Geographen-Weisheit

INFO

M4 Erkundung
Eine Erkundung hat das Ziel, außerhalb der Schule, das heißt vor Ort, möglichst viele Informationen zu einem Thema zu sammeln. Die Ergebnisse der Erkundung werden später zusammengetragen, geordnet, ausgewertet und eventuell auch präsentiert.

M5 **Du solltest bei einer Befragung immer genau erklären, warum und mit welchem Ziel du sie durchführst.**

Was kann man erkunden?
ein Naturschutzgebiet, ein Gewerbegebiet, eine Fußgängerzone, ein Stadtviertel, eine Forschungseinrichtung, einen Wirtschaftsbetrieb, ein Tierheim, eine Gemeideverwaltung ...

M6

Was sind die Themen?
die ökologischen Bedingungen in einem Gebiet, die Standortfaktoren von Gewerbebetrieben, das Einkaufsverhalten in einer Fußgängerzone, die Arbeitsabläufe in einer Verwaltung, die Produktionsabläufe in einem Industriebetrieb ...

M7

Welche Methoden kann man anwenden?
Befragung, Kartierung, Zählung, Fotoaufnahme

M8

➡ Übungsmöglichkeiten findest du auf den Seiten 123 und 190

1 2 3 4 5 6

Eine Erkundung durchführen

① Planung
- Was? – Was soll untersucht werden?
- Wo? – An welchem Ort soll die Erkundung durchgeführt werden?
- Wann? – Wann ist ein sinnvoller Zeitpunkt?

② Vorbereitung
- Stellt die benötigten Materialien zusammen: zum Beispiel Schreibblock, Schreibstift, Buntstifte, Kamera, Stadtplan, topographische Karte (evtl. ein Mitbringsel für eine Führerin/Führer).
- Berechnet mögliche entstehende Kosten und einigt euch, von wem diese bezahlt, wie diese aufgeteilt werden.
- Vereinbart rechtzeitig Termine mit beteiligten Personen (z.B. Beschäftigte in Betrieben, verantwortliche Försterin im Naturschutzgebiet).
- Bereitet die Fragen vor, die gestellt werden.

③ Durchführung der Erkundung
- Seid freundlich und höflich gegenüber den beteiligten Personen.
- Versucht, die Ergebnisse möglichst genau zu dokumentieren.

④ Auswertung der Ergebnisse
- Tragt die Ergebnisse zusammen.
- Wertet die Ergebnisse nach den vereinbarten Fragestellungen/Themen aus.
- Diskutiert die Ergebnisse, sodass alle über das Wesentliche Bescheid wissen.

⑤ Präsentation der Ergebnisse
Bereitet eine Präsentation vor. 218
Präsentiert eure Ergebnisse entweder gegenseitig oder einer interessierten Gruppe.

⑥ Reflexion der Vorgehensweise bei der Erkundung
Überlegt und diskutiert:
Was hätte man besser machen können?
- Waren der ausgewählte Ort und der ausgewählte Zeitraum sinnvoll?
- Waren die gewählten (Teil-)Themen und Fragen sinnvoll?
- Führten die Erkundungsmethoden zu den gewünschten Ergebnissen?

Kartierung, Interview, Meinungsumfrage – wie gehe ich vor?

INFO

M1 Kartierung

Bei einer Kartierung wird eine thematische Karte zu einem bestimmten Raum erstellt (Straße, Stadtviertel). Die Legende richtet sich dabei nach dem gewählten Thema.

Was ist bei einer Kartierung zu beachten?

- Den ausgewählten Raum genau abgrenzen, evtl. in Gruppen aufteilen.
- Die Legende lieber zu detailliert als zu oberflächlich planen. Man kann später immer noch verschiedene Signaturen zusammenfassen, aber nicht noch einmal kartieren ...

M2

INFO

M3 Interview

Um Informationen zu bekommen, bietet es sich an, Menschen zu interviewen, die sich in der Thematik auskennen (Experteninterview).

Was ist beim Interview zu beachten?

Du solltest dich schon vorher gut über das Thema informieren.
Erstelle einen Gesprächsleitfaden, in dem du die Abfolge deiner Fragen notierst.

M4

INFO

M5 Meinungsumfrage

In einer Meinungsumfrage wird eine größere Gruppe von Menschen zu einem Thema befragt. Das geschieht entweder schriftlich mit einem Fragebogen oder mündlich. (Die Antworten werden dann in einem Fragebogen notiert.)

Was ist bei den Fragen zu beachten?

Man unterscheidet
- gebundene Fragen – hier kann man zwischen vorgegebenen Antworten auswählen. Sie sind leicht auszuwerten (z. B. mit Strichlisten).
- ungebundene/freie Fragen – dabei können die Befragten frei antworten (sie sind aufwendig mitzuschreiben und auszuwerten).

M6

M7

1. Wie groß ist die Entfernung zwischen Ihrem Wohn- und Ihrem Arbeitsort? ________ km

2. Welches Verkehrsmittel nutzen Sie für den Arbeitsweg am häufigsten? (Bitte nur ein Kreuz!)

zu Fuß	Fahrrad	Motorrad	Pkw	ÖPNV
☐$_1$	☐$_2$	☐$_3$	☐$_4$	☐$_5$

3. Aus welchem Grund nutzen Sie dieses Verkehrsmittel am häufigsten?

____

4. Bitte bewerten Sie die Qualität der öffentlichen Verkehrsmittel in Ihrer Region/Stadt (1 = sehr gut bis 5 = unzureichend)

	1	2	3	4	5
$_1$Nähe zum Wohnort	☐	☐	☐	☐	☐
$_2$Nähe zum Arbeitsort	☐	☐	☐	☐	☐
$_3$Fahrthäufigkeit	☐	☐	☐	☐	☐
$_4$Anschlussmöglichkeiten	☐	☐	☐	☐	☐
$_5$Bequemlichkeit/Komfort	☐	☐	☐	☐	☐
$_6$Preis-Leistungs-Verhältnis	☐	☐	☐	☐	☐

M8 Ein Fragebogen mit unterschiedlichen Arten von Fragen

INTERNET

M9 **Grafstat** ist ein Programm zur Erstellung und Auswertung von Umfragen. Die Entwicklung des Programms wird durch die *Bundeszentrale für politische Bildung (bpb)* gefördert. Der Download ist für Schülerinnen und Schüler kostenlos.

Wie nehme ich einen Raum genauer unter die Lupe? – Die Raumanalyse

INFO

M10 Betrachtungsweisen der Geographie

Das Fach Erdkunde hat zwei unterschiedliche Betrachtungsweisen:

- Zum einen stehen geographische Themen im Vordergrund. Dies ist meistens der Fall. Das Thema lautet zum Beispiel „Wie wirtschaften Menschen in anderen Landschaftszonen?“ oder „Weltbevölkerung – Wachstum und Migration“. Dabei geht man Fragestellungen nach, die auf unterschiedlichste Räume der Erde anwendbar sind.
- Bei der zweiten Betrachtungsweise steht ein Raum im Vordergrund, z. B. „Sibirien – wie verändert der Klimawandel den Lebensraum?“ oder „Der Aralsee – wie konnte ein See zur Wüste werden?“

Dabei beschäftigt man sich unter einer bestimmten Fragestellung intensiv mit einem Raum und den Faktoren, die in ihm zusammenwirken.

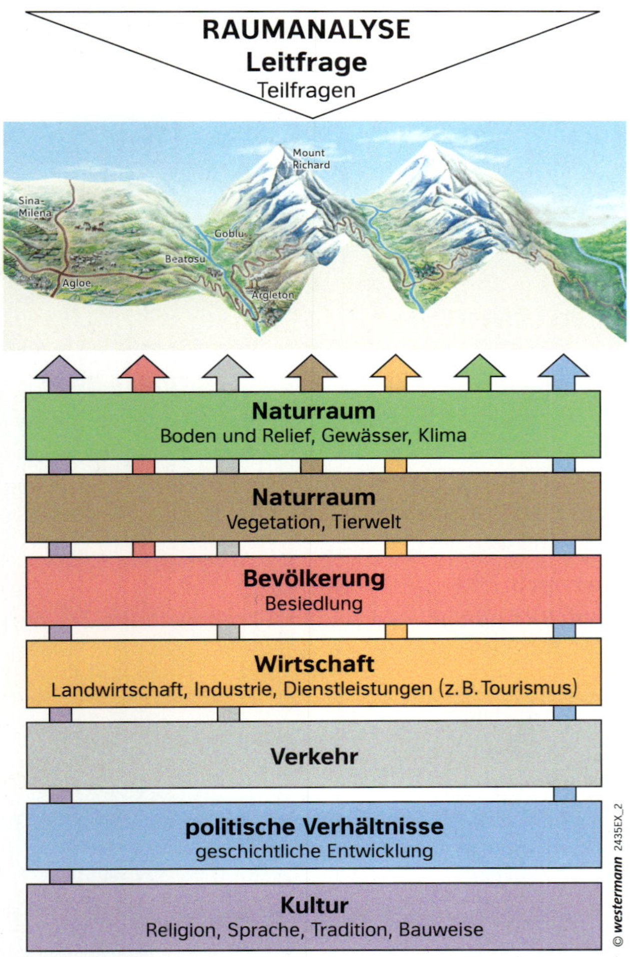

M11 Unterschiedliche Faktoren prägen einen Raum.

Im Fokus: ein Raum

Will man einer bestimmten Fragestellung zu einer Stadt, einer Landschaft oder einem Land nachgehen, dann ist es sinnvoll, diese Frage zunächst genau auszuformulieren, zum Beispiel:
Wie verändert sich der Lebensraum in Russland und seinen Nachbarstaaten durch den Einfluss der Menschen?

Nun überlegt man, ob sich diese Leitfrage in unterschiedliche Teilfragen aufgliedern lässt, zum Beispiel:
„Sibirien – wie verändert der Klimawandel den Lebensraum?“ oder *„Der Aralsee – wie konnte ein See zur Wüste werden?“*

Jetzt kann man sich auf die Suche nach geeigneten Informationsquellen machen: Zeitungen, (Fach-)Zeitschriften, Lexika, Atlanten, Fernsehfilme, Veröffentlichungen von Behörden und Organisationen, Informationen von Bekannten und natürlich das Internet. 213

Nachdem man sich einen ersten Überblick über die Informationen verschafft hat, prüft man, ob auch genügend (unterschiedliches) Material zur Beantwortung aller Teilfragen vorhanden ist.

Nun folgt die Auswertung.
Dabei wird deutlich: Einerseits ist vieles in dem bisher unbekannten Raum vergleichbar mit anderen, schon bekannten Räumen. Andererseits gleicht kein Raum dem anderen. Immer sind einzelne Faktoren (z. B. die Landwirtschaft) für die Fragestellung von großer Bedeutung und andere sind eher unwichtig. Bei einer anderen Fragestellung kann das anders sein.

Am Ende der Raumanalyse sind die Teilfragen und die Leitfrage beantwortet.

BEMERKENSWERT

M12 Jeder Raum auf unserer Erde ist unverwechselbar und einzigartig – jedoch in einzelnen Bereichen auch immer vergleichbar. Gerade das macht es so interessant, sich immer wieder mit unbekannten Ländern, ihren Landschaften und Menschen zu beschäftigen.

➡ Übungsmöglichkeiten zur Raumanalyse: Seite 192, 194, 196

Wie kann ich Arbeitsaufträge richtig verstehen und lösen?

Wir geben etwas wieder:

name – etwas aufzählen, ohne Erklärung wiedergeben, oft einzelne Begriffe.
(*Name five cities that are located in the tropical rainforest of Brazil.*)

list – Informationen in einer Liste (oder einer Tabelle) ohne Kommentierung geordnet wiedergeben

outline / point out – Informationen oder Zusammenhänge (aus dem Material oder dem Unterricht) geordnet in eigenen Worten wiedergeben

describe – Den Inhalt von Materialien zusammenhängend wiedergeben. (*Describe the satellite image of the Vesuvius area.*)

Wir erklären etwas:

explain – etwas so erklären, dass die Zusammenhänge deutlich werden, vor allem Gründe und Folgen (*Explain why some areas have a bigger risk of earthquakes than others.*)

illustrate – Zusammenhänge mithilfe ergänzender Informationen oder Beispiele erklären. (*Using the example of Vesuvius, illustrate why people live near volcanoes.*)

compare – verschiedene Dinge vergleichen – dabei Gemeinsamkeiten und Unterschiede herausarbeiten. Dabei sollte man vorher überlegen, was man miteinander vergleicht und ob sich das überhaupt vergleichen lässt. Häufig gibt die Aufgabe schon Hinweise darauf, anhand welcher Merkmale man etwas vergleichen soll. (*Compare the areas oft he volcanoes Vesuvius and Merapi in terms of [in Bezug auf] their population density and land use.*)

apply – etwas (beispielsweise ein Modell oder eine These) auf ein Beispiel anwenden. (*Apply the theory of plate tectonics to the situation in the Istanbul region.*)

analyse – komplexe Sachverhalte so darstellen, dass die Zusammenhänge zwischen den Einzelaspekten klar werden und man eine Schlussfolgerung ziehen kann. (*Analyse the Japan's vulnerability at the example of the Pacific coast.*)

Wir beurteilen etwas:

evaluate – sich eine begründete Meinung bilden. Vorher muss man jedoch möglichst viele Informationen bearbeitet und gegeneinander abgewogen haben. (*Evaluate the importance of protective measures for earthquakes for Istanbul.*)

discuss – die Gründe für (Pro-Argumente) und gegen (Kontra-Argumente) einen Sachverhalt gegenüberstellen. Dies geschieht meist mündlich – in einer Diskussion in der Klasse. Ziel ist es, Meinungen auszutauschen und zu einem begründeten Ergebnis zu kommen. (*Discuss if a new ski resort should be built at the slopes of the Vesuvius.*)

comment on – zu etwas begründet, nach Abwägung von verschiedenen Argumenten, seine eigene Meinung äußern (*Comment on the statement: „We shouldn't spend money on protective measures because they cannot stop natural disasters like earthquakes.".*)

Wie können die Arbeitsergebnisse aussehen?

Brief/Beschwerdebrief
Argumente werden aus der persönlichen Sicht vorgetragen. Anrede und Gruß am Ende sachlich halten.

Diagramm: 207, 208

Erklärvideo
Kurze Erklärung der wichtigsten Zusammenhänge mithilfe filmerischer Mittel (z. B. Spielszenen, Zeichnungen, Zeichentrick). Dauer: ca. 3 Minuten 219

Erlebnisaufsatz
Aus der eigenen Sicht wird ein Vorgang interessant und spannend dargestellt. Du kannst eigene Empfindungen und Gefühle äußern.

Informationsblatt
Kurze Hinweise als Tabelle mit Spiegelstrichen formulieren. Du kannst zusätzlich Zeichnungen verwenden.

Interview
Notiere zielgerichtete Fragen, die du einer Interviewpartnerin oder einem Interviewpartner stellen könntest und evtl. auch die möglichen Antworten. Schreibe so, wie man spricht.

Kurznachricht
Alles Wichtige im Telegrammstil mit wenigen Zeichen (z. B. 280) formulieren.

Lernplakat 218

Lexikoneintrag
Sehr sachlich und allgemein verständlich. Die Zusammenhänge werden genau erklärt.

Mind map, causal chain, causal network and concept map: 199

Präsentation 218

Radionachricht
Es gibt keine Überschrift. Der Text ist sachlich wie die Zeitungsnachricht. Er hat keine langen Sätze und wird verlesen oder aufgenommen.

Steckbrief
Kurze, listenartige Darstellung, die die wichtigsten Informationen zu Personen, Gegenständen oder Themen enthält.

Tabelle 207

Telegrammstil
Sehr kurze Darstellung ohne vollständige Sätze, keine Füllwörter verwenden.

Tweet
→ Kurznachricht

Werbetext/Werbeflyer mit Abbildungen
Einen Text mit Übertreibungen und vielen ausschmückenden Adjektiven formulieren, um eine Sache anzupreisen.

Zeitstrahl
Darstellung zeitlicher Abfolgen von Ereignissen an einer Achse (waagerecht oder senkrecht).

Zeitungsnachricht
In der Überschrift steht schon die wichtigste Information. Danach wird das Thema sehr sachlich und klar dargestellt.

Maße und Gewichte

Bei deiner Arbeit mit dem Buch wirst du häufig auf Gewichte und Maße stoßen, die dir unbekannt sind oder bei denen du gerade nicht weißt, wie du sie umrechnen kannst. Hier findest du eine Übersicht:

Längenmaße:

1 m (metre) = 10 dm (decimetre)
= 100 cm (centimetre)
= 1000 mm (millimetre)

1 km (kilometre) = 1000 m (metre)

1 mi (mile – Meile) ≈ 1,6 km

1 nmi (nautical mile) ≈ 1,85 km

Flächenmaße:

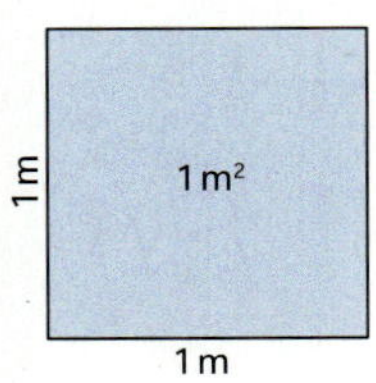

1 m² (square metre) = 1 m • 1 m

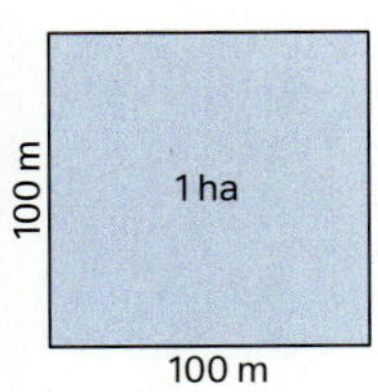

1 ha (hectare) = 100 m • 100 m
= 10 000 m²

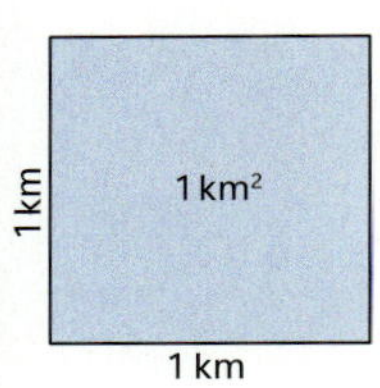

1 km² (square kilometre)
= 1 km • 1 km

100 ha (hectare) = 1 km²
= 1 000 000 m²

Raummaße:

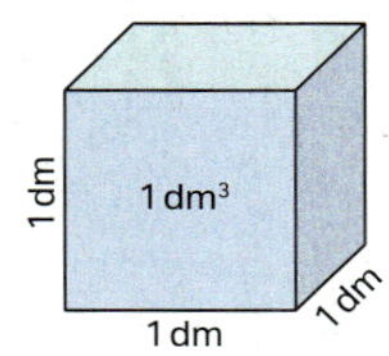

1 l (litre) = 1 dm³ (cubic decimetre)
= 1 dm • 1 dm • 1 dm

100 l (Liter) = 1 hl (hectolitres)

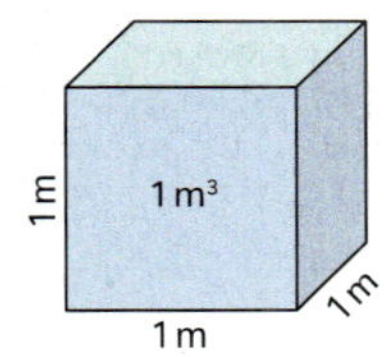

1 m³ (cubic metre) = 1 m • 1 m • 1 m

1 bbl (barrel oil) ≈ 159 l (litre oil)

Gewichte:

1 kg (kilogram) = 1000 g (gram)

1 t (tonne) = 1000 kg

1000 t (tonnes) = 1 kt (kiloton)

Energieeinheiten:

1 kWh (kilowatts per hour) = 1000 Wh (watts per hour)
→ damit kann man etwa 20 Stunden lang einen Laptop nutzen oder 25 Minuten staubsaugen.

Zählung und Dezimalzahlen:

1 million = 1,000,000 = 1 000 000
→ Zahlen ab Tausend werden in Englisch mit Komma oder Leerzeichen getrennt
3.1415 (three point one four one five)
→ Dezimaltrenner ist ein Punkt

Starthilfen

S. 27
M8 Lesebeispiel: Die 2060 geborenen Menschen werden nach einer Prognose der UNO in den Industrieländern durchschnittlich 85 Jahre alt, in den Entwicklungsländern 78 und weltweit im Durchschnitt 79 Jahre alt.

S. 31
M7 Lesebeispiel: 22 von 100 (= 22 %) Ehen, die zwischen 1910 und 1912 geschlossen wurden, hatten zwei Kinder, 32 hatten vier und mehr Kinder.

S. 45
M6 In seiner ersten Antwort spricht Abdelkader Chacha von legalen und illegalen Einwanderern. Illegal bedeutet, dass die Migranten ohne gültige Einreisepapiere eingewandert sind oder keine gültige Aufenthaltserlaubnis besitzen. Statt des Begriffs „illegal" wird heute oftmals der Begriff „irregulär" oder „undokumentiert" verwendet, da der Ausdruck „illegal" häufig mit Kriminalität verknüpft wird. Es soll nicht der Migrant an sich als „illegaler Mensch" dargestellt werden, sondern lediglich seine Einreise oder sein Aufenthalt als irregulär und undokumentiert.

S. 55
M10 Der Begriff „Migrationshintergrund" problematisch, weil er einige Ungenauigkeiten und Setzungen beeinhaltet, die man auch als rassistisch betrachten kann, etwa die Einteilung Deutscher nach der Herkunft ihrer Vorfahren. Begriff wird aber nach wie vor als eine statistische Kategorie verwendet.

S. 59
M4 Siehe Starthilfe zu Seite 45 **M6**.

S. 65
M3 Wirkungsgefüge, in denen Folgen wieder zum Ursprung zurückführen, in denen sich die Wirkungen selbst verstärken, werden auch als „Teufelskreise" bezeichnet.

S. 67
M7 Mögliche Bestandteile des Wirkungsgefüges (Auswahl): Ausbau der Infrastruktur/des Gesundheitswesens/des Bildungswesens, unzureichende Ernährung, Kinderarbeit, gesundheitliche Probleme, geringes Steueraufkommen, geringe Wirtschaftskraft

S. 81
M7 In den Bürohochhäusern sind vor allem Büros global tätiger Unternehmen untergebracht. Die Shoppingmalls und Vergnügungszentren werden nahezu ausschließlich von der wohlhabenden Bevölkerungsschicht besucht.

S. 89
M6 „It's just like Bangalore" ist nicht zu verwechseln mit „bangalored". Dieser Begriff beschreibt im Englischen den Arbeitsplatzverlust aufgrund von Outsourcing von Tätigkeiten in eine andere Region. Der Begriff ist von der Stadt Bangalore abgeleitet, in die zu Beginn des Outsourcings verstärkt US-amerikanische IT-Arbeitsplätze outgesourct wurden.

S. 92
M1 Dienstleistungsexporte umfassen alle von Einheimischen an Gebietsfremde erbrachten Dienstleistungen. Tourismusexporte geben den Wert der Einnahmen durch den ausländischen Tourismus an.

S. 100
M1 Im Quellentext findet sich die Schreibweise „Schüler*innen" mit *. Die Schreibweise bemüht sich um die Sichtbarmachung geschlechtergerechter Sprache. Alternativ wird häufig auch die Schreibweise „Schüler_innen" mit einem _ verwendet. Diese beiden Formen sollen Frauen und Männer sowie Geschlechter zwischen Frau und Mann (wie Transgender) berücksichtigen. Für die Ansprache von Frauen und Männern gibt es weitere Formen:
„SchülerInnen", „Schüler/-innen" oder „Schüler(-innen)". Im Alltag ist heute noch die Paarform „Schülerinnen und Schüler" die gebräuchlichste Variante. In deinem Schulbuch bemühen wir uns um eine gerechte und ausgewogene Sprache, das gelingt aufgrund von Platzmangel leider nur unzureichend.

S. 102
M5 Denke bei der Auswertung des Liniendiagramms daran, dass der Weltexport (Exporte aller Länder der Erde) seit 1980 stark zugenommen hat (1980: rund 2000 Mrd. US-$, 2015: rund 15 600 Mrd. US-$).

S. 110
M1 Verwechslungsgefahr! Der Begriff City bezeichnet im Deutschen das Zentrum einer größeren Stadt, das durch eine Vielzahl von typischen Merkmalen gekennzeichnet ist z. B. hohe Dichte an zentralen Einrichtungen und Funktionen wie Verwaltung und Einzelhandel, hohe Arbeitsplatzdichte, hoher Anteil der Tagbevölkerung im Vergleich zur Nachtbevölkerung und hohe Bebauungsdichte. Der englische Begriff bezeichnet hingegen eine Großstadt (allgemein), eine Stadt mit Bischofssitz und Kathedrale oder eine Stadt mit königlicher Urkunde.
Im Englischen bezeichnet man den Innenstadtbereich als city centre oder (down-)town – CBD bezeichnet hingegen den Central Business District, was in etwa dem deutschen Begriff Geschäfts- oder Bankenviertel entspricht

S. 119
M7 Hier geht es darum wie man „gute Adressen" schafft. Analysiere, welche Maßnahmen dazu im Text angesprochen werden.
„ethnische Segregation" = Aufteilung nach der Abstammung
„integration trap (Integrationsfalle)" – da steckt jemand fest, sitzt in der Falle, Integration findet daher nicht statt.

S. 147
M5 Lesehilfe: Die Entwicklung eines Smartphones erfolgt in den USA, die Produktion geschieht über viele globale Zulieferer von Rohstoffen, z. B. Kupfer aus Chile, und außerdem noch über viele Produktionsstätten von Bauteilen, z. B. Akkus aus Mexiko.

S. 157
M4 Ein Auto entsteht in den folgenden Produktionsschritten.
Press shop (Presswerk): liefert die Blechteile für die Karosserie.
Body shell (Rohbau): Hier werden die Karosserieteile zu einer fertigen Karosserie eines Fahrzeuges zusammengefügt.
Painting (Lackiererei): Hier wird die Karosserie lackiert.
Assembly (Montage): In diesem Bereich wird die lackierte Karosserie um noch fehlende Teile ergänzt und es wird die „Hochzeit" vollzogen. In der Automobilproduktion nennt man das Zusammenfügen von Karosserie und Motor mit Antriebsstrang „Marriage (Hochzeit)".

S. 166
M3 Die blaue Kurve (West-EU) bezieht sich auf die linke Y-Achse und die rote Kurve (Ost-EU) auf die rechte Y-Achse.

S. 170/171
M1, M6 Bei der Analyse helfen dir deine Kenntnisse zu Entwicklungsländern aus dem sechsten Kapitel des Buches ab Seite 62.

S. 179
M9 Darknet: Parallelwelt im Internet, wo die Verbindungswege der Daten samt Sender und Empfänger völlig verschleiert sind, im Dunkeln liegen. Es wird oft mit Illegalem in Verbindung gebracht. In erster Linie steht der Schutz der eigenen Privatsphäre im Vordergrund. Die Bewegungen im Darknet werden nicht, wie im regulären Netz dokumentiert. Diese Daten werden von Behörden zur Beobachtung und Verfolgung benutzt, Kriminelle nutzen sie für sog. Scamming (Betrug) und Marketing- und Analysefirmen um Kauf- oder Wahlentscheidungen zu analysieren und auch zu beeinflussen.
Die Anonymität im Darknet machen sich hingegen auch Kriminelle zu Nutzen. So werden auf einschlägigen Seiten Drogen, Waffen, verbotene Musik oder Propagandamaterial, und Kinderpornos getauscht oder gehandelt.

S. 186
M1 Das regionale BIP pro Kopf liegt beispielsweise in Deutschland in der Region Mecklenburg-Vorpommern bei 84 % des Durchschnitts der Europäischen Union (EU) und in der Region Hamburg bei 197 % des Durchschnitts der EU.

S. 187
M4 Entlang von Entwicklungsachsen (z. B. Bahnlinien, Autobahnen) soll eine räumliche Entwicklung vorangetrieben werden. In der Regionalplanung wurden sogenannte Entwicklungsachsen verschiedener Ordnungen ausgewiesen, die in der Regel zentrale Orte miteinander verbinden und den Raum entlang der Achsen als vorrangige Entwicklungsgebiete ausweisen.

S. 220
M2 Siehe Starthilfe zu Seite 100 **M1**

Europe

a) Llanfairpwllgwyngyll-gogerychwyrndrobwll-llantysiliogogogoch is considered the longest official place name in the world. 20 km south of the village is the mountain Snowdon. How high is it?

b) View from Euromast observation tower ussichtsturm over Rotterdam. Name the capital and seat of government of the Netherlands.

c) Magnitogorsk (Russia). Which mountain range is considered the border between Europe and Asia?

h) Val d'Isere in the French Alps. Name eight alpine countries.

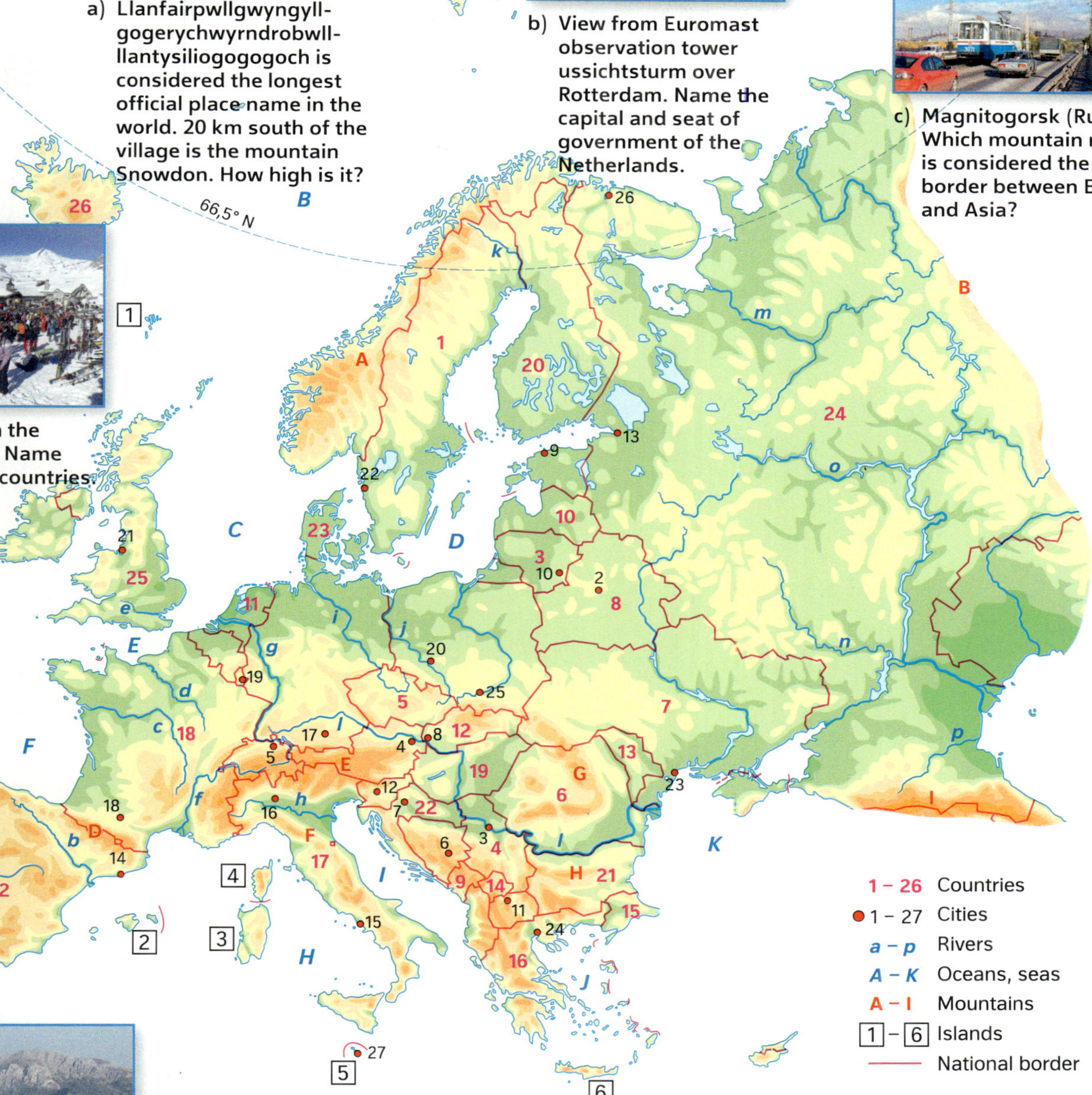

g) The southernmost place on the European mainland. What is its name, in which country and on which latitude is it located, and which country do you see in the background of the picture?

f) Vienna* – the capital of Austria. Which river flows through it? Which countries does the river flow through besides Austria?

*If you visit Vienna on a trip, the Natural History Museum is worth a visit: it is one of the most impressive natural history museums in the world.

e) Particularly large specimens of the European river catfish (Waller) live in the river Po. In which country does the Po River and into which sea does it flow?

d) Sarajevo – the venue of the 1984 Winter Olympics, the city was besieged in the Bosnian War from 1992 to 1996. The capital of which country is it today?

Africa

h) The port of Tangier-Med in Morocco.* Which other North African countries border the Mediterranean Sea?

*You can see the Rock of Gibraltar in the left background.

a) The Guelta d'Archei on the Ennedi Plateau. Name the capital of the country where it is located.

b) Geologists at the Erta Ale volcano. In which depression is it located? How deep is the depression at its deepest point?

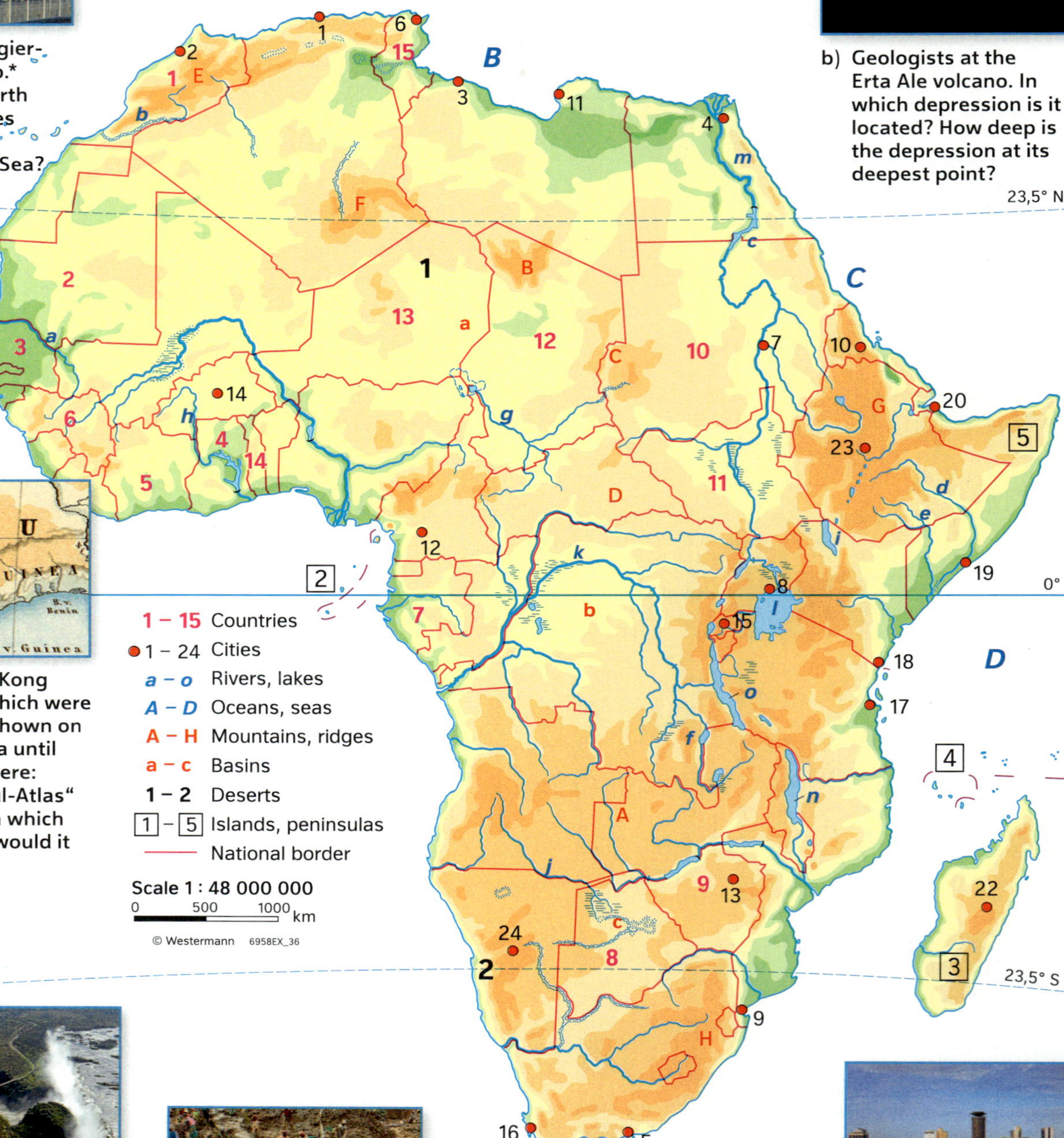

g) The fictitious Kong Mountains, which were erroneously shown on maps of Africa until about 1900 (here: „Diercke Schul-Atlas" from 1883). In which states today would it rise?

f) The Victoria Falls. Which river forms the waterfall and in which countries is it located?

e) Gold mining in the northeast of the DR Congo. What was the country called until 1997 and which countries does it border?

d) Winter landscape in Lesotho. Name the capital of Lesotho. Which neighboring state completely surrounds the country?

c) Skyline of Nairobi, Kenya. Name two national parks in the country.

Asia, Australia und Oceania

g) The summit of the second highest mountain on earth. What is its name and how high is it?

a) In winter, the frozen Lake Baikal serves as a road. How deep is it at its deepest point and in which country is it located?

b) The border between North and South Korea. Name the capitals of both countries?

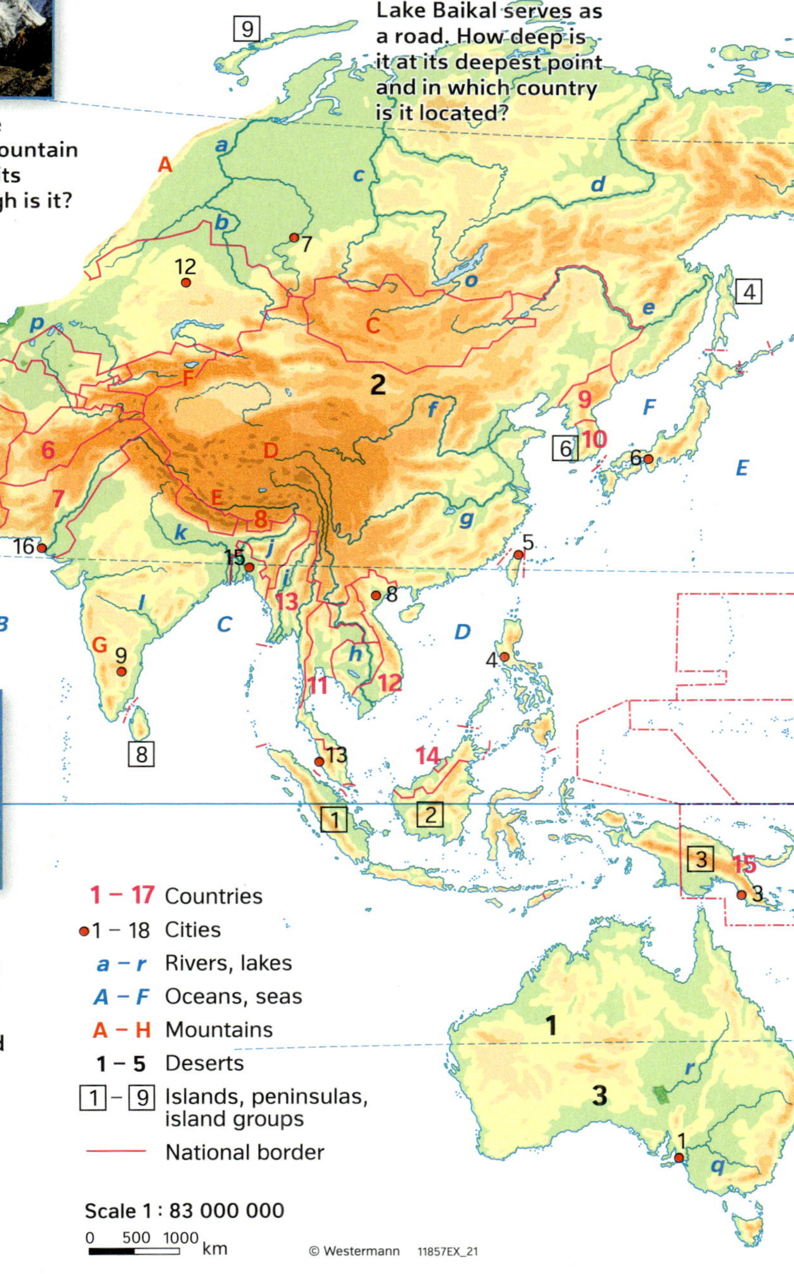

c) First atomic bomb test in Bikini Atoll in 1946. In which ocean is the atoll located and to which country does it belong?

f) The shore of the Dead Sea - the deepest part of the land surface that is not covered by water or ice. How deep is it? Name the countries and territories that border the Dead Sea.

e) Mumbai. Name two other Indian cities with more than five million inhabitants.

d) A Kakapo. A mating attempt with a photographer in the documentary „Last Chance to See“ (BBC) helped the endangered parrot species to international fame. What is the name of the home country of the kakapo, which is home to other unusual birds such as the kiwi and kea?

North America

a) Alert is the northernmost continuously inhabited place in the world (82° 28′ N). On which Canadian island is it located?

b) Toronto is the largest city in Canada, but not the capital. What is the name of the capital of Canada?

i) Denali (formerly: Mount McKinley) - the highest mountain in North America. How high is it and in which country is it located?

h) West of Ontario, the border between Canada and the United States follows which latitude?

c) Detroit: Name the river next to the city and the name of the city across this river?

g) Silicon Valley - the largest high-tech cluster in the world. On which bay is it located?

1 – 2 Countries
● 1 – 16 Cities
a – g Rivers, lakes
A – E Oceans, seas
A – B Mountains
1 – 2 Deserts
[1] – [5] Islands, peninsulas, island groups
National border

Scale 1 : 35 000 000
0 500 1000 km

f) Los Angeles has a population of just under four million, and as many as 17 million in the metropolitan area. Name three other US cities with more than one million inhabitants.

e) Bingham Mine südlich von Salt Lake City. In welchem US-amerikanischen Bundesstaat liegt der Tagebau?

d) The Everglades in southern Florida. Name three metropolitan areas with more than 500,000 residents in the state?

Central and South America

a) Port-au-Prince, Haiti after a heavy earthquake in 2010. On which island is Haiti located?

b) Torre de David in Caracas – the high-rise building was never completed. Until its eviction in 2015, the shell was used for years as an informal settlement. Caracas is the capital of which country?

h) A sloth in the tropical rainforest of Costa Rica: The country is known for its biodiversity: Which states and oceans/seas does the country border?

c) Teatro Amazonas – the opera is a landmark of the largest city in the Amazon basin. What is the name of the city?

g) Naturalists Alexander von Humboldt (standing) and Aimé Bonpland (seated) before their attempt to climb Chimborazo volcano (left) in 1802: How high is the volcano and in which country is it located?

e) Lake Titicaca. The border of which two states runs through the lake? Name their capitals?

d) Stanley – capital of the Falkland Islands. In 1982, a war broke out over the archipelago. It is an overseas territory of which state? Name the ocean where it is located?

f) Funicular (cable railway) in Valparaíso, Chile. On which ocean is the city located?

Minilexikon

A

advantages of agglomeration – Agglomerationsvorteil
(Seite 116)
Standortvorteil, der sich aus der Lage eines Betriebes in der Nähe anderer Betriebe ergibt, z. B. niedrige Transportkosten, großer Absatzmarkt und viele gut ausgebildete Arbeitskräfte.

Agenda 2030
(Seite 15)
Die Agenda 2030 der Vereinten Nationen (→ UNO) umfasst 17 Ziele für die nachhaltige Entwicklung der Erde, der Menschheit (→ sustainable development goals, SDGs). Sie traten am 01.01.2016 in Kraft und sollen bis 2030 erreicht sein.

age-gender pyramid – Bevölkerungspyramide
(Seite 33)
Die Bevölkerungspyramide (auch Alterspyramide) stellt die Altersstruktur (→ age structure) der Bevölkerung eines Raumes, wie die einer Stadt oder eines Landes, in einem doppelten Säulendiagramm da. Dabei wird der Altersaufbau der Bevölkerung nach Altersgruppen und nach Geschlecht getrennt.

ageing population – Überalterung
(Seite 31)
Überalterung herrscht dann vor, wenn sich eine Bevölkerung aus wenigen Kindern und Jugendlichen sowie Menschen im arbeitsfähigen Alter von 15 bis 65 Jahren zusammensetzt, jedoch aus vielen älteren Menschen, die über 65 Jahre alt sind.

age structure – Altersstruktur
(Seite 33)
Die Altersstruktur ist die Zusammensetzung einer Bevölkerung im Hinblick auf das Alter der Menschen, die sie bilden. Häufig wird die Bevölkerung in drei Gruppen gegliedert: Kinder und Jugendliche (0 bis 15 Jahre), Menschen im erwerbsfähigen Alter (15 bis 65 Jahre) und nicht mehr erwerbsfähige Personen (über 65 Jahre). Deutlich ablesbar ist die Altersstruktur in einer Bevölkerungspyramide (→ age-gender pyramid).

agglomeration – Agglomerationsraum
(Seite 48)
(auch Ballungsraum, Verdichtungsraum)
In einem Agglomerationsraum ballen sich Bevölkerung und Wirtschaftsbetriebe. Oft handelt es sich dabei auch um eine Ansammlung von Städten, die eng miteinander verknüpft sind.

Aktivraum (Seite 187)
Gebiet mit hoher wirtschaftlicher Leistung und daher ein Zuwanderungsgebiet. Menschen ziehen hierher, weil in modernen Industrien gut bezahlte Arbeitsplätze angeboten werden. Anspruchsvolle Dienstleistungsberufe wie z. B. Rechtsanwälte oder Architekten sind häufig vertreten. Der Lebensstandard der Bevölkerung ist höher als in anderen Gebieten. (→ passive space)

aquaculture – Aquakultur
(Seite 21)
In verschiedenen Formen intensiver Haltung (Teiche, Netze, Käfige) werden Süßwasserfische, Salzwasserfische und Schalentiere gezüchtet, gemästet und anschließend vermarktet. Wie bei der Massentierhaltung gibt es auch hier Vorgaben, wie viele Tiere in einem Netz oder Käfig gehalten werden dürfen. In diesem Rahmen gewinnt heute auch der intensive Anbau von Meeresgemüse (z. B. Algen) immer mehr an Bedeutung.

artificial intelligence (AI) – künstliche Intelligenz (KI)
(Seite 156)
Teilgebiet der Informatik. Ihr Ziel ist es, menschliche Intelligenz nachzuahmen und sogar zu verbessern. KI-Systeme sind Software, Maschinen oder Roboter, die Aufgaben eigenständig erledigen, ohne dass jeder Schritt vom Menschen programmiert wird. Sie zeichnen sich dadurch aus, dass sie lernen und sich selbst und andere programmieren und weiterentwickeln können.

B

bad governance
(Seite 77)
schlechte Regierungsführung, gekennzeichnet durch fehlende demokratische Beteiligungsmöglichkeiten der Bevölkerung, Korruption und Günstlingswirtschaft. (→ good governance)

basic need – Grundbedürfnis
(Seite 65)
Mindestbedarf, den ein Mensch zum Leben braucht. Die wichtigsten Grundbedürfnisse sind Nahrung, Trinkwasser, Kleidung, Unterkunft sowie Bildung, Arbeit, ärztliche Versorgung und politische Mitbestimmung.

basic parts of living – Grunddaseinsfunktion
(Seite 128)
(auch Daseinsgrundfunktion) Dieser Begriff aus der Raumplanung bezeichnet die Mindestanforderungen an das Leben in der Gesellschaft: wohnen, arbeiten, sich versorgen, sich bilden, sich erholen, in Gemeinschaft leben und am Verkehr teilnehmen. → M2 Seite 138

birth rate – Geburtenrate
(Seite 25)
Die Geburtenrate ist die Anzahl der Lebendgeborenen pro 1000 Einwohner, gemessen für ein bestimmtes Gebiet und für einen bestimmten Zeitraum.

brain drain – Braindrain
(Seite 43)
Braindrain heißt wörtlich übersetzt „Gehirn-Abfluss" und meint den Verlust von hoch qualifizierten Arbeitskräften durch Abwanderung. Häufig tritt dieses Phänomen in den → Entwicklungsländern auf, wo sich zum Beispiel Ärzte, Techniker oder Wissenschaftler für einen Fortgang in ein → Industrieland entscheiden.

C

car-friendly city – autogerechte Stadt
(Seite 126)
Die autogerechte Stadt ist ein städtebauliches Leitbild der 60er- und 70er-Jahre. Man wollte dem schnell wachsenden Pkw- und Lkw-Verkehr einen möglichst problemlosen Verkehrsfluss gewährleisten: durch mehrspurige Straßen, Fußgängerunterführungen, Parkhäuser und später elektronische Leitsysteme. Heute gilt dieses Leitbild als weitestgehend überholt. Ziel ist nun eher eine „menschengerechte" Stadt.

carrying capacity – Tragfähigkeit
(S. 12)
Die Menge an Menschen, die unter Befriedigung aller Grundbedürfnisse (→ basic needs) auf längere Sicht in einem Raum (Region, Kontinent, Erde) leben kann. Dabei dürfen die Eingriffe der Menschen in den Raum nur so groß sein, dass dieser sich wieder davon erholen kann. Nachhaltigkeit (→ sustainability)

child labour – Kinderarbeit
(Seite 25)
Kinder unter 15 Jahre, die arbeiten müssen, statt in die Schule zu gehen, weil sie zum Lebensunterhalt ihrer Familie beitragen müssen. Dies ist in Entwicklungsländern (→ developing country) häufig der Fall. Diese Arbeit gefährdet die körperliche und geistige Gesundheit der Kinder und verletzt ihre Rechte. Unterschieden werden müssen Kinderarbeiter von Kindern, die z. B. im elterlichen Haushalt helfen oder von Jugendlichen, die einer legalen Arbeit wie dem Austragen von Zeitungen nachgehen.

city centre – City
(Seite 25)
Kern einer Stadt, in dem wichtige Einrichtungen und Funktionen angesiedelt sind. Die City hat so auch wichtige Funktion für das Umland der Stadt.

cluster – Cluster
(Seite 165)
Eine Art Netzwerk bei der Herstellung von Produkten und Verfahren, in dem verschiedene Betriebe bzw. Organisationen zusammenarbeiten. Einzelne Unternehmen stehen in Teilbereichen durchaus in Konkurrenz zueinander. Die Beteiligten siedeln sich dabei in räumlicher Nähe zueinander an. Ein Cluster können z. B. Industriebetriebe, Hochschulen, Forschungseinrichtungen, zukunfts- und technologieorientierte Unternehmen, spezialisierte Dienstleistungsbetriebe oder Einrichtungen der Verwaltung und Wirtschaftsförderung bilden.

coal crisis – Kohlekrise
(Seite 161)
Rückgang des Absatzes von Steinkohle Ende der 1950er-Jahre, weil Erdöl preiswerter angeboten wurde. Daraufhin wurden Zechen stillgelegt und Bergleute entlassen. Die Steinkohlenförderung wurde gedrosselt.

colony – Kolonie
(Seite 78)
Gebiet ohne eigenes Hoheitsrecht in Besitz und Abhängigkeit eines anderen Staates. Viele Entwicklungsländer (→ developing country) waren bis in das 20. Jahrhundert Kolonien europäischer Staaten.

common market – gemeinsamer Binnenmarkt
(Seite 182)
Binnenmarkt der EU (→ European Union) mit freiem Waren-, Dienstleistungs- und Kapitalverkehr zwischen den Mitgliedstaaten, freie Wahl des Arbeitsplatzes für Arbeitnehmer und freie Standortwahl für Unternehmen.

countries of the global south – Länder des globalen Südens
(Seite 65)
Als „Länder des globalen Südens“ werden sehr unterschiedliche Staaten gefasst, auch solche, die uns als entwickelt erscheinen. Sie sind zwar in vieler Hinsicht sehr modern und scheinen wohlhabend, dennoch sind die Grundbedürfnisse vieler Menschen dort nicht befriedigt. Auch in den sogenannten Industrieländern gibt es Elend und Armut. Dennoch sind hier die Grundbedürfnisse der meisten Menschen erfüllt.

D

death rate – Sterberate
(Seite 26)
Die Sterberate beschreibt die Anzahl der Gestorbenen pro 1000 Einwohner in einem bestimmten Gebiet und in einem bestimmten Zeitraum. Geburtenrate (→ birth rate)

demographic transition – demographischer Übergang
(Seite 29)
Allgemeines Modell der Bevölkerungsentwicklung im Rahmen der wirtschaftlichen Entwicklung, welches die Veränderung der Geburtenrate, Sterberate und Wachstumsrate beschreibt.

developed country – Industrieland
(Seite 65)
Im Vergleich zu einem Entwicklungsland (→ developing country) weit entwickeltes Land mit einem entsprechend hohen Bruttonationaleinkommen (→ gross national income – GNI) pro Einwohner. Weitere Merkmale von Industrieländern sind ein hoher Anteil an Beschäftigten in der Industrie und im Dienstleistungssektor sowie eine gut ausgebaute Infrastruktur (→ infrastructure). Die Grundbedürfnisse (→ basic needs) der meisten Menschen sind hier befriedigt.

developing country – Entwicklungsland
(Seite 11, 65)
Land, das im Vergleich zu einem Industrieland (→ developed country), weniger entwickelt ist. Entwicklungsländer weisen typische Merkmale auf, z. B. ein hohes Bevölkerungswachstum, viele Analphabeten und Elendssiedlungen. Die Grundbedürfnisse (→ basic needs) der meisten Menschen sind hier nicht befriedigt.

development cooperation –Entwicklungszusammenarbeit
(Seite 84)
Sie umfasst Maßnahmen zur Unterstützung des wirtschaftlichen Wachstums und der sozialen Entwicklung in den Entwicklungsländern (→ developing country). Dieser Begriff wird in der deutschen Entwicklungspolitik anstelle von „Entwicklungshilfe“ benutzt.

digital infrastructure – digitale Infrastruktur
(Seite 149)
Die digitale Infrastruktur umfasst alle technischen Einrichtungen, die eine reibungslose digitale Kommunikation und Abwicklung von Geschäften (z. B. im → Onlinehandel) ermöglichen. Sie ist heute ein wesentlicher → Standortfaktor. Zu einer leistungsfähigen digitalen Infrastruktur gehören z. B. ein flächendeckender leistungsfähiger Breitbandzugang, Glasfasernetze, Mobilfunknetze, Internetknoten, öffentliches WLAN.

disaster relief – Katastrophenhilfe
(Seite 85)
Kurz- und mittelfristige Sofort- und Nothilfe, die darauf abzielt, die Grundbedürfnisse von Menschen zu sichern, die von Krieg, Naturkatastrophen oder Epidemien betroffen sind.

diversification – Diversifizierung
(Seite 161)
Verringerung der Abhängigkeit von einem Produkt oder einer Branche durch Ausweitung des Sortiments oder der Branchenstruktur. Ziel ist es, → Monostrukturen abzubauen.

E

economically underdeveloped region – strukturschwacher Raum / strukturschwache Region
(Seite 75, 183)
Gebiet, in dem sich wegen schlechter Standortfaktoren (→ location factor), wie z. B. unzureichender Infrastruktur (→ infrastructure), kaum Wirtschaftsunternehmen ansiedeln. Daher gibt es dort nur wenige Arbeitsplätze und zahlreiche Arbeitslose. Die Abwanderung ist hoch.

economic structure – Wirtschaftsstruktur
(Seite 161)
Gliederung der Wirtschaft einer Region. Dabei ist insbesondere die Gliederung in die drei Wirtschaftssektoren, der Anteil des informellen Sektors (→ informal sector) und eine mögliche Monostruktur (→ monostructure) von Interesse.

European Union (EU)– Europäische Union (EU)
(Seite 182)
Staatenverbund aus 27 europäischen Ländern, durch den eine möglichst enge, wirtschaftliche und politische Zusammenarbeit erreicht werden soll.

F

fair trade – fairer Handel
(Seite 97)
Gemeint ist damit ein Handel ohne benachteiligende Handelsbeschränkungen (Quoten, Zölle, Einfuhrbestimmungen) oder übervorteilende Preise. Die Erzeuger erhalten so einen angemessenen Preis, der ihnen menschenwürdige Lebensbedingungen sichert.

family planning – Familienplanung
(Seite 34)
Familienplanung wird als die bewusste Anwendung von Maßnahmen von Paaren bezeichnet, die Zahl und den Zeitpunkt der Geburt von Kindern zu planen. Bei Programmen zur Familienplanung in Entwicklungsländern (→ developing countries) steht die Senkung der Geburtenrate (→ birth rate) im Vordergrund.

foreign debt – Auslandsverschuldung
(Seite 79)
Verschuldung eines gesamten Landes gegenüber anderen Ländern. Die → Entwicklungsländer mussten sich in den letzten Jahren bei den Industrieländern (→ developed country) immer mehr Geld leihen, um die teuren Fertigwaren bezahlen zu können.

foreign direct investment (FDI) – ausländische Direktinvestition (ADI)
(Seite 89, 91)
Investition eines Unternehmens in einem anderen Land. Dies geschieht oft in der Form eines → Joint Ventures. Je höher die ausländischen Direktinvestitionen sind, desto attraktiver ist der Wirtschaftsstandort. Die Zahl der ausländischen Direktinvestitionen ist im Zuge der Globalisierung (→ globalisation) stark angestiegen und macht so die zunehmende wirtschaftliche Vernetzung deutlich.

foreign exchange – Devisen
(Seite 92)
Ausländisches Geld (z. B. als Bargeld oder als Scheck)

foreign national – Ausländer/in
(Seite 44)
Ausländer/innen in Deutschland haben nicht die deutsche Staatsangehörigkeit, sind also einem anderen Staat zugeordnet. Mit der Staatsangehörigkeit sind Rechte, wie das Wahlrecht, und Pflichten verbunden.

fragmentation – Fragmentierung
(Seite 81, 170)
Aufsplitterung einer Stadt in Gebiete, die global integriert sind und solche, die das nicht sind. Ein direktes Nebeneinander von armen und reichen Gebieten. Typisch ist, dass diese sehr unterschiedlichen Gebiete unmittelbar aneinandergrenzen (z. B. Elendssiedlungen grenzen unmittelbar an wohlhabende Wohngebiete oder gar → gated communities).

G

gated community – Gated Community
(Seite 51)
Eine Gated Community ist eine geschlossene Wohnanlage von wohlhabenden Menschen innerhalb einer Großstadt, die durch zahlreiche Maßnahmen wie Sperrzäune und Wachdienste vom Rest der Siedlung abgetrennt ist. Der Zutritt ist nur bestimmten Personen möglich. Man findet Gated Communities häufig in den großen Städten der Entwicklungsländer (→developing countries), aber auch in denen der Industrieländer (→ developed country).

Global City
(Seite 173)
Stadt wie New York, London oder Tokio, die weltweit von besonderer kultureller, politischer und wirtschaftlicher Bedeutung ist, z. B. als globales Finanzzentrum oder als Hauptsitz zahlreicher global tätiger Unternehmen.

globalisation – Globalisierung
(Seite 142)
Zunahme der weltweiten, vor allem wirtschaftlichen Verflechtung der Staaten der Erde, die durch moderne Kommunikationsmittel und Transportsysteme gefördert wird.

Global Player
(Seite 150)
Weltweit agierendes, oft multinationales Unternehmen, das Produktionsstandorte in der ganzen Welt hat und den Weltmarkt beliefert. Beispiele: IBM, Shell, Daimler.

good governance
(Seite 77)
Gute Regierungsführung, die sich durch Transparenz des Regierungshandelns und durch Mitbestimmungsmöglichkeiten auszeichnet, häufig Vorbedingung für die Gewährung von Entwicklungszusammenarbeit (→ development cooperation).
(→ bad governance)

gross national income (GNI) – Bruttonationaleinkommen (BNE)
(Seite 66)
(früher Bruttosozialprodukt, BSP) Summe des Wertes aller während eines Jahres von den Bewohnern eines Raumes (z. B. Staat) im Inland oder Ausland produzierten Güter und erbrachten Dienstleistungen. Im Unterschied dazu enthält die Angabe des Bruttoinlandsprodukts (BIP) die Werte der Waren und Dienstleistungen, die innerhalb eines Jahres in einem Raum (z. B. Staat) erzeugt werden.

growth industry – Wachstumsindustrie, Wachstumsbranche
(Seiten 88, 90)
Branche, bei der auch in Zukunft überdurchschnittliches wirtschaftliches Wachstum mit positiven Impulsen für die übrige Wirtschaft zu erwarten ist (z. B. in Industrieländern (→ developed country) die Freizeitbranche, in Entwicklungsländern (→ developing country) die Bauindustrie).

growth rate – Wachstumsrate
(Seite 26)
Die Wachstumsrate einer Bevölkerung ergibt sich aus der Differenz zwischen Geburtenrate (→ birth rate) und Sterberate (→ death rate). Eine hohe Wachstumsrate einer Bevölkerung ist kennzeichnend für ein Entwicklungsland (→ developing country)

H

Hidden Champion
(Seite 150)
Unternehmen kleiner und mittlerer Größe mit bedeutsamer, internationaler Marktposition.

Human Development Index (HDI)
(Seite 69)
Maßeinheit, mit der die → UNO seit 1992 den Entwicklungsstand der Länder bestimmt. Dabei werden die Lebenserwartung, der Anteil der Analphabeten, die durchschnittliche Dauer des Schulbesuchs und das Bruttonationaleinkommen (→ gross national income – GNI) berücksichtigt. Resultat ist eine Rangfolge der Länder der Erde vom höchsten bis zum niedrigsten HDI.

hybrid variaty – Hybridsorte
(Seite 17)
Aus Kreuzung verwandter Individuen hervorgegangene Pflanze, die höhere Erträge liefert oder andere gewünschte Eigenschaften hat.

I

illiterate – Analphabet
(Seite 25)
Ein Analphabet ist ein erwachsener Mensch, der weder lesen noch schreiben kann.

indigenous people – indigene Bevölkerung
(Seite 76)
Nachfahren der ersten Bewohner einer Region. Der Begriff wird heute verwendet anstelle der früher gebräuchlichen Begriffe „Ureinwohner", „Indianer" oder „Eingeborene".

Industry 4.0 – Industrie 4.0
(Seite 156)
In der Industrie 4.0 werden industrielle Produktionsprozesse mit modernster Informations- und Kommunikationstechnik vernetzt. So können die Wirtschaftlichkeit der Produktion und die Flexibilität erhöht werden. Im Mittelpunkt der Industrie 4.0 steht die Smart Factory (→ smart factory), die intelligente Fabrik.

informal sector – informeller Sektor
(Seite 51, 66)
Für Entwicklungsländer (→developing countries) typischer, offiziell nicht erfasster Bereich des Klein- und Dienstleistungsgewerbes (z. B. Straßenhandel, Schuhputzer). Obwohl keine Steuern an den Staat gezahlt werden, wird der informelle Sektor von den Regierungen geduldet, da er für große Teile der Bevölkerung ein Existenzminimum ermöglicht.

infrastructure – Infrastruktur
(Seite 75)
Alle Einrichtungen wie Verkehrswege, Energieversorgung, Kommunikationsnetze, aber auch Krankenhäuser, Schulen, öffentliche Verwaltung, Justizeinrichtungen. Die Infrastruktur bildet die Grundlage für gute Wohnbedingungen und wirtschaftliche Leistungsfähigkeit.

integration – Integration
(Seite 46)
Integration bedeutet allgemein Eingliederung. Hier ist die Einbindung einer Minderheit von Menschen (z. B. Ausländer/innen → foreign nationals) in eine vorherrschende Gesellschaft gemeint, von der sie sich unterscheiden, beispielsweise weil sie nicht dieselbe Sprache sprechen.

J

Joint Venture – Gemeinschaftsunternehmen
(Seiten 88, 90, 167)
Zusammenarbeit von Unternehmen, um ein gemeinsames Projekt durchzuführen. Häufig kommen diese Unternehmen aus verschiedenen Ländern, z. B. aus einem Industrie- und einem → Entwicklungsland (→ developed country, → developing country).

Just in Time – Just-in-time
(Seite 157)
Organisation der Zulieferung von Teilen, die bei der Produktion gebraucht werden. Der Hersteller bestellt aus Kostengründen erst dann bei den einzelnen Zulieferern, wenn der Lagerbestand zur Neige geht.

K

L

Least Developed Country (LDC) (Seite 69)
Land in der Gruppe von Entwicklungsländern (→ developing country), die laut Beschluss der → UNO am wenigsten entwickelt sind. Die Zuordnung erfolgt anhand der Werte Bruttonationaleinkommen pro Kopf (→ gross national income → per capita) und weiterer wirtschaftlicher Indikatoren wie Anteil der industriellen Produktion und Dienstleistungen am BIP sowie sozialen Merkmalen wie Alphabetisierungsgrad oder Kalorienzufuhr.

life expectancy – Lebenserwartung
(Seite 27)
Die durchschnittlich zu erwartende Lebensdauer eines Menschen zum Zeitpunkt seiner Geburt. Einfluss auf die Lebenserwartung haben unter anderem die Art der Ernährung und die medizinische Versorgung. Auch das Geschlecht kann entscheidend sein, denn im Durchschnitt werden Frauen älter als Männer.

location factor – Standortfaktor (weich/hart)
(Seite 165)
Faktor, der einen Betrieb dazu veranlasst, sich an einem bestimmten Standort niederzulassen, z. B. Infrastruktur (→ infrastructure), Subventionen, Steuern, vorhandene Arbeitskräfte. Weiche Standortfaktoren sind nur schwer messbar. Sie liegen oft im sozialen oder psychischen Bereich (z. B. günstige Wohnsituation, kulturelles Angebot innerhalb eines Raumes). Harte Standortfaktoren sind im Gegensatz dazu genau finanziell messbar (z. B. Arbeitskosten, Infrastrukturanbindung).

logistics – Logistik
(Seite 147)
Die Planung, Steuerung, Abwicklung und Kontrolle des Transportes, des Umschlags und der Lagerung von Gütern. Logistikzentren sind z. B. Häfen, Güterbahnhöfe und Flugplätze.

M

mass tourism – Massentourismus
(Seite 92)
Massentourismus ist eine Form des Fremdenverkehrs, an der eine große Zahl von Menschen teilnimmt. Neben dem finanziellen Gewinn für die Menschen, die in den Tourismusgebieten leben, bedeutet der Massentourismus oft eine Gefahr für Natur und Landschaft. Der Begriff wird daher häufig abwertend im Sinne einer Kritik an den Auswüchsen des Tourismus gebraucht.

megacity – Megastadt
(Seite 38, 49)
Sehr große Städte mit mehr als zehn Millionen Einwohnern. Meist sind damit gerade die rasant wachsenden Großstädte in den Entwicklungsländern gemeint. Mit dem Anstieg der Bevölkerungszahl dieser Städte ist auch ihr flächenhaftes Wachstum verbunden. Sie breiten sich immer weiter in ihr Umland aus.

metropolis – Metropole
(Seite 49)
Eine Metropole ist eine Großstadt, die der überragende politische, gesellschaftliche und wirtschaftliche Mittelpunkt eines Landes ist. Oft sind Metropolen die Hauptstädte der Länder.

migration – Migration
(Seite 11, 40)
Wanderung von Menschen mit dem Ziel, ihren Wohnort zu wechseln. Emigration bezeichnet die Auswanderung, Immigration die Einwanderung in ein Land oder eine Region. Heute leben mehr als 230 Millionen Menschen außerhalb ihres Heimatlandes. Noch viel mehr Menschen sind innerhalb von Staaten unterwegs. Binnenmigration nennt man die Migration innerhalb eines Landes, internationale Migration die über Grenzen hinweg.

monostructure – Monostruktur
(Seite 161)
Wirtschaft, die durch einen Industriezweig oder eine Branche bestimmt wird.

N

newly industrialising country (NIC) – Schwellenland
(Seite 68)
Land, das sich im Übergang vom Entwicklungsland (→ developing country) zum Industrieland (→ developed country) befindet.

non-governmental organisation (NGO) – Nichtregierungsorganisation (NRO)
(Seite 85)
Freiwillige Zusammenschlüsse einzelner Bürger mit gleichen oder ähnlichen Interessen auf lokaler, regionaler, nationaler oder internationaler Ebene.

O

online trade – Onlinehandel
(Seite 152)
Der Onlinehandel umfasst unterschiedliche Formen von Geschäften, die über das Internet abgewickelt werden, z. B. der Kauf von Waren oder Dienstleistungen bei Firmen durch Privatpersonen, der Handel von Firmen untereinander oder der Handel mit Aktien. Voraussetzung für den Onlinehandel ist eine funktionierende digitale Infrastruktur (→ digital infrastructure).

outsourcing – Outsourcing
(Seite 157)
Auslagerung von Teilen der Produktion und der Dienstleistungen mit dem Ziel, Kosten zu sparen und sich auf seine Kernkompetenz zu konzentrieren.

P

passive space – Passivraum
(Seite 187)
Ein Passivraum ist ein Gebiet mit geringer wirtschaftlicher Leistung. Vor allem junge Menschen wandern aus Passivräumen ab, weil diese kaum Arbeitsplätze bieten. (→ active space)

per capita income – Pro-Kopf-Einkommen
(Seite 67)
Das statistische Durchschnittseinkommen jedes Einwohners eines Landes. Es wird errechnet, indem man das Bruttoinlandsprodukt (→ gross national product – GNP) oder das Bruttonationaleinkommen (→ gross national income – GNI) durch die Bevölkerungszahl dividiert.

population pressure – Bevölkerungsdruck
(Seite 13)
Bevölkerungszahl, die im Verhältnis zu den wirtschaftlichen Möglichkeiten eines Raumes zu hoch ist, um eine ausreichende Ernährung aller Menschen zu gewährleisten.

poverty line – Armutsgrenze
(Seite 67)
Existenzminimum zur Deckung der lebensnotwendigen Grundbedürfnisse (→ basic needs). Unterhalb dieser Grenze hat ein Mensch zu wenig Mittel, um seinen Lebensunterhalt bestreiten zu können. Die Weltbank sieht alle diejenigen Menschen als arm an, die pro Tag weniger als 1,90 US-$ zur Verfügung haben. Um internationale Vergleichbarkeit zu gewährleisten, wird die lokale Kaufkraft der Währung einberechnet. 2015 lebten laut Weltbank weltweit 702 Mio. Menschen unterhalb der Armutsgrenze.

promotion of women – Frauenförderung
(Seite 87)
Schwerpunkt innerhalb der → Entwicklungszusammenarbeit mit dem Ziel, in den Entwicklungsländern (→ developing country) den in der Regel stark benachteiligten Frauen die Befriedigung ihrer Grundbedürfnisse zu ermöglichen (→ basic needs – z. B. Bildung, politische Mitbestimmung). Die Frauenförderung wird als Basis für die nachhaltige Entwicklung (→ sustainable development) auch in anderen Bereichen (z. B. Bildung, Ernährung, Gesundheit) gesehen.

pull factor – Pull-Faktor
(Seite 42)
Pull-Faktoren beeinflussen wie die Push-Faktoren (→ push factores) das → Migrationsverhalten (→ migration) einer Bevölkerung. Die Gründe dieser Wanderungsbewegung sind attraktiver erscheinende Lebensbedingungen des Zielgebietes.

push factor – Push-Faktor
(Seite 42)
Push-Faktoren beeinflussen wie die Pull-Faktoren (→ pull factores) das → Migrationsverhalten (→ migration) einer Bevölkerung. Die Gründe dieser Wanderungsbewegung sind die schlechten Lebensbedingungen im Herkunftsgebiet.

Q

R

refugee – Flüchtling
(Seite 40)
Menschen, die zum Beispiel aufgrund von Krieg, politischer oder religiöser Verfolgung ihr Land oder ihre Region verlassen mussten. Flüchtlinge genießen besonderen internationalen Schutz. Im Allgemeinen versteht man unter dem Begriff Menschen, die gezwungen wurden, ihre Heimat zu verlassen.

regional policy – Regionalpolitik
(Seite 187)
Der Teil der Politik, der sich verstärkt um regionalspezifische Belange kümmert. Bei der Regionalpolitik der → EU geht es vor allem darum, die großen räumlichen Disparitäten (→ spatial disparity) innerhalb Europas abzubauen.

remittance – Rücküberweisung
(Seite 42)
Geldzahlungen von Migranten etwa an die Familie in der Heimat. Die meist kleineren Beträge machen in der Summe riesige Kapitalströme aus.

reurbanisation – Reurbanisierung
(Seite 116)
Bevölkerungszunahme innerhalb der Stadt, oft ausgelöst durch Maßnahmen der Stadterneuerung, durch gesunkene Mietpreise infolge einer vorhergehenden Suburbanisierung (→ suburbanisation) oder durch steigende Wertschätzung des Wohnens in der Stadt (statt am Stadtrand oder auf dem Land).

rural-urban migration – Landflucht
(Seite 48)
Unter Landflucht versteht man die aufgrund von Push- und Pull-Faktoren (→ push/pull factores) ausgelösten Abwanderungen vom Land in die Stadt.

S

scenario – Szenario
(Seite 11)
In einem Szenario wird generell ein mögliches Bild von zukünftigen Ereignissen oder Zuständen beschrieben. Für die Voraussage werden vergangene und gegenwärtige Entwicklungen beachtet. Bei den getroffenen Voraussagen handelt sich sozusagen um eine „Was-wäre-wenn-Prognose". Die → UNO erstellt beispielsweise Szenarien zur weltweiten Bevölkerungsentwicklung.

segregation – Segregation
(Seite 119)
Der Prozess oder der Zustand der räumlichen Trennung von ethnischen oder sozialen Bevölkerungsgruppen, vor allem innerhalb einer Siedlungseinheit. Die sich trennenden Gruppen haben jeweils Gemeinsamkeiten, die sie als Gruppe definieren (z. B. Sprache, Religion, soziale Schicht). Segregation kann erzwungen sein und äußert sich räumlich in der Entstehung von Gettos, sie kann aber auch erwünscht und freiwillig sein, was oftmals zur Bildung von → gated communities führt.

slum – Slum
(Seite 50)
Ein Slum ist ein dicht besiedeltes Wohnviertel innerhalb einer Stadt, in dem sozial schwach gestellte Menschen in Behausungen mit einer heruntergekommenen Bausubstanz leben. Es wird auch als Elendsviertel bezeichnet. Der Wohnungsstandard ist niedrig und die Ausstattung mit Infrastruktur gering. Der Begriff „Slum" wird umgangssprachlich aber häufig auch als Sammelbezeichnung für Elendssiedlungen wie z. B. Hüttensiedlungen (→ squatter settlements) verwendet.

smart city – intelligente Stadt
(Seite 129, 130)
Sammelbegriff für Konzepte der Stadtentwicklung, die zum Ziel haben, Städte technologisch fortschrittlicher (z. B. digitale Verwaltung), umweltfreundlicher und sozial inklusiver zu gestalten.

spatial disparities – räumliche Disparitäten
(Seite 80, 187)
Unausgeglichenheit in der Ausstattung von Regionen. Diese zeigt sich in einem unterschiedlichen Angebot an Arbeitsplätzen oder an unterschiedlichen Lebensbedingungen bzw. ungleichen Entwicklungsmöglichkeiten. strukturschwache Region (→ economically underdeveloped region)

structurally weak region – strukturschwache Region
(Seite 187)
Gebiet, in dem sich wegen schlechter Standortfaktoren (→ location factor), wie z. B. unzureichender Infrastruktur (→ infrastructure), kaum Wirtschaftsunternehmen ansiedeln. Daher gibt es dort nur wenige Arbeitsplätze und zahlreiche Arbeitslose. Die Abwanderung ist hoch.

special economic zone – Sonderwirtschaftszone
(Seite 175)
Abgegrenzte Gebiete, wie z. B. in China, in denen ausländische Unternehmen unter anderen gesetzlichen Regelungen als im Rest des Landes zu besonders günstigen Bedingungen (Zollfreiheit, Steuererleichterungen, geringe Baukosten, billige Arbeitskräfte) produzieren können.

squatter settlement – Hüttensiedlung
(Seite 51)
Im Gegensatz zu einem → Slum sind Hüttensiedlungen Elendsviertel, die sich hauptsächlich in Randgebieten von Städten bilden. Dort leben Menschen in einfachen Behausungen aus Stroh, Wellblech oder Plastikplanen und haben häufig keinen Zugang zu Trinkwasser und zur Kanalisation. Weiterhin fehlen Einrichtungen wie Schulen, eine Müllabfuhr oder Krankenhäuser.

start-up – Start-up
(Seite 163)
Ein neu gegründetes, noch kleines Unternehmen mit einer innovativen Geschäftsidee. Wegen der großen Hoffnungen auf schnelles Wachstum ist die Ansiedlung von Start-ups sehr begehrt.

steel crisis – Stahlkrise
(Seite 161)
Krise der Eisen- und Stahlindustrie in den 1970er-Jahren. Gründe für die Krise waren ein weltweites Überangebot von Eisen und Stahl sowie ein Rückgang der Nachfrage nach Eisen- und Stahlprodukten in Deutschland. Es mussten Hüttenwerke stillgelegt werden. Tausende von Arbeitnehmern wurden entlassen.

structural change – Strukturwandel
(Seite 160)
Tiefgreifende Veränderung in der Struktur eines Raumes (z. B. Wirtschaftsstruktur – → economic structure), eines Wirtschaftsbereichs oder eines Betriebes (Betriebsstruktur). Von Bedeutung ist vor allem der Wandel der Wirtschaftsstruktur, z. B. von einer Monostruktur (→ monostructure) hin zu einer Diversifizierung (→ diversification) oder dem Wechsel von einer Branche zu einer anderen.

suburbanisation – Suburbanisierung
(Seite 116)
Unter diesem Begriff versteht man die Ausdehnung einer Stadt in ihr Umland durch Wohnbebauung, Industrie- und Dienstleistungsbetriebe. Die Gegenbewegung wird als Reurbanisierung (→ reurbanisation) bezeichnet.

supply chain – Lieferkette
(Seite 97, 146)
Alle Stationen, die bei der Erzeugung eines Produktes durchlaufen werden. Die Steuerung der Zu- und Weiterlieferungen ist Aufgabe der Logistik (→ logistics).

sustainability – Nachhaltigkeit
(Seite 15)
Nachhaltigkeit ist ein Leitbild für das Leben, Wirtschaften und politische Handeln. Ziel ist es, hinsichtlich der Umwelt, der Wirtschaft und der Gesellschaft so zu handeln, dass für möglichst alle ein menschenwürdiges Leben nicht nur jetzt, sondern auch in Zukunft möglich ist. Es darf jetzt nichts unternommen werden, was zukünftigen Generationen das Leben in Umwelt, Wirtschaft und Gesellschaft erschwert.

sustainable development – nachhaltige Entwicklung
(Seite 86)
Entwicklung, deren Ziel es ist, dass nicht nur eine kurzfristige, sondern eine langfristig anhaltende Verbesserung der Lebensbedingungen der Menschen erzielt wird. Dabei wird vor allem auf die Schonung der Umwelt geachtet. Nachhaltige Entwicklung beinhaltet auch eine nachhaltige Verbesserung der Grundbedürfnisse (→ basic needs) oft durch Hilfe zur Selbsthilfe (→ supporting self-help).

Sustainable Development Goals (SDG) – Nachhaltigkeitsziele
(Seite 15)
Im Rahmen der → Agenda 2030 erklärten über 180 Staaten Nachhaltigkeit (→ sustainability) zu einem wichtigen Leitziel ihrer Entwicklung. Die 17 von der → UNO formulierten Sustainable Development Goals fordern Nachhaltigkeit im Handeln in allen Bereichen des Lebens und Wirtschaftens weltweit, das heißt in allen Ländern der Erde.

T

Terms of Trade – Handelsbedingungen
(Seite 79)
Verhältnis zwischen Exportpreisen und Importpreisen. Das Verhältnis verschlechtert sich für ein Land, wenn die Exportpreise fallen und die Importpreise steigen oder die Exportpreise langsamer steigen als die Importpreise.

U

United Nations Organization (UNO) – Vereinte Nationen
(Seite 11)
Auch United Nations UN; Die Vereinten Nationen wurden 1945 gegründet. Ihr Hauptsitz ist in New York. Heute gehören ihr nahezu alle Staaten der Erde an. Die wichtigsten Aufgaben der UNO sind die Sicherung des Weltfriedens, die Förderung der Entwicklungszusammenarbeit (→ development cooperation) und der Schutz der Menschenrechte. Daneben kümmert sie sich um weitere globale Probleme wie den Klimawandel (→ climate change), die Desertifikation (→ desertification) oder die Bevölkerungsentwicklung.

urban farming – Urban Farming
(Seite 19)
Landwirtschaft und Tierhaltung innerhalb von Städten zur Versorgung der Region. Dabei können die Nutzflächen sowohl Gartenanlagen (z. B. Schrebergärten) als auch die Flachdächer von Gebäuden umfassen. Auch intensive Nutzungen wie der großflächige Anbau in Hochhausetagen oder Aquakulturen (→ aquaculture) gehören dazu. → Vertical Farming

urban heat island – Wärmeinsel
(Seite 130)
Phänomen der oft erhöhten Lufttemperatur in Städten gegenüber dem ländlichen Umland.

urbanisation – Verstädterung
(Seite 48)
Das Phänomen der Verstädterung beschreibt, dass ein immer größerer Anteil der Gesamtbevölkerung in Städten lebt. Diese dehnen sich flächenhaft aus. Zu beobachten ist die Verstädterung in fast allen Teilen der Erde. Hauptursache ist die Wanderung der ländlichen Bevölkerung in die Städte.

V

vertical farming – Vertical Farming
(Seite 19)
Innovatives Konzept für eine zukünftige Landwirtschaft, in der pflanzliche und tierische Erzeugnisse in Ballungsgebieten in mehrstöckigen Gebäuden (sogenannten Farmscrapern) erzeugt werden sollen.

vulnerability – Verletzlichkeit / Vulnerabilität
(Seite 75)
Die Anfälligkeit von Menschen, Gesellschaft, Infrastruktur (→ infrastructure) und Sachwerten eines Lebensraumes gegenüber unterschiedlichen Arten von Katastrophen (z. B. Naturkatastrophen, soziale Unruhen, politische Unruhen).

W

World Trade – Welthandel
(Seite 144)
Importe und Exporte aller Länder der Erde, d. h. der Austausch von Gütern und Dienstleistungen über die jeweiligen Staatsgrenzen hinweg.

World Trade Organization (WTO) – Welthandelsorganisation
(Seite 96, 182)
Die WTO ist eine internationale Organisation, die seit 1994 den Welthandel regelt, koordiniert und bei Streitigkeiten eingreift. Sie hat 164 Mitgliedstaaten.

X / Y

Z

Klimadaten

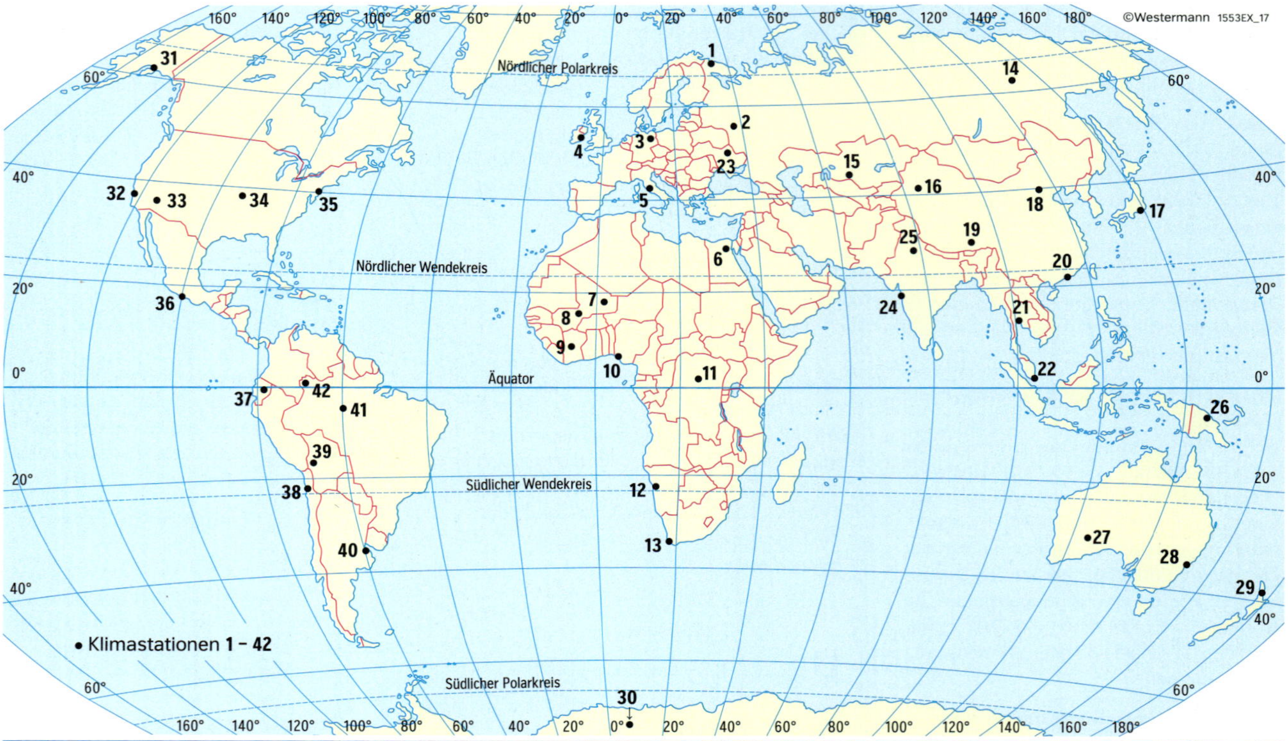

M1 Lage der Klimastationen

Europa

				J	F	M	A	M	J	J	A	S	O	N	D	Jahr
1	Murmansk	46 m ü. M.	°C	-10,9	-11,4	-8,1	-1,4	3,9	10,0	13,4	11,1	6,9	0,9	-3,8	-7,9	0,2
	(Russland)	68° 58′ N / 33° 03′ O	mm	19	16	18	19	25	40	54	60	44	30	28	33	386
2	Moskau	156 m ü. M.	°C	-10,3	-9,7	-5,0	3,7	11,7	15,4	17,8	15,8	10,4	4,1	-2,3	-8,0	3,6
	(Russland)	55° 45′ N / 37° 34′ O	mm	31	28	33	35	52	67	74	74	58	51	36	36	575
3	Berlin	51 m ü. M.	°C	-0,6	-0,3	3,6	8,7	13,8	17,0	18,5	17,7	13,9	8,9	4,5	1,1	8,9
	(Deutschland)	52° 28′ N / 13° 18′ O	mm	43	40	31	41	46	62	70	68	46	47	46	41	581
4	Dublin	68 m ü. M.	°C	4,5	4,8	6,5	8,4	10,5	13,5	15,0	14,8	13,1	10,5	7,2	5,8	9,6
	(Irland)	53° 26′ N / 6° 15′ W	mm	71	52	51	43	62	55	66	80	77	68	67	77	769
5	Rom	46 m ü. M.	°C	6,9	7,7	10,8	13,9	18,1	22,1	24,7	24,5	21,1	16,4	11,7	8,5	15,5
	(Italien)	41° 54′ N / 12° 29′ O	mm	76	88	77	72	63	48	14	22	70	128	116	106	880

Afrika

				J	F	M	A	M	J	J	A	S	O	N	D	Jahr
6	Kairo	95 m ü. M.	°C	13,3	14,7	17,5	21,1	25,0	27,5	28,3	28,3	26,1	24,1	20,0	15,0	21,7
	(Ägypten)	30° 08′ N / 31° 34′ O	mm	4	5	3	1	1	0	0	0	0	1	1	8	24
7	Gao	270 m ü. M.	°C	22,0	25,0	28,8	32,4	34,6	34,5	32,3	29,8	31,8	31,9	28,4	23,3	29,6
	(Mali)	16° 16′ N / 0° 03′ W	mm	<1	0	<1	<1	8	23	71	127	38	3	<1	<1	270
8	Mopti	280 m ü. M.	°C	22,6	25,2	29,0	31,6	32,8	31,2	28,6	27,3	28,3	28,8	26,8	23,1	27,9
	(Mali)	14° 30′ N / 4° 12′ W	mm	<1	<1	1	5	23	56	147	198	94	18	1	<1	543
9	Bouaké	365 m ü. M.	°C	27,1	28,0	28,4	27,9	27,2	26,1	24,8	24,5	25,5	26,1	26,7	26,7	26,6
	(Elfenbeinküste)	7° 42′ N / 5° 00′ W	mm	13	46	92	140	154	135	99	108	225	140	35	23	1210
10	Douala	11 m ü. M.	°C	26,7	27,0	26,8	26,6	26,3	25,4	24,3	24,1	24,7	25,0	26,0	26,4	25,8
	(Kamerun)	4° 01′ N / 9° 43′ O	mm	57	82	216	243	337	486	725	776	638	388	150	52	4150
11	Yangambi	487 m ü. M.	°C	24,7	25,3	25,5	25,2	24,9	24,5	23,6	23,9	24,3	24,5	24,3	24,3	24,6
	(D. R. Kongo)	0° 49′ N / 24° 29′ O	mm	85	99	148	150	177	126	146	170	180	241	180	126	1828
12	Swakopmund	12 m ü. M.	°C	17,2	18,1	17,5	15,7	15,1	15,2	13,0	12,1	12,6	13,7	14,9	16,4	15,1
	(Namibia)	22° 41′ S / 14° 31′ O	mm	1	2	2	2	0	0	0	0	0	0	1	0	8
13	Kapstadt	17 m ü. M.	°C	21,2	21,5	20,3	17,5	15,1	13,4	12,6	13,2	14,5	16,3	18,3	20,1	17,0
	(Südafrika)	33° 54′ S / 18° 32′ O	mm	12	8	17	84	82	85	85	71	43	29	17	11	506

Asien

	Station	Lage		J	F	M	A	M	J	J	A	S	O	N	D	Jahr
14	Jakutsk	100 m ü. M.	°C	-43,2	-35,8	-22,0	-7,4	5,6	15,4	18,8	14,8	6,2	-7,8	-27,7	-39,6	-10,2
	(Russland)	62° 05′ N / 129° 45′ O	mm	7	6	5	7	16	31	43	38	22	16	13	9	213
15	Kysylorda	129 m ü. M.	°C	-9,6	-7,5	0,5	11,2	18,8	23,6	24,6	22,5	15,8	7,8	-0,6	-7,0	8,3
	(Kasachstan)	44° 46′ N / 65° 32′ O	mm	13	15	14	14	11	5	4	3	4	7	10	14	114
16	Kuqa	970 m ü. M.	°C	-12,6	-3,6	7,2	13,1	18,1	21,4	24,1	22,5	18,1	12,2	1,7	-7,2	9,6
	(China)	41° 40′ N / 83° 06′ O	mm	3	3	5	3	3	33	18	8	5	0	2	8	91
17	Tokio	4 m ü. M.	°C	3,7	4,3	7,6	13,1	17,6	21,1	25,1	26,4	22,8	16,7	11,3	6,1	14,7
	(Japan)	35° 41′ N / 139° 46′ O	mm	48	73	101	135	131	182	146	147	217	220	101	61	1562
18	Peking	52 m ü. M.	°C	-4,7	-1,9	4,8	13,7	20,1	24,7	26,1	24,9	19,9	12,8	3,8	-2,7	11,8
	(China)	39° 57′ N / 116° 19′ O	mm	4	5	8	17	35	78	243	141	58	16	11	3	619
19	Lhasa	3685 m ü. M.	°C	-1,7	1,1	4,7	8,1	12,2	16,7	16,4	15,6	14,2	8,9	3,9	0,0	8,3
	(China)	29° 40′ N / 91° 07′ O	mm	2	13	8	5	25	64	122	89	66	13	3	0	410
20	Hongkong	33 m ü. M.	°C	15,6	15,0	17,5	21,7	25,6	27,5	28,1	28,1	27,2	25,0	20,9	17,5	22,5
	(China)	22° 18′ N / 114° 10′ O	mm	33	46	74	292	394	381	394	361	247	114	43	30	2409
21	Bangkok	2 m ü. M.	°C	26,0	27,8	29,2	30,1	29,7	28,9	28,5	28,4	28,0	27,7	27,0	25,7	28,1
	(Thailand)	13° 45′ N / 100° 28′ O	mm	9	30	36	82	165	153	168	183	310	239	55	8	1438
22	Singapur	10 m ü. M.	°C	26,4	27,0	27,5	27,5	27,8	27,5	27,5	27,2	27,2	27,0	27,0	27,0	27,2
	(Singapur)	1° 18′ N / 103° 50′ O	mm	251	173	193	188	173	173	170	196	178	208	254	256	2413
23	Charkow	152 m ü. M.	°C	-7,1	-6,5	-1,6	7,8	15,0	18,9	21,1	19,9	14,1	7,3	0,7	-4,0	7,1
	(Ukraine)	49° 56′ N / 36° 17′ O	mm	36	33	32	33	50	60	75	48	34	42	39	37	519
24	Mumbai (Bombay)	11 m ü. M.	°C	23,9	23,9	26,1	28,1	29,7	28,9	27,2	27,0	27,0	28,1	27,2	25,6	26,9
	(Indien)	18° 54′ N / 72° 49′ O	mm	3	3	3	2	18	485	617	340	264	64	13	3	1815
25	Neu-Delhi	218 m ü. M.	°C	13,9	16,7	22,5	28,1	33,3	33,6	31,4	30,0	28,9	26,1	20,0	15,3	25,0
	(Indien)	28° 35′ N / 77° 12′ O	mm	23	18	13	8	13	74	180	173	117	10	3	10	642
26	Madang	6 m ü. M.	°C	27,3	27,0	27,3	27,2	27,5	27,2	27,2	27,2	27,2	27,5	27,5	27,5	27,3
	(Papua-Neuguinea)	5° 14′ S / 145° 45′ O	mm	307	302	378	429	384	274	193	122	135	254	338	368	3484

Australien

	Station	Lage		J	F	M	A	M	J	J	A	S	O	N	D	Jahr
27	Kalgoorlie	380 m ü. M.	°C	25,7	24,9	23,0	18,7	14,7	12,0	10,8	12,3	15,3	18,2	21,4	24,3	18,4
	(Australien)	30° 45′ S / 121° 30′ O	mm	24	27	24	18	22	25	24	23	13	14	15	13	244
28	Sydney	42 m ü. M.	°C	22,0	21,9	20,8	18,3	15,1	12,8	11,8	13,0	15,2	17,6	19,5	21,1	17,4
	(Australien)	33° 51′ S / 151° 31′ O	mm	104	125	129	101	115	141	94	83	72	80	77	86	1207
29	Auckland	49 m ü. M.	°C	19,2	19,6	18,4	16,4	13,8	11,8	10,8	11,3	12,6	14,3	15,9	17,7	15,2
	(Neuseeland)	36° 51′ S / 174° 46′ O	mm	84	104	71	109	122	140	140	109	97	107	81	79	1243

Antarktika

	Station	Lage		J	F	M	A	M	J	J	A	S	O	N	D	Jahr
30	Südpol	2800 m ü. M.	°C	-28,8	-40,1	-54,4	-58,5	-57,4	-56,5	-59,2	-58,9	-59,0	-51,3	-38,9	-28,1	-49,3
		90° S	mm	0	0	0	0	0	0	1	1	0	0	0	0	2

Amerika

	Station	Lage		J	F	M	A	M	J	J	A	S	O	N	D	Jahr
31	Anchorage	27 m ü. M.	°C	-10,9	-7,8	-4,8	2,1	7,7	12,5	13,9	13,1	8,8	1,7	-5,4	-9,8	1,8
	(USA)	61° 10′ N / 149° 59′ W	mm	20	18	13	11	13	25	47	65	64	47	26	24	373
32	San Francisco	16 m ü. M.	°C	10,4	11,7	12,6	13,2	14,1	15,1	14,9	15,2	16,7	16,3	14,1	11,4	13,8
	(USA)	37° 47′ N / 122° 25′ W	mm	116	93	74	37	16	4	0	1	6	23	51	108	529
33	Death Valley	56 m ü. M.	°C	11,1	14,4	18,9	23,9	28,9	34,2	38,6	37,2	32,0	23,9	16,1	11,4	24,2
	(USA)	36° 28′ N / 116° 51′ W	mm	3	1	3	3	5	5	8	8	5	1	3	1	46
34	Kansas City	226 m ü. M.	°C	-0,7	1,6	6,0	12,9	18,4	24,1	27,2	26,3	21,6	15,4	6,7	1,6	13,4
	(USA)	39° 07′ N / 94° 35′ W	mm	36	32	63	90	112	116	81	96	83	73	46	39	867
35	New York	96 m ü. M.	°C	0,7	0,8	4,7	10,8	16,9	21,9	24,9	23,9	20,3	14,6	8,3	2,2	12,5
	(USA)	40° 47′ N / 73° 58′ W	mm	84	72	102	87	93	84	94	113	98	80	86	83	1076
36	Acapulco	3 m ü. M.	°C	26,7	26,5	26,7	27,5	28,5	28,6	28,7	28,8	28,1	28,1	27,7	26,7	27,7
	(Mexiko)	16° 50′ N / 99° 56′ W	mm	6	1	<1	1	36	281	256	252	349	159	28	8	1377
37	Quito	2818 m ü. M.	°C	13,0	13,0	12,9	13,0	13,1	13,0	12,9	13,1	13,2	12,9	12,8	13,0	13,0
	(Ecuador)	0° 13′ S / 78° 30′ W	mm	124	135	159	180	130	49	18	22	83	133	110	107	1250
38	Antofagasta	119 m ü. M.	°C	19,9	20,1	18,5	16,2	15,1	13,5	13,1	13,5	14,5	15,2	16,8	18,4	16,2
	(Chile)	23° 26′ S / 70° 28′ W	mm	0	0	0	0,3	0,1	1,7	2,4	1,2	1,0	0,8	0,2	0	7,7
39	La Paz	3632 m ü. M.	°C	17,5	16,2	15,5	14,1	11,7	10,1	9,8	10,9	14,4	15,5	17,5	17,9	14,3
	(Bolivien)	16° 30′ S / 68° 08′ W	mm	92	89	62	26	11	2	4	7	34	28	48	85	488
40	Buenos Aires	25 m ü. M.	°C	23,7	23,0	20,7	16,6	13,7	11,1	10,5	11,5	13,6	16,5	19,5	22,1	16,9
	(Argentinien)	34° 35′ S / 58° 29′ W	mm	104	82	122	90	79	68	61	68	80	100	90	83	1027
41	Manaus	48 m ü. M.	°C	26,2	26,2	26,4	26,2	26,3	26,6	26,8	27,5	27,9	27,8	27,6	26,8	26,9
	(Brasilien)	3° 08′ S / 60° 01′ W	mm	266	247	269	267	194	100	64	38	60	124	152	216	1897
42	Uaupés	85 m ü. M.	°C	25,5	25,8	25,6	25,4	25,1	24,7	24,3	25,0	25,5	25,8	25,1	25,7	25,4
	(Brasilien)	0° 08′ S / 67° 05′ W	mm	284	261	284	263	329	244	234	186	160	164	190	270	2869

Länderdaten

Land	Fläche (in km²)	Einwohner (in Mio.) 2020	Städtische Bevölkerung (in %) 2020	Geburtenrate (in ‰)	Sterberate (in ‰)	Lebenserwartung in Jahren männl.	weibl.
Europa (EU 27)							
Belgien	30528	11,6	98	10	10	79	84
Bulgarien	110879	6,9	76	9	16	71	78
Dänemark	43094	5,8	88	11	9	79	83
Deutschland	357022	83,8	78	10	12	78	83
Estland	45228	1,3	69	11	12	74	82
Finnland	338145	5,5	86	8	10	79	84
Frankreich	551500	65,3	81	11	9	80	86
Griechenland	131957	10,4	80	8	11	79	84
Irland	70273	4,9	64	12	6	78	83
Italien	301340	60,5	71	7	11	81	85
Kroatien	56594	4,1	58	9	13	75	81
Lettland	64589	1,9	68	10	14	70	80
Litauen	65300	2,7	68	10	14	71	81
Luxemburg	2586	0,6	92	10	7	80	85
Malta	316	0,4	95	9	8	80	85
Niederlande	41543	17,1	92	10	9	80	83
Österreich	83880	9,0	59	10	9	79	84
Polen	312685	37,8	60	10	11	74	82
Portugal	92090	10,2	66	9	11	78	83
Rumänien	238391	19,2	54	11	14	72	79
Schweden	450295	10,1	88	11	9	81	85
Slowakei	49035	5,4	54	10	10	74	80
Slowenien	2273	2,1	55	10	10	78	84
Spanien	505370	46,8	81	8	9	80	86
Tschechien	78867	10,7	74	11	11	76	82
Ungarn	93028	9,7	72	9	13	73	79
Zypern	5370	1,2	67	11	7	79	83
weitere europäische Staaten							
Großbritannien und Nordirland	243610	67,9	84	11	9	79	83
Russland	17098242	145,9	75	11	13	68	78
Ukraine	603550	43,7	70	8	14	67	77
Afrika							
Ägypten	1001450	102,3	43	23	6	73	75
Äthiopien	1104300	115,0	22	33	6	65	69
Kenia	580140	53,8	28	28	5	67	70
Libyen	1759540	6,9	81	24	4	74	79
Nigeria	923768	206,1	52	37	12	54	56
Senegal	196722	16,7	48	34	6	66	70
Südafrika	1219090	59,3	67	20	9	62	68
Nordamerika							
Kanada	9984670	37,7	82	10	8	80	84
USA	9833517	331,0	83	12	9	76	81
Mittel- und Südamerika							
Argentinien	2780400	45,2	92	17	8	74	81
Bolivien	1098581	11,7	70	22	6	70	77
Brasilien	8515770	212,6	87	14	6	72	80
Mexiko	1964375	128,9	81	17	6	72	78
Peru	1285216	33,0	78	19	5	74	80
Asien							
China, Volksrepublik	9596960	1439,3	61	10	7	75	79
Indien	3287263	1380,0	35	20	6	68	70
Israel	20937	8,7	93	20	5	81	85
Japan	377915	126,5	92	7	11	81	87
Pakistan	796095	220,9	37	28	6	67	71
Saudi-Arabien	2150000	34,8	84	15	3	74	78
Singapur	719	5,9	100	9	5	81	85
Südkorea	99720	51,3	81	6	6	80	86
Türkei	783562	84,3	76	18	5	76	81
Vietnam	331210	97,3	37	16	6	71	76
Australien und Ozeanien							
Australien	7741220	25,5	86	13	6	83	86
Neuseeland	268838	4,8	87	12	7	80	84

Quellen: CIA World Factbook 2020, DSW/PRB 2020, UN DESA 2020

Land	Alphabetisierungsrate 2017/2018 in % männl.	weibl.	Direkter Zugang zu sauberem Trinkwasser 2017 in %	CO_2-Emissionen pro Kopf 2019 in t	BNE (KKP*) in US-$ pro Einwohner 2019	Auslandsverschuldung 2019 in Mio. US-$	Beschäftigte nach Sektoren 2020 (I. / II./ III.) in %**
Europa (EU 27)							
Belgien	>98	>98	>99	8,6	55 590	1 298 500	1/ 21/ 78
Bulgarien	>98	>98	97	6,0	24 390	40 400	6/ 30/ 64
Dänemark	>98	>98	97	5,6	61 960	501 600	2/ 18/ 80
Deutschland	>98	>98	>99	8,4	57 810	5 588 100	1/ 27/ 72
Estland	>98	>98	93	10,5	38 030	23 300	3/ 29/ 68
Finnland	>98	>98	>99	7,5	48 360	603 700	4/ 22/ 74
Frankreich	>98	>98	98	5,0	51 670	6 268 400	2/ 20/ 78
Griechenland	98,5	97,4	>99	6,4	30 470	500 200	12/ 15/ 73
Irland	>98	>98	97	7,6	70 010	2 852 000	4/ 19/ 77
Italien	99,4	99,0	95	5,6	44 630	2 503 000	4/ 26/ 71
Kroatien	>98	>98	90	4,3	29 680	45 900	6/ 27/ 67
Lettland	99,9	99,9	95	4,3	31 740	40 100	7/ 24/ 70
Litauen	>98	>98	92	4,9	37 170	37 100	7/ 26/ 67
Luxemburg	>98	>98	>99	15,9	77 570	4 090 600	1/ 12/ 87
Malta	93,0	96,0	>99	3,5	42 770	98 400	1/ 18/ 81
Niederlande	>98	>98	>99	9,1	59 790	4 311 000	2/ 16/ 82
Österreich	>98	>98	99	7,6	59 240	686 400	3/ 25/ 71
Polen	>98	>98	>99	8,5	33 060	354 000	9/ 32/ 59
Portugal	97,4	95,1	95	4,8	35 790	460 400	6/ 25/ 70
Rumänien	99,1	98,6	82	3,9	31 840	123 300	21/ 30/ 49
Schweden	>98	>98	>99	4,3	57 220	888 800	2/ 18/ 81
Slowakei	>98	>98	>99	6,1	33 470	118 500	2/ 36/ 62
Slowenien	>98	>98	98	6,6	40 330	49 200	5/ 33/ 61
Spanien	98,9	98,0	98	5,4	42 260	2 371 800	4/ 20/ 76
Tschechien	>98	>98	98	9,4	40 640	193 800	3/ 37/ 60
Ungarn	>98	>98	90	5,1	33 580	147 300	5/ 33/ 63
Zypern	>98	>98	>99	6,1	39 830	212 600	2/ 16/ 82
weitere europäische Staaten							
Großbritannien und Nordirland	>98	>98	>99	5,5	47 880	8 768 200	1/ 18/ 81
Russland	99,7	99,7	76	11,5	28 270	491 400	6/ 27/ 68
Ukraine	>98	>98	92	5,1	13 750	121 700	14/ 25/ 61
Afrika							
Ägypten	76,5	65,5	–	2,5	11 840	112 700	23/ 28/ 49
Äthiopien	59,2	44,4	11	0,1	2310	28 300	66/ 10/ 24
Kenia	85,0	78,2	–	0,3	4430	43 200	54/ 7/ 39
Libyen	–	–	–	6,9	16 130	–	19/ 22/ 60
Nigeria	71,3	52,7	20	0,7	5190	54 800	35/ 12/ 53
Senegal	64,8	39,8	–	0,6	3470	13 600	29/ 14/ 57
Südafrika	87,7	86,5	–	8,2	12 670	185 400	5/ 23/ 72
Nordamerika							
Kanada	>98	>98	99	15,4	50 810	2 110 100	1/ 19/ 79
USA	>98	>98	>99	16,1	66 080	20 600 700	1/ 20/ 79
Mittel- und Südamerika							
Argentinien	98,9	99,1	–	4,0	22 120	278 500	0/ 21/ 79
Bolivien	–	–	–	2,0	8930	14 300	30/ 19/ 50
Brasilien	93,0	93,4	–	2,2	14 890	675 800	9/ 20/ 71
Mexiko	96,2	94,6	43	3,4	19 990	464 300	12/ 26/ 61
Peru	97,1	91,7	50	1,7	12 790	64 200	27/ 15/ 58
Asien							
China, Volksrepublik	98,5	95,2	–	7,1	16 790	2 057 300	25/ 28/ 47
Indien	82,4	65,8	–	1,9	6920	563 900	41/ 26/ 32
Israel	>98	>98	>99	7,5	41 950	105 000	1/ 17/ 82
Japan	>98	>98	98	8,7	44 810	4 239 200	3/ 24/ 73
Pakistan	71,1	46,5	35	1,1	4800	100 900	36/ 26/ 38
Saudi-Arabien	97,1	92,7	–	17,0	49 520	183 800	2/ 25/ 73
Singapur	98,9	95,9	>99	6,7	92 270	1 576 000	1/ 15/ 84
Südkorea	>98	>98	98	11,9	43 520	457 745	5/ 25/ 70
Türkei	98,8	93,5	–	4,9	27 660	434 200	18/ 26/ 56
Vietnam	96,5	93,6	–	2,6	7910	–	36/ 28/ 35
Australien und Ozeanien							
Australien	>98	>98	–	16,3	51 760	1 539 600	3/ 20/ 78
Neuseeland	>98	>98	>99	7,6	42 710	193 000	6/ 19/ 75

* Das BNE in KKP (Kaufkraftparitäten; Parität = Gleichheit) gibt das Einkommen an, das für den Konsum von Waren und Dienstleistungen (z. B. Nahrungsmittel, Bekleidung, Gesundheitsdienstleistungen) tatsächlich zur Verfügung steht. Es ist ein Maßstab für den Lebensstandard der Bevölkerung eines Landes.
** rundungsbedingte Abweichungen
Quellen: Global Carbon Atlas 2020, ILO 2020, UNESCO UIS 2020, UNICEF 2019, Weltbank 2020, WHO 2020

Vokabeln

A

abort – abtreiben
abortion – Schwangerschaftsabbruch
abroad – im Ausland
absolute/relative number – absolute/relative Werte/Zahlen
abundance of fish – Fischreichtum
abuse of power – Machtmissbrauch
accelerate – beschleunigen
accelerated designation – beschleunigte Ausweisung/Kennzeichnung
accessibility – Erreichbarkeit
accumulate waste – Abfälle anhäufen
achieve – erreichen
acquisition – (Unternehmens-)Kauf
adapt – anpassen
adaptation – Anpassung
adequate payment – angemessene Zahlungen
advantages of agglomeration – Agglomerationsvorteile
affected – betroffen
affluence – Wohlstand
affordable housing – bezahlbarer Wohnraum
agglomeration – Agglomeration/Ballungsraum/Ballungszentrum
agreement – Vertrag
agricultural policy – Landwirtschaftspolitik
agricultural scientist – Agrarwissenschaftler
agricultural workers' union – Landarbeitergewerkschaft
agriculture limit – Anbaugrenze
air pollution – Luftverschmutzung
algae – Algen
amenity – Annehmlichkeit
ancestors – Vorfahren
appeal – Attraktivität
applicable – anwendbar
aquaculture – Aquakultur
arable farming – Ackerbau
arable land – landwirtschaftliche Nutzfläche
aridity – Trockenheit
armed conflict – bewaffneter Konflikt / Krieg
artificial – künstlich
arts and crafts – Kunsthandwerk
assembly plant – Montagewerk
assembly – Montage
assets – Werte/materielle Güter
asylum – Asyl
at the seams – aus den Nähten
automobile sales – Automobilabsatz
automotive industry – Automobilindustrie
average – Durchschnitt

B

B2B – business-to-business (Unternehmen zu Unternehmen-Beziehung)
backward – rückständig
ban – Verbot
barley – Gerste
basic parts of living – Grunddaseinsfunktionen
be dependent on – abhängig sein von
be on the cutting edge – auf der Höhe der Zeit sein
be underway – im Gange sein
beacon – Leuchtfeuer
beef – Rinfleisch
bell – Glocke
biodiversity – Artenvielfalt
blessing – Segen
body shell – Rohkarosserie
body – hier: Leiche
bottle feeding – Flaschennahrung
bourgeois – bürgerlich
brain drain – Braindrain (wörtlich: Gehirn-Abfluss)
branch – Zweig, hier: Niederlassung
breast milk – Muttermilch
breeding – Zucht
building site – Baugebiet
bureaucracy – Bürokratie
burial site – Begräbnisstätte
bury – bestatten
business model – Geschäftsmodell
business promotion – Wirtschaftsförderung
business – Wirtschaftsbetrieb
by-catch – Beifang
bypass motorway – Umgehungsstraße

C

carbon footprint – CO2-Fußabdruck
carbon monoxide – Kohlenmonoxid
car-friendly city – autogerechte Stadt
cargo turnover – Warenumschlag
carrying capacity – Tragfähigkeit
catch rate – Fangquote
catchment area – Einzugsbereich
cattle – Rinder
cause – Ursache
central warehouse – Zentrallager
certification – Zertifizierung
charging station – Ladesäule
child abuse – Kindesmissbrauch
child mortality – Kindersterblichkeit
childcare facilities – Kinderbetreuungsstätten
childcare – Kinderbetreuung
cholera – Cholera
chronically malnourished – chronisch unterernährt
cinnamon – Zimt
cisterne – Zisterne
city dweller – Stadtbewohner/-in
civil servant – Beamter/Beamtin
civil war – Bürgerkrieg
climate-related hazard – klimabedingte Gefährdung
clog up – verstopfen
coastal – küstennah
coca cultivation – Cocaanbau
cohort – Kohorte, hier: Jahrgang/Alterskohorte
cold air corridor – Kaltluftschneise
colliery site – Zechengelände
colonial era – Kolonialzeit
colonial power – Kolonialmacht
combustion – Verbrennung
commission – Provision
commodity prices – Rohstoffpreise
commuter – Pendler/Pendlerin
company premise – Firmengelände
compatibility of famliy and work – Vereinbarkeit von Familie und Beruf
competition – Wettbewerb
competitive – wettbewerbsfähig
competitiveness – Wettbewerbsfähigkeit
component – Bauteil
Computer Science – Informatik
confectionery – Süßwaren
conqueror – Eroberer/-in
consequence – Folge
construction sector – Bausektor
construction – Baugewerbe
consumer protection rule – Verbraucherschutzvorschrift
consumer – Verbraucher/-in
consumption – Verbrauch
contraception – Verhütung
contraceptive – Schwangerschaftsverhütungsmittel
contrast – hier: gegenüberstellen
control system – Steuerungssystem
conurbation – Ballungsgebiet/Städteballung

cook-shop – Garküche
co-owner – Miteigentümer
cost of living – Lebenshaltungskosten
council housing unit – Sozialwohnung
country of origin – Herkunftsland
courtyard – Innenhof
crate – Kiste
creditor – Gläubiger/-in
crop diversity – Nutzpflanzenvielfalt
crop rotation – Fruchtfolge
crop – Anbauprodukt
cross-border flow – grenzenlose Ströme
customised – angepasst
cut flowers – Schnittblumen

D

dam – Damm, hier: Staumauer
dashboard – Armaturenbrett
data centre – Rechenzentrum
data throughput – Datendurchsatz
decisive role – entscheidende Rolle
decline – Niedergang
deficiency disease – Mangelkrankheit
deficiency – Defizit
degree – Grad, hier: (Schul-)Abschluss
demand – Nachfrage
demographer – Bevölkerungswissenschaftler/-in
demographic development – Bevölkerungsentwicklung
demographic transition model (DTM) – Modell des demographischen Übergangs
denounce – anprangern
densely populated – dicht besiedelt
dependence – Abhängigkeit
destination – Zielregion
deterioration – Verschlechterung
development – Entwicklung
development assistance – Entwicklungshilfe
development cooperation – Entwicklungszusammenarbeit (EZ)
development index – Entwicklungsindex
development strategies – Entwicklungsstrategien
digital map service – digitaler Kartendienst
direct/indirect employment – direkte/indirekte Arbeitsplätze
disaster relief – Katastrophenhilfe
disease – Krankheit
distribute – verteilen
distribution – Verteilung
distribution centre – Verteilzentrum
district – Stadtteil
diversification – Diversifizierung
division of labour – Arbeitsteilung
domestic – inländisch
downside – Schattenseite
downstream – flussabwärts
driver of development – Entwicklungsmotor
dustbin – Mülltonne
dwarf – Zwerg

E

ecological footprint – ökologischer Fußabdruck
economic crisis – Wirtschaftskrise
economic development – Wirtschaftsförderung
economic indicators – ökonomische/wirtschaftliche Indikatoren
economic performance – Wirtschaftsleistung
economic sector – Wirtschaftszweig
economic zone – Wirtschaftszone
economical – wirtschaftlich
edible – essbar
educational facility – Bildungseinrichtung
egress – Ausgang
electric vehicles – Elektrofahrzeuge
electricity meter – Stromzähler
electronic components – elektronische Bauteile
emergency shelter – Notunterkunft
emigrant – Auswanderer/in
employee – Angestellte/r
employment contract – Arbeitsvertrag
energy supplier – Energieversorger
enterprise – Unternehmen
entry regulations – Einreisebestimmungen
environmental and animal welfare requirements – Umwelt- und Tierschutzauflagen
environmental regulations – Umweltschutzbestimmungen
environmentally friendly – umweltfreundlich
epidemic – Seuche / Epidemie
equal – gleich
establish – einrichten / ansiedeln
estimate – Schätzung
exchange earner – Devisenbringer
executive board – Konzernleitung
expected years of schooling – erwartete Schulbesuchsdauer in Jahren
exploit – ausbeuten, ausnutzen, ausschöpfen
exploitative – ausbeuterisch
exposure – Belastung

F

factory vessel – Fabrikschiff
famine relief – Hungerhilfe
famine – Hungernot
farmscraper – Farmhochhaus
fatal – tödlich
fattening – Mast
favela – Favela / Marginalsiedlung in Brasilien
fence – Zaun
fence in – einzäunen
fertile – fruchtbar
fertiliser – Dünger
fibre – Fasern
fibre optic cable – Glasfaserkabel
final assembly – Endmontage
financial ressources – finanzielle Mittel
financial support – finanzielle Unterstützung
fine – Strafe/Strafgebühr
fish consumption – Fischkonsum
fish meal – Fischmehl
fish oil – Fischöl
fish production – Fischproduktion
fish stocks – Fischbestände
fishing fleet – Fangflotte
fishing vessel – Fangschiff/Fischerboot
flooding – Überschwemmung
florish – florieren / blühen
flow of goods – Warenstrom
fluctuate – schwanken
fluctuating prices – schwankende Preise
fodder – Futtermittel
foetus – Fötus
food security – Ernährungssicherung
forced recruitment – Zwangsrekrutierung
forecast – Vorhersage/Prognose
foreign – ausländisch
foreign currency/ foreign exchange – Fremdwährung/Devisen
foreign trade volume – Außenhandelsvolumen
foreign trade – Außenhandel
fortress – Festung
fossil fuel – fossile Brennstoffe
founder – Gründer/in
fragmentation – Fragmentierung
freight traffic – Frachtaufkommen
freight transport – Gütertransport
fruit pulp – Fruchtfleisch

G

garden hose – Gartenschlauch
garlic – Knoblauch
general conditions – Rahmenbedingungen
genetically modified – gentechnisch verändert
geriatric nurse – Altenpfleger/in
German Empire – Deutsches Kaiserreich
global economic crisis – Weltwirtschaftskrise
global economy – Weltwirtschaft
global hunger – Welthunger
global population – Weltbevölkerung
governance – Regierungsführung
graduation – Studienabschluss
grain production – Getreideproduction
greenhouse – Gewächshaus
gross domestic product (GDP) – Bruttoinlandsprodukt (BIP)
gross national income (GNI) – Bruttonationaleinkommen (BNE)
growth – Wachstum
growth period – Wachstumsperiode

H

hard coal mining – Steinkohleförderung
hard location factor – harter Standortfaktor
harden – aushärten
harvest – Ernte
headline – Schlagzeile
headquarters (HQ) – Hauptsitz
health insurance – Krankenversicherung
high wage sector – Hochlohnsektor
high-elevation-railway – Hochgebirgsbahn
higher professional education – höherer Berufsabschluss
high-street retailing – stationärer Einzelhandel
hoe – Hacke
homeland – Heimatland
horticulture – Gartenbau
hotel chain – Hotelkette
housing – hier: Wohnen
hub – Drehscheibe
human rights – Menschenrechte
humanitarian relief – humanitäre Hilfe
hunger – Unterernährung
hybrid variety – Hybridsorte

I

ice cream parlour – Eisdiele
illiteracy rate – Analphatebenrate
immigrant – Einwanderer/in
impeached by – abgesetzt von
in descending order – in absteigender Reihenfolge
incentive – Anreiz
income tax – Einkommenssteuer
in-commuter – Einpendler
increase – steigern/Steigerung
indicator of development – Entwicklungsstand
indigenous – indigen/einheimisch
industrial fishing – industrielle Fischerei
industrial robots – Industrieroboter
inexorable – unaufhaltsam
infant mortality rate – Säuglingssterblichkeit
informal sector – informeller Sektor
information technology (IT) – Informationstechnologie
informative value – Aussagekraft
inland port – Binnenhafen
inland water – Binnengewässer
innovation capacity – Innovationsfähigkeit
integration trap – Integrationsfalle
internal migration – Binnenmigration
internet exchange – Internetknoten
investigation – Ermittlungsverfahren
investment grant – Investitionszuschuss
iron and steel works – Hüttenwerk
iron ore – Eisenerz
irrigation – Bewässerung
irrigation agriculture – Bewässerungslandwirtschaft
irrigation techniques – Bewässerungstechniken
isolated – abgeschottet

jobseekers – Arbeitssuchende
joint effort – gemeinsame Anstrengung
joint project – Gemeinschaftsprojekt

K

key industry – Schlüsselindustrie
kidnapping – Entführung
kitchen garden – Nutzgarten

L

land of unlimited opportunity – Land der unbegrenzten Möglichkeiten
land use – Flächennutzung
landlocked country – Binnenstaat
land-use plan – Flächennutzungsplan
large cable railway – Großkabinenbahn
large estates – Großgrundbesitz
large landowners – Großgrundbesitzer/in
large scale – großmaßstäblich
large-scale retailer – Großhandelsunternehmen
layout – hier: Gliederung
leakage effect – Sickerrate
legume – Hülsenfrucht
lend money – Geld verleihen
lentils – Linsen
level of development – Entwicklungsstand
life expectancy – Lebenserwartung
life-insurance – Lebensversicherung
life-threatening – lebensbedrohend
liquidate stocks – Vorräte auflösen
livable future – lebenswerte Zukunft
living conditions – Lebensbedingungen
locally – local / vor Ort
location factors – Standortfaktoren
logistics centre – Logistikzentrum
low wage sector – Niedriglohnsektor
lung cancer – Lungenkrebs
lurk – lauern

M

maintain – aufrecht erhalten
maize – Mais
majority – Mehrheit
malnutrition – Mangelernährung
manufacturing industry – verarbeitendes Gewerbe
manufacturing – Herstellung
margin – hier: Rand
market for housing – Wohnungsmarkt
marketing manager – Marketingleiter/in
marmorated – marmoriert
maternal mortality – Müttersterblichkeit
mayor – Bürgermeister/in
mean years of schooling – durchschnittliche Schulbesuchsdauer in Jahren
meaningful – aussagekräftig
means of communication – Kommunikationsmittel
measure – messen, Maßnahme
mechanical loom – mechanischer Webstuhl
medical care – medizinische Versorgung
medical research – medizinische Forschung
megacity – Megastadt
melting pot – Schmelztiegel

merchandise – Handelsware
merging – Verschmelzung
metropolis – Metropole
metropolitan region – Stadtregion
migrant – Migrant/-in
millet – Hirse
mine – Bergwerk
mineral resources – Bodenschätze
minerals – Mineralstoffe
misery – Elend
mislead – irreführen/ täuschen
mismanagement – Misswirtschaft
mitigate – entschärfen
model village – hier: Arbeiterkolonie
MP/member of parliament – Abgeordnete/r
multistory – mehrstöckig
municipal house building – kommunaler Wohnungsbau
municipality – Kommune

N

nanny – Kindermädchen
nationality – Staatsangehörigkeit
navigable river – schiffbarer Fluss
nepotism – Vetternwirtschaft
net pen – Netzgehege
net-migration – Wanderungssaldo
network of streets – Straßennetz
networking impulse – Vernetzungsimpuls
newly industrialised country – Schwellenland
niche product – Nischenprodukt
nitrogen oxides – Stickoxide
Nobel Peace Prize – Friedensnobelpreis
non-compliance – nicht-Einhalten
nourishment – Nährstoffe
nursing staff – Pflegepersonal
nutrition – Ernährung
nutritional information – Angaben zum Nährwert

O

obstacle – Hindernis
ocean-going vessel – hochseetüchtiges Schiff
odd job – Gelegenheitsarbeit
official development assistance (ODA) – öffentliche Entwicklungszusammenarbeit
old-age poverty – Altersarmut
on the internet – im Internet
on the verge – an der Schwelle
one-child-policy – Ein-Kind-Politik
one-off product – Einzelstück/Unikat
one-sided orientation – einseitige Ausrichtung
origin – Herkunft
outsource – auslagern
overexploitation – Übernutzung/Ausbeutung
overfishing – Überfischung

P

pace – Tempo
parcel shipment – Paketsendung
participations – (Unternehmens-)Beteiligungen
passenger traffic – Passagieraufkommen
pasture – Weideland
payment services – Zahlungsdienste
peace and order – Ruhe und Ordnung
peaty soil – Torfmoorboden
peel shrimps – Garnelen pulen
per capita – pro Kopf
permeate – durchdringen
persecution – Verfolgung
pest infestation – Schädlingsbefall
polio – Kinderlähmung
pollutant – Schadstoff
pollutant emission – Schadstoffemission
population distribution – Bevölkerungsverteilung
population policy – Bevölkerungspolitik
population pressure – Bevölkerungsdruck
postpone – verschieben
postwar period – Nachkriegszeit
poverty – Armut
poverty line – Armutsgrenze
power cut – Stromausfall
power grid – Stromnetz
power network – power network
precondition – Voraussetzung
predecessor – Nachfolger/in
predominant – überwiegend
pregnancy – Schwangerschaft
premium payment – Prämie
press plant – Presswerk
previously – vorher / vormals
primarily – vor allem
primary school – Grundschule
processing – (Weiter)-verarbeitung
procurement – Beschaffung
production line – Produktionsband
production site – Produktionsstätte
profits – Gewinne
programmable logic controller – speicherprogammierbare Steuerung
promote – fördern
promotion – Förderung
prosperity – Wohlstand
prosperous – wohlhabend
protected area – Schutzgebiet
protection – Schutz
protective gloves – Schutzhandschuhe
protein – Eiweiß
provision of services – Erbringung von Dienstleistungen
proximity – Nähe
public participation – Bürgerbeteiligung
publicly awarded – öffentlich ausgezeichnet
punitive tariff – Strafzoll
purchase price – Kaufpreis
purchasing power – Kaufkraft
pursue a goal – ein Ziel verfolgen

Q

qualitative leap – Qualitätssprung

R

rape – Vergewaltigung
rapeseed – Raps
rare earths – seltene Erden
ratio – Verhältnis
raw material supplier – Rohstofflieferant
reduction of poverty – Armutsbekämpfung
refugee – Flüchtling
refugee camp – Flüchtlingslager
regional disparities – regionale Disparitäten
remittance – Rücküberweisung
remote region – abgelegene Region
remote – abgelegen
rental price – Mietpreis
research and development (R&D) – Forschung und Entwicklung (FuE)
research institution – Forschungseinrichtung
reservation – hier: Reservat
resettler – Aussiedler/in
residential area – Wohngebiet
resin – Harz
restore confidence – Vertrauen wiederherstellen
restricted – begrenzt
restriction – Beschränkung
retail – Einzelhandel
retailer – Einzelhändler/-in
reurbanisation – Reurbanisierung
revenue groth – Umsatzwachstum
revitalise – wiederbeleben
river oasis – Flussoase
robbery – Überfall
roof garden – Dachgarten
roof tile – Dachziegel
rubber – hier: Kautschuk
rubbish disposal – Müllentsorgung
rubbish truck – Müllwagen
run-off – Abfluss
rural areas – ländliche Gebiete
rural-urban migration – Landflucht
salary – Gehalt

S

sanitation – Abwassersystem
school attendance – Schulbesuch/Anwesenheit in der Schule
scientific facility – Wissenschaftseinrichtung
scientific publications – wissenschaftliche Veröffentlichungen
seafood – Meeresfrüchte
sealed – versiegelt
sealing – Versiegelung
seaport – Seehafen
season ticket – Dauerkarte
seasonality – Saisonalität
security guard – Wache
seed bank – Saatgutbank
segregation – Segregation
semi-desert – Halbwüste
sequester – hier: binden
services/service industry – Dienstleistungen/Dienstleistungsbranche
settle – (sich) niederlassen
sewage system – Abwasserentsorgung
sex education – Aufklärung
sexually transmitted disease (STD) – sexuell übertragbare Krankheit
shaft – Stab, hier: Schacht
shanty – Baracke
share – teilen, Anteil
shared accomodation – Gemeinschaftsunterkunft
shareholder – Gesellschafter/in, Anteilseigner/in
shares – Anteile
shipbuilding – Schiffbau
shrinking population – schrumpfende Bevölkerung
sights – Sehenswürdigkeiten
significance – Aussagekraft, Bedeutung
significant – nennenswert, wichtig, bedeutungsvoll
single internal market – gemeinsamer Binnenmarkt
skilled workers/professionals – Fachkräfte
skin cancer – Hautkrebs
slaughterhouse – Schlachthof
sluice – Schleuse
small scale agriculture – kleinbäuerliche Landwirtschaft
smallholder – Kleinbauer/-bäuerin
smart factory – intelligente Fabrik
SME – small/medium-sized enterprise – KMU - Klein-/Mittelstandsunternehmen
smuggler – Schleuser/Schmuggler
social disparities – soziale Disparitäten
social indicators – soziale Indikatoren
social scientist – Sozialwissenschaftler/in
social standards – gesellschaftliche Normen
social-spatial – sozialräumlich
soft location factor – weicher Standortfaktor
soy – Soja
spacial proximity – räumliche Nähe
spatial distribution – räumliche Verteilung
special economic zone (SEZ) – Sonderwirtschaftszone
squatter settlement – Hüttensiedlung
stackable – stapelbar
stagnate – stagnieren
stakeholder – Interessenvertreter/-in
state welfare/social welfare – staatliche/soziale Wohlfahrt
state-of-the-art – neuester Stand der Technik
stationary retail trade – stationärer Einzelhandel
stem cell – Stammzelle
sterilise – sterilisieren
stink bug – Baumwanze
stock exchange – Börse, Wertpapierbörse
storage space – Speicherplatz
stowaway – blinde/r Passagier/-in
street vendor – Straßenhändler/in
structural change – Strukturwandel
structurally strong – strukturstark
structurally weak – strukturschwach
subsidies – Fördermittel
subsidised – subventioniert
suburb – Vorort
suburbanisation – Suburbanisierung
superfluous – überflüssig
supply and demand – Angebot und Nachfrage
supply chain – Lieferkette
supply industry – Zulieferindustrie
support measures – Fördermaßnahmen
support – tragen / unterstützen, Unterstützung
supression – Unterdrückung
surplus – Überschuss
surplus of importance – Bedeutungsüberschuss
sustain – ertragen
sustainability – Nachhaltigkeit
sustainable development goal – Nachhaltigkeitsziel
sustainable development – nachhaltige Entwicklung
sustainable goals – nachhaltige Ziele
sworn in – vereidigt
synchronised traffic lights – grüne Welle

T

tank – hier: Becken
tariff – Zoll / Abgabe
tax – Steuer / Abgabe
tax advantage – Steuervergünstigung
tax allowences – Steuervergünstigungen
tax reduction – Steuerentlastung
technology park – Technologiepark
tenement block – Mietskaserne
tertiary sector – tertiärer Sektor/ Dienstleistungssektor
theft – Diebstahl
theme park – Vergnügungspark
think tank – Denkfabrik
ticket machine – Fahrkartenautomat
timber – Holz / Bauholz
timeline – Zeitstrahl
tissue sample – Gewebeprobe
top-selling – umsatzstark
torture – Folter
tourism revenues – Tourismuseinnahmen
trace elements – Spurenelemente
trade agreement – Handelsabkommen
trade barrier – Handelsbarriere
trade flows – Handelsströme
trade market – Absatzmarkt
trade policy – Handelspolitik
trade relations – Handelsbeziehungen
traffic jam – Verkehrsstau
traffic management – Verkehrssteuerung
trafficking – (illegaler) Handel
training – hier: Ausbildung
transmission – Übertragung
transport link – Verkehrsanbindung
transshipment point – Umschlagplatz
transshipment – Umschlag (im Hafen), Warenumschlag
treat – behandeln
trigger – auslösen, Auslöser
tropical cyclones – tropische Wirbelstürme
tuber – Knollenfrucht
turnover – Umsatz
two-family dwelling – Zweifamilienhaus
typhoid – Typhus

U

unanimously adopted – einstimmig verabschiedet
uncertain – unsicher
undernourishment – Unterernährung
underwater cable – Unterseekabel
unemployed – arbeitslos/Arbeitslose/r
unemployment insurance – Arbeitslosenversicherung
unemployment – Arbeitslosigkeit
unfavourable – ungünstig
University of Applied Sciences – Fachhochschule
unscathed – schadlos
unskilled – ungelernt/unqualifiziert
upstream – flussaufwärts
upstream/downstream industry – vor- und nachgelagerte Branche
urban dweller – Großstadtbewohner/-bewohnerin
urban farming – urbane Landwirtschaft
urban growth – Städtewachstum
urban heat island – Wärmeinsel in der Stadt
urban hinterland – städtisches Umland
urban poor – städtische Arme
urban population – Stadtbevölkerung
urban sprawl – Zersiedelung
urbanisation – Verstädterung
urn – Urne
utopian – utopisch

V

vacancy rate – Leerstandsquote
value added – Wertschöpfung
vicious circle – Teufelskreis
video monitoring – Videoüberwachung

W

wages – Löhne
warehousing – Lagerei
waste management – Abfallwirtschaft
water expanse – Wasserfläche
water leak – Wasserleck
water treatment plant – Wasseraufbereitungsanlage
wealthy – wohlhabend
weather conditions – Witterung
well-funded – kaufkräftig
wide base – breite Basis
wide-body aircraft – Großraumflugzeug
wire resistor – Drahtwiderstand
womens' cooperative – Frauen-Kooperative
wordle (tag cloud) – Wortwolke
workers' settlement – Werkssiedlung
working age – arbeitsfähiges Alter
working conditions – Arbeitsbedingungen
World Heritage Site – Weltkulturerbe
world market leader – Weltmarktführer
world market share – Weltmarktanteil

X

Y

Z

Bildquellenverzeichnis

|123RF.com, Hong Kong: 229.6; eugenesergeev 228.2; lowthian 230.3. |action press, Hamburg: ZUMA Wire / Zuma Press 154.1. |Airbus Defence and Space GmbH, München: 183.3. |akg-images GmbH, Berlin: 31.1; Bildarchiv Steffens 78.2; Ziese/ Stefan 113.2. |Alamy Stock Photo, Abingdon/Oxfordshire: BIOSPHOTO 99.2; Blossey, Hans 115.2; Boethling, Joerg 18.1; Brown, Ed 142.4; Cristofori, Marco 89.1; Daemmrich, Bob 55.2; Dagnall, Ian 198.1; Doering, Ulrich 35.2; FLHC A23 169.1; imageBROKER 85.4; Imaginechina Limited 153.1; Oberhauser, Rupert 119.2; Otto, Werner 110.2; Pavone, Sean 216.1; Pavone, Sean/Prisma by Dukas Presseagentur GmbH 216.2; Prisma by Dukas Presseagentur GmbH 216.3; public domain sourced / access rights from Yogi Black 16.1; Souza, Joas 188.1; Stark, Friedrich 149.2; Stout, Mark 178.1; ton koene 97.1; vant Hoenderdaal, Robert 138.3; volkerpreusser 185.2; Welsh, Ken 177.4; YOORAN PARK 197.2. |Alamy Stock Photo (RMB), Abingdon/Oxfordshire: 32.2; Aerial Archives 230.6; Asia File 173.3; bamboofox 171.4; BIOSPHOTO 227.7; Boethling, Joerg 18.3, 87.4, 180.2; Charles O. Cecil 145.2; dave stamboulis 66.5; Greatstock 99.1; guy oliver 228.6; Hoferichter, Bert 66.3; Image Source 41.8; imageBROKER 191.2; JLImages 230.8; John Warburton-Lee Photography 228.5; Kesseler, Marco 180.1; Kublins, Ikars 182.2; Laurance, Robin 57.1; Lyell, Jake 91.1; Miller, Kevin 230.4; Mullins, William 193.1; Probst, Peter 65.3; robertharding 42.1; Rye, Tony 228.1; Sackermann, Joern 111.2; SOPA Images Limited 143.2; Tozer, Christopher 104.1; Uhlman, Tom 230.7; Ulrich Doering 177.7; WENN Ltd 231.1; Werner Otto 119.1; Willie, Des 143.7. |AUDI AG, Ingolstadt: 159.1. |Baaske Cartoons, Müllheim: Plaßmann, Thomas 39.1, 122.1; Stauber, Jules 10.1. |Bauer, Jürgen, Nierderrimsingen: 220.1. |Berghahn, Matthias, Bielefeld: 75.1, 82.4, 232.1. |Bezirksregierung Köln, Bonn: Geoportal.NRW 122.2, 122.3, 123.1, 123.2, 123.3, 123.4, 123.5. |BMW AG, München: 157.2. |Bolivien-Brücke e.V., Aachen: Los Masis, Sucre/Bolivien 101.1. |bpk-Bildagentur, Berlin: Stiftung Preussische Schlösser und Gärten Berlin-Brandenburg /J. P. Anders 231.5. |Braune, Barbara, Peine: 221.2. |Brilliance China Automotive Holdings Ltd., Beijing: 167.1. |Brot für die Welt, Berlin: 101.3. |Bundesamt für Kartographie und Geodäsie (BKG), Frankfurt: Geoportal.rlp.de 179.2. |Colourbox.com, Odense: 197.1. |DE-CIX Management GmbH, Köln: 148.3. |DER SPIEGEL, Hamburg: Cover DER SPIEGEL 36/2016 169.3. |Deutsche Stiftung Weltbevölkerung (DSW), Hannover: 37.4; www.dsw.org 101.8. |Deutsche Welthungerhilfe e.V., Bonn: © Welthungerhilfe 101.6. |DLR Deutsches Zentrum für Luft- und Raumfahrt, Weßling, OT Oberpfaffenhofen: contains modified Copernicus Sentinel-2 data (2016) 206.1. |dreamstime.com, Brentwood: Aleksandar Todorovic 227.8; Antikainen 227.3. |eBay Deutschland, Kleinmachnow: 128.1. |Ecosia GmbH, Berlin: 213.2. |ESA - European Space Agency, Frascati (Roma): 230.1. |Eurostat - Statistisches Amt der Europäischen Union, Luxemburg: Quelle: Eurostat, © Europäische Union, http://ec.europa.eu/eurostat/web/main/home, 1995 - heute 73.2, 212.2. |Fairtrade Deutschland e.V., Köln: Kaliszewski, Jakub 98.1. |Feldkamp, Friedhelm, Celle: http://hannovers.brot-fuer-die-welt.de/uploads/tx_templavoila/Karikatur_Tourismus.jpg 95.3. |Fischer Kinder- und Jugendbuch Verlag GmbH, Frankfurt/Main: Frankfurt am Main 2021/Jörg Müller, Anita Siegfried, Jürg E. Schneider, Auf der Gasse und hinter dem Ofen. Eine Stadt im Spätmittelalter. Erstmals erschienen 1995 im Sauerländer Verlag./ Müller, Jörg: Stadt im Spätmittelalter © VG Bild-Kunst, Bonn 2021 114.1. |Focus Photo- u. Presseagentur GmbH, Hamburg: Peter Menzel 4.1, 4.2, 4.3, 4.4, 62.1, 62.2, 63.1, 63.2, 72.1. |Ford-Werke GmbH, Köln: 194.1. |fotolia.com, New York: A. Edelmann 228.4; D. Kohn 190.12; Etzold, Daniel 19.6; froxx 190.7; janossygergely 231.3; Javen 230.2; Jeanette Dietl 34.2; Kushnirov Avraham 229.5; photo 5000 227.6; RCH 231.6; Reitz-Hofmann, Birgit 190.11; Schwier, Christian 211.1, 211.2, 211.3; stockphoto-graf 190.3. |FrankenAir, Weismain: Baeuerlein, Ingo 121.2. |Getty Images, München: AFP/MAINA, SIMON 91.3; AFP/Stringer 95.2; Frederic J. Brown/AFP 59.2; Gallo Images (PTY) LTD/Foto24 97.2; Jose Cendon/Bloomberg 180.3; Karumba, Tony 19.2; Stringer/Allison Joyce 25.1; Torgovnik, Jonathan 35.1, 37.1; Zakir Hossain Chowdhury / Barcroft Media 181.2. |Getty Images (RF), München: Stocktrek Images 229.4. |Google Earth: 203.1, 203.2, 203.3, 203.4. |Güttler, Peter - Freier Redaktions-Dienst (GEO), Berlin: 184.1. |Gyarmaty, Jens, Hannover: 22.1. |Haitzinger, Horst, München: 30.1, 73.1, 100.1. |Horsch, Wolfgang, Niedernhall: 209.1. |Human Help Network e.V., Mainz: 101.2. |Imago, Berlin: Blossey, Hans 161.2; Cord, Anja 163.2; imago/Xinhua 106.1; ITAR-TASS 85.2; Shotshop 70.1; Tang Ke 146.2; VWPics/Vallecillos, Lucas 132.1; Wölk, Rüdiger 127.1. |International Animal Rescue Indonesia: Alejo Sabugo 176.8. |iStockphoto.com, Calgary: 1001nights 21.1; 1001slide 93.1; africa924 99.3; alvarez 173.1; Beeldbewerking 138.4; bkindler 116.1; blackred 208.1; ByeByeTokyo 40.1; C_Fernandes 61.1; Chan, Jui-Chi 173.2; Dewald, Claudia 231.2; Ernst, Daniel 90.1; fbxx 64.1; Fiammenghi, Davide 131.1; fotoVoyager 186.1; futureimage 190.13; Gaul, Pawel 230.5; Guenter Guni 228.3; hipproductions 227.1; hudiemm 190.2; Hussain, Ghulam 229.3; justhavealook 169.2; kruwt 136.1; lucky-photographer 197.3; Madmaxer 149.1; McDonald, Joshua 176.4; Mercedes Rancaño Otero 186.2; Mette, Holger 66.2; Mike_Sheridan 227.4; Naylor, Matt 230.9; ollo 183.5; Onfokus 190.14; OSTILL 77.1; OY Titel; pagadesign 190.6; Prostock-Studio 148.2; quangpraha 175.1; Radist 41.5; seewhatmitchsee 190.4; skynesher 29.2; szymanskim 29.1; tirc83 66.6; triloks Titel; vinhdav 170.1, 171.2; Wenjie Dong 177.3; xeni4ka 192.1; znm 93.2. |Karto-Grafik Heidolph, Dachau: 16.2, 130.1, 160.3, 200.1. |Kartographie Michael Hermes, Hardegsen Hevensen: 69.1, 117.1, 117.2, 124.1, 124.2, 126.1, 127.2, 136.2, 137.2, 137.4, 137.5, 138.5, 138.6, 139.3, 139.4, 139.5, 144.2, 145.1, 149.3, 150.2, 152.2, 153.3, 164.1, 167.2, 167.3, 179.4, 182.3, 183.1, 183.2, 189.1, 198.2, 207.1. |KRAH Elektronische Bauelemente GmbH, Drolshagen: 150.3. |laif, Köln: Alex MacLean/NYT/Redux 176.5; Jungeblodt 45.1; Loeffelbein, Kai 142.2; Modrow, Jörg 6.1, 140.1; Paola Di Bella/Redux 20.2; Zackary Canepari/The New York Times/Redux 51.1. |Latinomedia - Büro für Journalismus, Tübingen: Cecibel Romero 59.1. |Liebe deine Stadt e.V., Köln: Merlin Bauer / www.liebedeinestadt.org /© VG Bild-Kunst Bonn, 2021 135.1. |Lopez, Pablo, Polanco: 3.1, 8.1. |Lung, Lucien, Paris: Photography Lucien Lung 103.1. |LVR-Zentrum für Medien und Bildung, Düsseldorf: BIPARCOURS 191.4. |Mark Carwardine, Bristol: 229.7. |Masterplan com GmbH, Berlin: 162.1. |mauritius images GmbH, Mittenwald: Alamy/McConville, Patti 142.1. |Microsoft Deutschland GmbH, München: 219.1, 219.2. |MISEREOR e.V., Aachen: 101.5. |Mithoff, Stephanie, Ahorn: 2.2, 19.3, 55.1, 60.1, 67.1, 74.1, 76.1, 78.1, 79.1, 80.1, 82.1, 82.2, 82.3, 82.5, 83.1, 83.2, 83.3, 83.4, 83.5, 83.6, 83.7, 83.8, 85.3, 88.2, 89.3, 90.2, 91.2, 92.1, 129.1, 132.3, 142.5, 146.3, 155.1, 156.2, 157.1, 158.1, 158.2, 160.2, 172.2, 200.2, 205.1, 206.2, 211.4, 218.1, 218.2. |Müller, Niels, Köln: www.ganzkoeln.de 111.1. |NASA, Washington: 176.1, 176.2. |Nascaprojekt – Verein Dr. Maria Reiche e.V. Dresden, Dresden: 176.7. |Naumann, Andrea, Aachen: 2.1, 11.1, 13.1, 14.1, 15.19, 18.2, 29.3, 33.1, 37.2, 65.2, 72.2, 76.2, 79.2, 84.1, 85.6, 86.1, 93.3, 94.1, 97.3, 145.3, 148.4, 159.4, 163.5, 166.2, 175.3, 177.8, 183.4, 184.2, 207.2, 208.2, 212.1. |nelcartoons.de, Erfurt: 36.1. |Philips GmbH, Hamburg: 190.9. |Picture-Alliance GmbH, Frankfurt a.M.: AP Images/ ITSUO INOUYE 216.4; AP Images/Abd, Rodrigo 176.9; AP Photo/Heri Juanda 23.1; AP Photo/Muhammad, Bazuki 177.1; AP Photo/Sladky, Lynne 143.9; Arco Images GmbH/Boensch, B. 163.4; augenklick/Minkoff, Sammy 181.1; Bratic, Hasan 155.3; Bruno Vedel/dpa 187.1; dpa-ZB/Patrick Pleul 47.1; dpa/Büttner, Jens 159.3; dpa/Christoph Schmidt 47.2; dpa/Hildenbrand, Karl-Josef 213.1; dpa/Imaginechina 35.3; dpa/Kaiser, Henning 161.1; dpa/Markus Scholz 47.3; dpa/Reinhardt, Daniel 191.1; dpa/Tirl, Marc 142.3; EPA/Nv, Jagadeesh 89.2; euroluftbild.de/Blossey, Hans 110.3; euroluftbild.de/Grahn, Robert 165.3; Everett Collection 179.3; imageBROKER/Schoenen, Daniel 132.2; landov/Acker, Daniel 194.2; Parry, David 19.5; Photoshot 105.1; REUTERS 19.1; REUTERS/Bizimana, Jean 99.4; REUTERS/KHAM 175.2; REUTERS/Rietschel, Matthias 165.4; REUTERS/Song, Aly 153.2; REUTERS/TYRONE SIU 125.1; Rothermel, Winfried 182.1; Shapran, Andrej 20.1; Westend61/Ingold, Daniel 156.1; Westend61/Trubavin, Konstantin 95.1; Xinhua News Agency/Kun, Liu 168.1. |Porsche Leipzig GmbH, Leipzig: Stand: 01/2021 165.1. |Prof. Dr. Stephan Lessenich, Muenchen: 181.4. |Sakurai, Heiko, Köln: 15.1. |Scholz, Ulrich, Linden: 177.5, 177.6. |Schwarzstein, Yaroslav, Hannover: 222.1. |Shutterstock.com, New York: A and N photography 26.1, 34.1; Aerovista Luchtfotografie 138.2; Akella Srinivas Ramalingaswami 52.1; Amit.pansuriya 146.4; Andrew V Marcus 170.2; Armyagov, Andrey 24.1; Asada Nami 17.1; Asianet-Pakistan 51.2; Asier Romero 41.1; BrianPlrwin 143.4; Carey, Rich 177.2; catwalker 181.5; Chernyshev, Kirill 139.1; DFLC Prints 81.2; Eagle9 32.1; erlucho 66.4; fizkes 143.6; Glazkov, Pavlo 137.3; GoneWithTheWind 41.2; ibrar.kunri 48.1; iofoto 87.8; Jess Kraft 81.1; Jirousek, Vladislav T. 231.7; Juilliart, Richard 85.1; junrong 179.1; Kaminer, Mattis 229.1; Katvic 229.2; Klein, Herb 66.1; kleinstofzuigertje 138.1; Lay, Steffen 215.1; Mangostar 88.1; Maschek, Victor 227.2; Melnik, Vladimir 27.1; MichaelJayBerlin 54.1, 218.3, 221.1; Miss Ty 143.8; Mumemories 171.3; naluwan 41.3; Osman, Marcin 176.6; Pantoja, Alvaro 231.4; photo.ua 139.2; r.classen 113.1; Samyan, Sek 181.3; SAPhotog 228.7; Serjio74 231.8; Simona 86 41.6; StanislavBeloglazov 65.1; testing 151.1; thi 171.1; think4photop 176.3; udeyismail 23.2; Westmacott, Edward 77.2; Yasonya 137.1; YewLoon Lam 50.1; Zhyhman, Mark 87.1; Zimniy 159.5. |Staatskanzlei Nordrhein-Westfalen, Düsseldorf: Zeichner: Jacques Tilly 5.1, 108.1. |Stadt Bochum, Kommunikationsservice - Presse und Informationsamt, Bochum: 160.1. |Stiftung DIE GRÜNE STADT, Berlin: 131.2, 131.3, 131.4. |Stiftung Schüler Helfen Leben b.R., Neumünster: 100.2. |stock.adobe.com, Dublin: 147.1; Antonov, Nikolay N. 136.3; asife 41.4; asnidamarwan 143.3; chee siong teh 144.1, 146.1, 148.1, 150.1, 152.1; Cline, Dusty 190.5; Daniel Smolcic 248.1; Fälchle, Jürgen 64.2; Gonzalez, Juanamari 227.5; ihnat 138.7; Isa ÖZDERE 143.5; M.Style 128.2, 128.3, 128.4, 128.5, 128.6, 128.7; mdworschak 217.1; Phimak 21.2; Rynio Productions 190.8; Scanrail 190.10; Successo images 19.4; tanyatorgonskaya 121.1; tibanna79 41.7; willyam 159.2; Yuuyen, Pawarit 172.1; zephyr_p 23.3. |Studienkreis GmbH, Bochum: Studienkreis - Die Nachhilfe /www.studienkreis.de/ Karikaturenwettbewerb „Umwelt? - Natürlich!“, Anna Hörner, 16 Jahre 155.2. |Sturz, Nicole: Schwentker, Björn 159.6. |TechnologieZentrumDortmund GmbH, Dortmund: Foto: Roland Kentrup / TZDO 163.1. |Tekülve, Rita, Essen: 144.3, 180.4, 190.1. |terre des hommes Deutschland e.V., Osnabrück: 101.9; Stolz, Iris 174.1. |The Club of Rome, Winterthur: 13.2. |TMD Friction Services GmbH, Leverkusen: 2017 191.5. |Tomasi, Juan Carlos, Barcelona: 85.5. |toonpool.com, Berlin, Castrop-Rauxel: Kamensky, Marian 99.5. |Trebels, Rüdiger, Düsseldorf: 12.1. |twitter.com: Fraunhofer-Institut für Software- und Systemtechnik ISST 163.3. |ullstein bild, Berlin: Boness/IPON 143.1. |UNICEF Deutschland, Köln: © UNICEF 101.7. |United Nations/Department of Global Communications, New York, NY: https://www.un.org/sustainabledevelopment/ „The content of this publication has not been approved by the United Nations and does not reflect the views of the United Nations or its officials or Member States“ 15.2, 15.3, 15.4, 15.5, 15.6, 15.7, 15.8, 15.9, 15.10, 15.11, 15.12, 15.13, 15.14, 15.15, 15.16, 15.17, 15.18, 87.2, 87.3, 87.5, 87.6, 87.7. |vario images, Bonn: euroluftbild.de 191.3. |von Wille, August: 115.1. |Wiedenroth, Götz/www.wiedenroth-karikatur.de, Flensburg: 166.1. |Wirtschaftsförderung Sachsen GmbH, Dresden: www.standort-sachsen.de 165.2, 185.1. |Zygomatic, WB Vianen: www.wortwolken.com 110.1, 134.1. |© Oxfam Deutschland e.V., Berlin: 101.4. |© Statistisches Bundesamt (Destatis), Wiesbaden: 37.3.

Earth from above 206

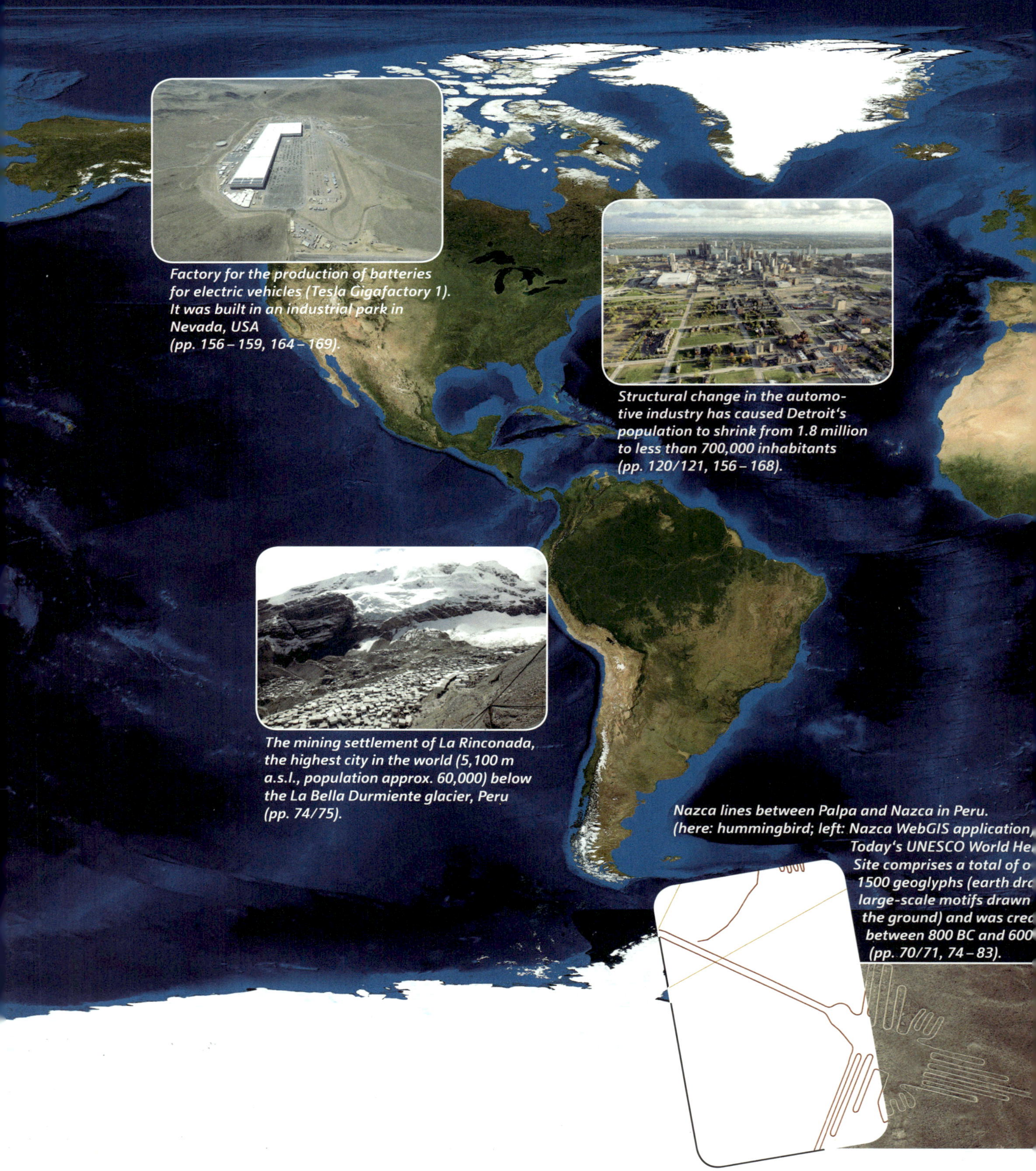

Factory for the production of batteries for electric vehicles (Tesla Gigafactory 1). It was built in an industrial park in Nevada, USA (pp. 156 – 159, 164 – 169).

Structural change in the automotive industry has caused Detroit's population to shrink from 1.8 million to less than 700,000 inhabitants (pp. 120/121, 156 – 168).

The mining settlement of La Rinconada, the highest city in the world (5,100 m a.s.l., population approx. 60,000) below the La Bella Durmiente glacier, Peru (pp. 74/75).

Nazca lines between Palpa and Nazca in Peru. (here: hummingbird; left: Nazca WebGIS application, Today's UNESCO World He Site comprises a total of o 1500 geoglyphs (earth dra large-scale motifs drawn the ground) and was crea between 800 BC and 600 (pp. 70/71, 74 – 83).